Handbook of
Electronics Tables and Formulas

Handbook of
Electronics Tables and Formulas

SIXTH EDITION

Compiled and Edited by
The Howard W. Sams Engineering Staff

Howard W. Sams & Co.
A Division of Macmillan, Inc.
4300 West 62nd Street, Indianapolis, IN 46268 USA

SIXTH EDITION
FIRST PRINTING—1986

International Standard Book Number: 0-672-22469-0
Library of Congress Catalog Card Number: 86-60032

Editor: *Sara Black*
Illustrator: *Ralph E. Lund*
Interior Design: *T. R. Emrick*
Cover Art: *Stephanie Ray*
Shirley Engraving Co., Inc.
James F. Mier, Keller, Mier, Inc.
Composition: *Photo Comp Corp.*

Printed in the United States of America

Contents

PREFACE

The electronics industry is rapidly changing. New developments require frequent updating of information if any handbook such as this is to remain a useful tool. With this thought in mind, each item in the sixth edition was reviewed. Where necessary, additions or changes were made.

In previous editions, we asked for recommendations of additional items to consider for inclusion in future editions. Many suggestions were received and considered; most of them are incorporated in this volume. Hence, this book contains the information that users of the first five editions—engineers, technicians, students, experimenters, and hobbyists—have told us they would like to have in a comprehensive, one-stop edition.

We have added new sections on resistor and capacitor color codes, laws of heat flow in transistors and heat sinks, operational amplifiers, and basic fiber optics. We also detail how to add, subtract, multiply, and divide vectors on a computer as well as work with natural logarithms in computer programs. Computer programs that calculate many of the electronics formulas that appear in the text are part of the two new appendices.

Throughout the text we have attempted to clarify many misconceptions. For example, we clearly distinguish between the physical movement of a free electron and the guided wave motion produced by the electron's field. In addition, we present the volt as a unit of work or energy rather than a unit of electrical pressure or force. We also make a distinction between formulas or mathematical concepts and physical objects or measurements.

In addition, we have retained our comprehensive coverage of the broad range of commonly used electronics formulas and mathematical tables from the fifth edition.

- *Chapter 1*—The basic formulas and laws, so important in all branches of electronics. Nomographs that speed up the solution of DC power, parallel resistance, and reactance. Dimensions of the electrical units are also discussed.
- *Chapter 2*—Useful, but hard-to-remember constants and government- and industry-established standards. The comprehensive table of conversion factors is especially helpful in electronics calculations.
- *Chapter 3*—Symbols and codes that have been adopted over the years. The latest semiconductor information is included.

- *Chapter 4*—Items of particular interest to electronics service technicians.
- *Chapter 5*—Data most often used in circuit design work. The filter and attenuator configurations and formulas are particularly useful to service technicians and design engineers.
- *Chapter 6*—Mathematical tables and formulas. The comprehensive table of powers, roots, and reciprocals is an important feature of this section.
- *Chapter 7*—Miscellaneous items such as measurement conversions, table of elements, and temperature scales.
- *Appendices*—Computer programs for basic electronics formulas.

No effort has been spared to make this handbook of maximum value to anyone, in any branch of electronics. Once again your comments, criticisms, and recommendations for any additional data you would like to see included in a future edition will be welcomed.

List of Tables

List of Tables

Chapter 1

Electronics Formulas and Laws

OHM'S LAW FOR DIRECT CURRENT

All substances offer some obstruction to the flow of current. According to Ohm's law, the current that flows is directly proportional to the applied voltage and inversely proportional to the resistance. Thus, referring to Fig. 1-1:

$$I = \frac{E}{R}$$

$$E = IR$$

$$R = \frac{E}{I}$$

where

I is the current, in amperes,
E is the voltage, in volts,
R is the resistance, in ohms.

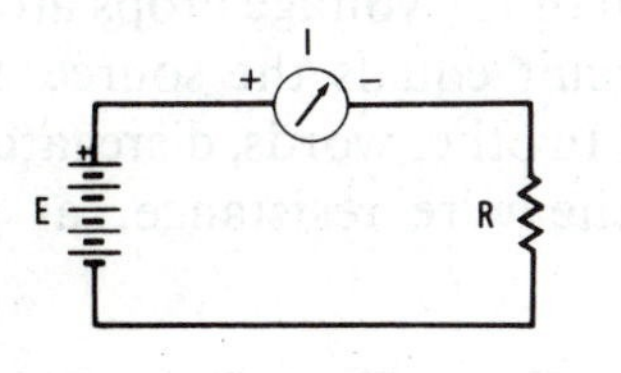

Fig. 1-1

Note. The volt is the work that is done by a battery or generator in separating unit charges through unit distance; the volt is the basic unit of potential energy per unit of charge flow.

DC POWER

The power P expended in load resistance R when current I flows under a voltage pressure E can be determined by the formulas:

$$P = EI$$

$$P = I^2R$$

$$P = \frac{E^2}{R}$$

where

P is the power, in watts,
E is the voltage, in volts,
I is the current, in amperes,
R is the resistance, in ohms.

OHM'S LAW FORMULAS

A composite of the electrical formulas that are based on Ohm's law is given in Fig.

1-2. These formulas are virtually indispensable for solving DC electronic circuit problems.

Unknown Value	Formulas		
E	E=IR	E=P/I	$E=\sqrt{PR}$
I	I=E/R	I=P/E	$I=\sqrt{P/R}$
P	P=EI	$P=E^2/R$	$P=I^2R$
R	R=E/I	$R=E^2/P$	$R=P/I^2$

Fig. 1-2

Free electrons travel slowly in conductors because there is an extremely large number of free electrons available to carry the charge flow (current). If a current of 1 A flows in ordinary bell wire (diameter about 0.04 in), the velocity of each free electron is approximately 0.001 in/s. Thus, if the wire were run 3000 mi across the country, it would take more than 6025 years for an electron entering the wire at San Francisco to emerge from the wire at New York. Nevertheless, because each free electron exerts a force on its adjacent electrons, the electrical impulse travels along the wire at the rate of 186,000 mi/s. Or, the electrical impulse would be evident at New York in less than 0.02 s.

Formulas are used to calculate unknown values from known values. For example, if it is known that $E=10$ and $R=2$, then the formula $I=E/R$ can be used to calculate that $I=5$ A. Similarly, the formula $P=EI$ can be used to calculate that $P=50$ W. Since $I=E/R$, the formula $P=EI$ can be used to calculate that $P=E^2$/R, or $100/2=50$ W. The same answer is obtained whether the formula $P=EI$ or the formula $P=E^2$/R is used.

Note, however, that E is physically real and that E^2 is physically unreal. In other words, E is both physically and mathematically real. On the other hand, E^2 is physically unreal, although E^2 is mathematically real. E^2 is a mathematical stepping-stone to go from one physical reality to another physical reality. Thus, the formula $P=E^2/$R states a mathematical reality, although this formula is a physical fiction. Such relations are summarized by the basic principle that states equations are mathematical models of electrical and electronic circuits.

OHM'S LAW NOMOGRAPH

The nomograph presented in Fig. 1-3 is a convenient way of solving most Ohm's law and DC power problems. If two values are known, the two unknown values can be determined by placing a straightedge across the two known values and reading the unknown values at the points where the straightedge crosses the appropriate scales. The figures in **boldface** (on the right-hand side of all scales) cover one range of given values, and the figures in lightface (on the left-hand side) cover another range. For a given problem, all values must be read in either the **bold-** or lightface figures.

Example. What is the value of a resistor if a 10-V drop is measured across it and a current of 500 mA (0.5 A) is flowing through it? What is the power dissipated by the resistor?

Answer. The value of the resistor is 20 Ω. The power dissipated in the resistor is 5 W.

KIRCHHOFF'S LAWS

According to Kirchhoff's voltage law, "The sum of the voltage drops around a DC series circuit equals the source or applied voltage." In other words, disregarding losses due to the wire resistance, as shown in Fig. 1-4:

$$E_T = E_1 + E_2 + E_3$$

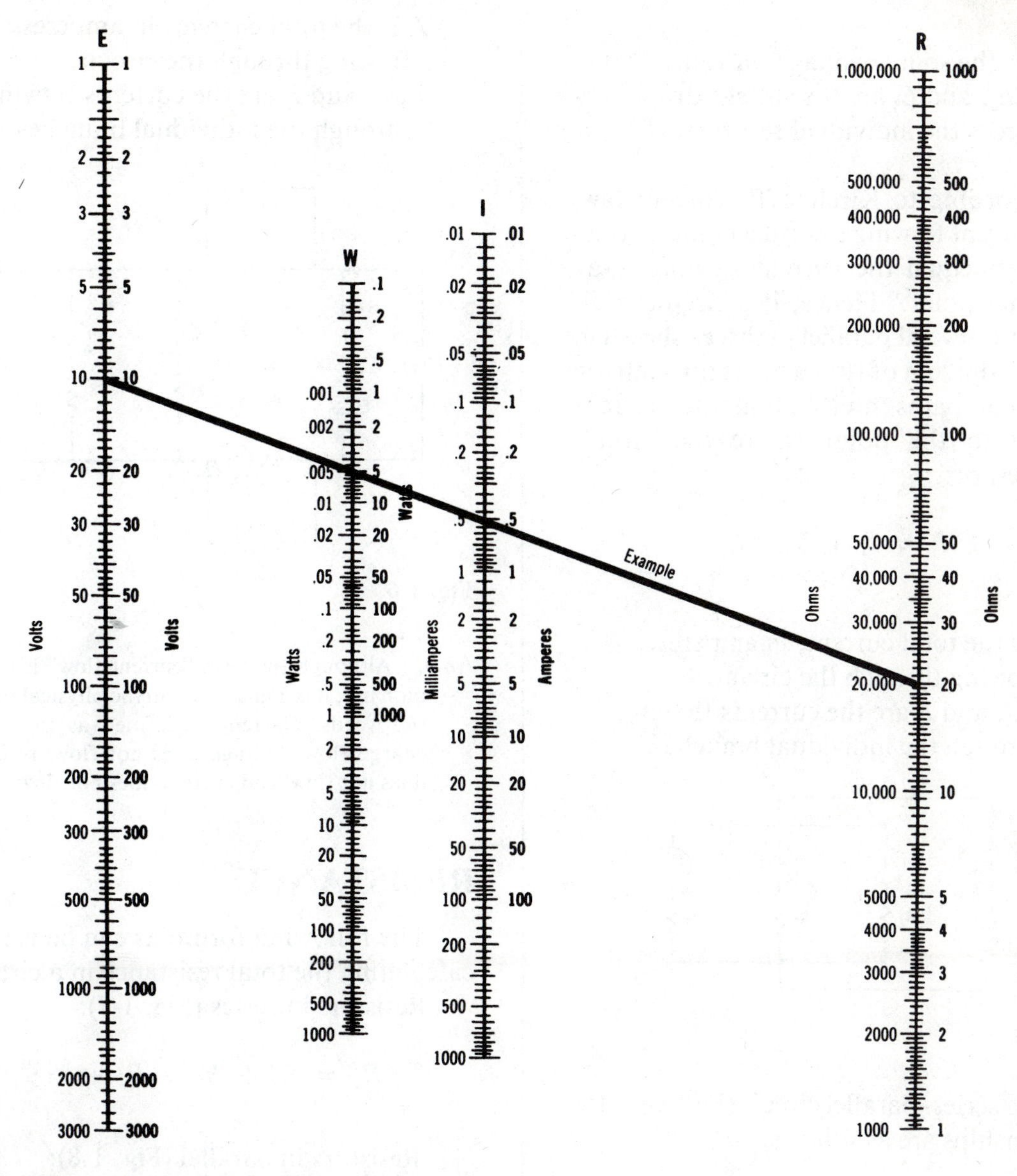

Fig. 1-3. Ohm's law and DC power nomograph.

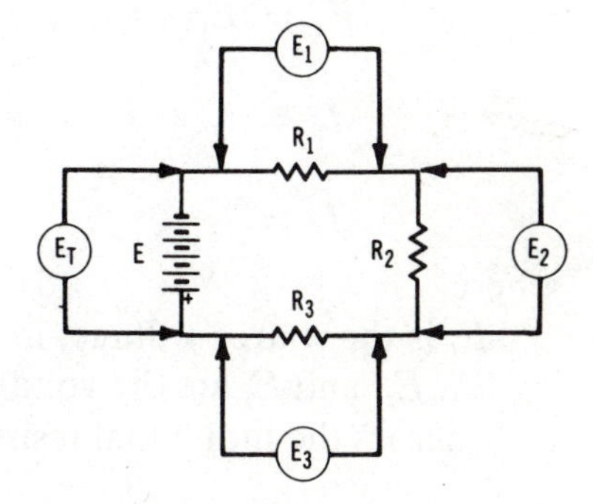

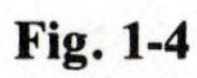
Fig. 1-4

where

E_T is the source voltage, in volts,
E_1, E_2, and E_3 are the voltage drops across the individual resistors.

According to Kirchhoff's current law, "The current flowing toward a point in a circuit must equal the current flowing away from that point." Hence, if a circuit is divided into several parallel paths, as shown in Fig. 1-5, the sum of the currents through the individual paths must equal the current flowing to the point where the circuit branches, or:

$$I_T = I_1 + I_2 + I_3$$

where

I_T is the total current, in amperes, flowing through the circuit,
I_1, I_2, and I_3 are the currents flowing through the individual branches.

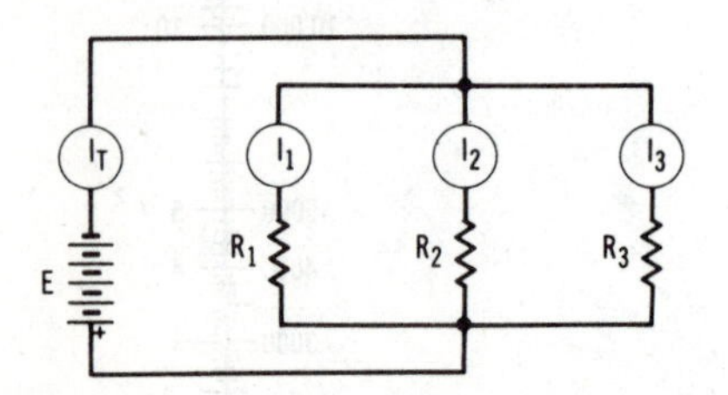

Fig. 1-5

In a series–parallel circuit (Fig. 1-6), the relationships are as follows:

$$E_T = E_1 + E_2 + E_3$$

$$I_T = I_1 + I_2$$

$$I_T = I_3$$

where

E_T is the source voltage, in volts,
E_1, E_2, and E_3 are the voltage drops across the individual resistors,
I_T is the total current, in amperes, flowing through the circuit,
I_1, I_2, and I_3 are the currents flowing through the individual branches.

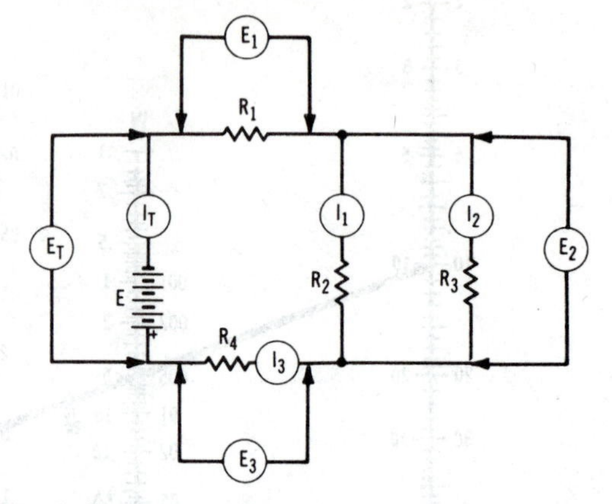

Fig. 1-6

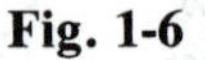

Note. Although the term "current flow" is in common use, it is a misnomer in the physical sense of the words. Current is defined as the rate of charge flow. Voltage does not flow, resistance does not flow, and current does not flow.

RESISTANCE

The following formulas can be used for calculating the total resistance in a circuit.

Resistors in series (Fig. 1-7):

$$R_T = R_1 + R_2 + R_3 + \cdots$$

Resistors in parallel (Fig. 1-8):

$$R = \frac{1}{\frac{1}{R_1} + \frac{1}{R_2} + \frac{1}{R_3} + \cdots}$$

Two resistors in parallel (Fig. 1-9):

$$R_T = \frac{R_1 \times R_2}{R_1 + R_2}$$

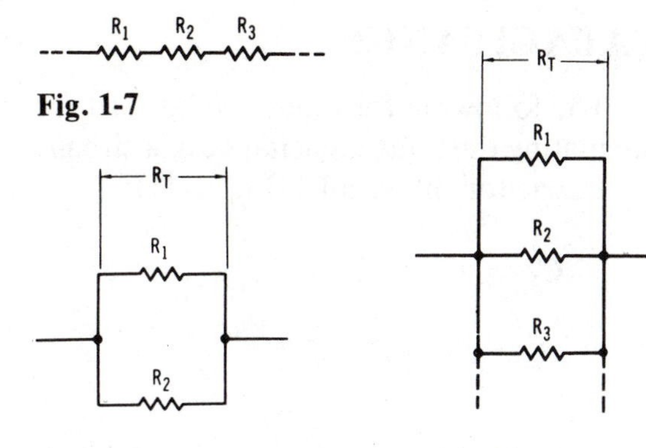

Fig. 1-7

Fig. 1-9

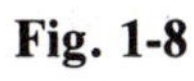

Fig. 1-8

where

R_T is the total resistance, in ohms, of the circuit,

R_1, R_2, and R_3 are the values of the individual resistors.

The equivalent value of resistors in parallel can be solved with the nomograph in Fig. 1-10. Place a straightedge across the points on scales R_1 and R_2 where the known value resistors fall. The point at which the

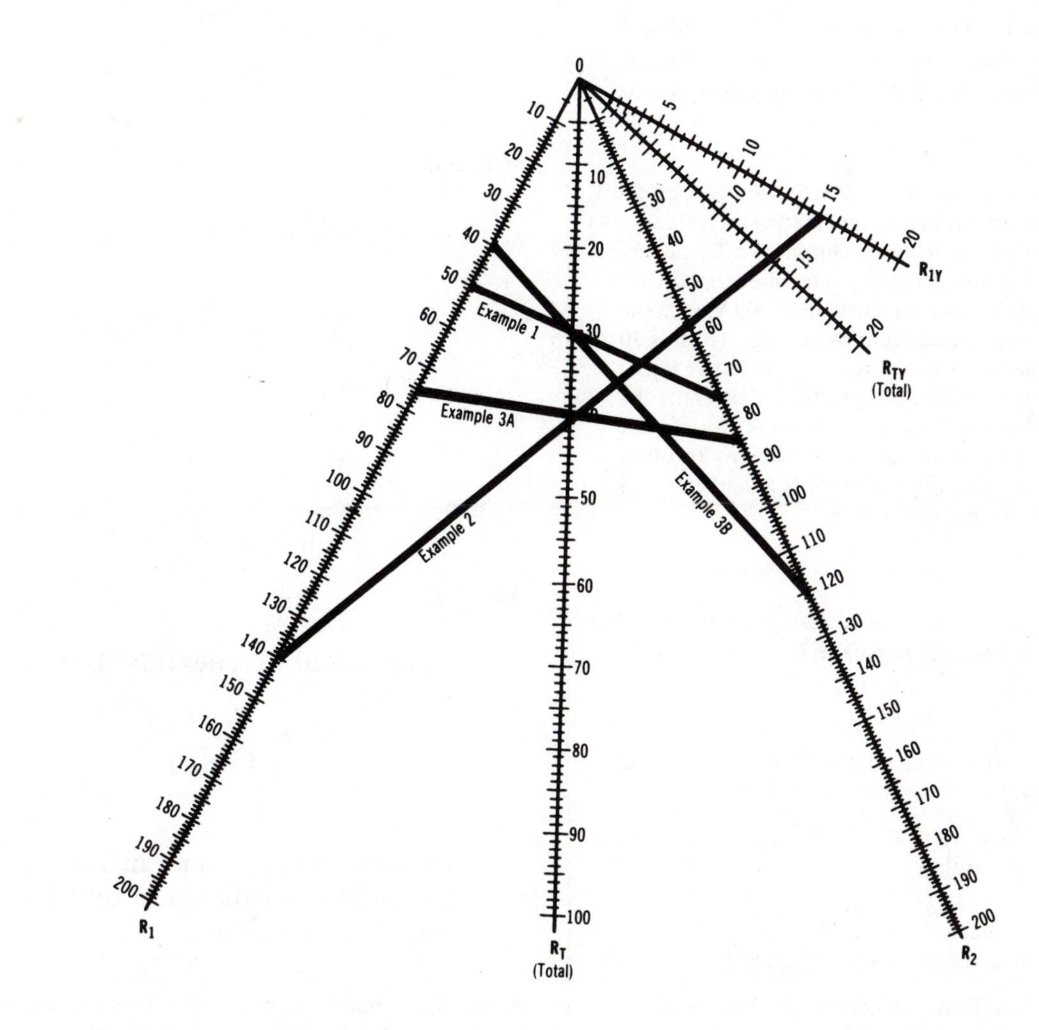

Fig. 1-10. Parallel-resistance nomograph.

straightedge crosses the R_T scale will show the total resistance of the two resistors in parallel. If three resistors are in parallel, first find the equivalent resistance of two of the resistors, then consider this value as being in parallel with the remaining resistor.

If the total resistance needed is known, the straightedge can be placed at this value on the R_T scale and rotated to find the various combinations of values on the R_1 and R_2 scales that will produce the needed value.

Scales R_{1Y} and R_{TY} are used with the B_1 scale when the values of the known resistors differ greatly. The range of the nomograph can be increased by multiplying the values of all scales by 10, 100, 1000, or more, as required.

Note. Ohm's law states that $R = E/I$. In turn, effective resistance is often calculated as an E/I ratio. For example, if the beam current in a TV picture tube is 0.5 mA, and the potential-energy difference from cathode to screen is 15,000 V, then the effective resistance from cathode to screen is 30 MΩ. The power dissipated in the effective resistance is 7.5 W. From a practical viewpoint, the physical power is dissipated by the screen and not in the space from cathode to screen. In other words, the effective resistance is a mathematical reality but a physical fiction.

Example 1. What is the total resistance of a 50-Ω and a 75-Ω resistor in parallel?

Answer. 30 Ω.

Example 2. What is the total resistance of a 1500-Ω and a 14,000-Ω resistor in parallel?

Answer. 1355 Ω. (Use R_1 and R_{TY} scales; read answer on R_{TY} scale.)

Example 3. What is the total resistance of a 75-Ω, an 85-Ω, and a 120-Ω resistor in parallel?

Answer. 30 Ω. (First, consider the 75-Ω and 85-Ω resistors, which will give 40 Ω; then consider this 40 Ω and the 120-Ω resistor, which will give 30 Ω.)

CAPACITANCE

The following formulas can be used for calculating the total capacitance in a circuit.

Capacitors in parallel (Fig. 1-11):

$$C_T = C_1 + C_2 + C_3 + \cdots$$

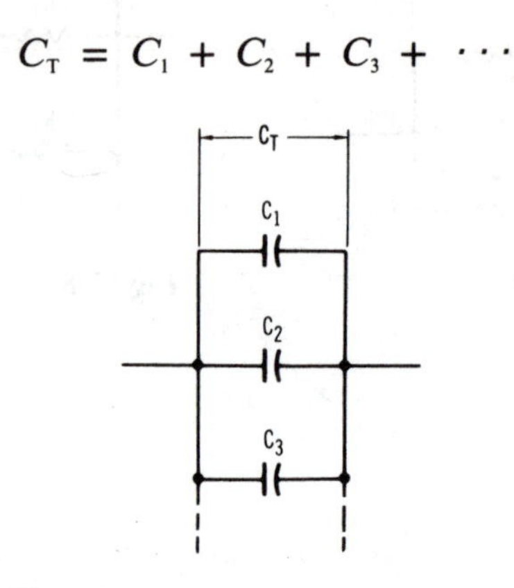

Fig. 1-11

Capacitors in series (Fig. 1-12):

$$C_T = \frac{1}{\frac{1}{C_1} + \frac{1}{C_2} + \frac{1}{C_3} + \cdots}$$

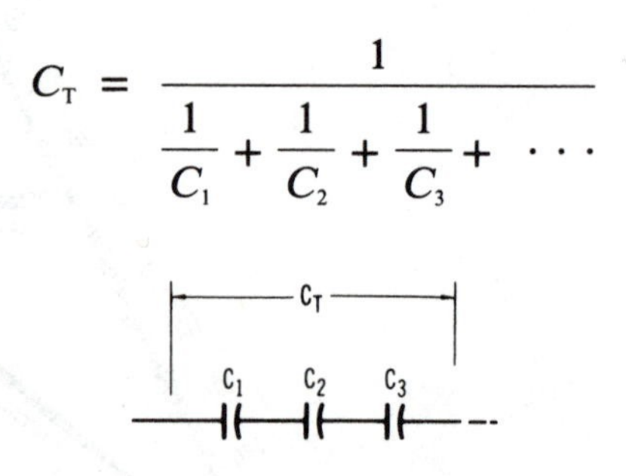

Fig. 1-12

Two capacitors in series (Fig. 1-13):

$$C_T = \frac{C_1 \times C_2}{C_1 + C_2}$$

where

C_T is the total capacitance in a circuit,

C_1, C_2, and C_3 are the values of the individual capacitors.

Note. C_1,C_2 may be in any unit of measurement as long as all are in the same unit. C_T will be in this same unit.

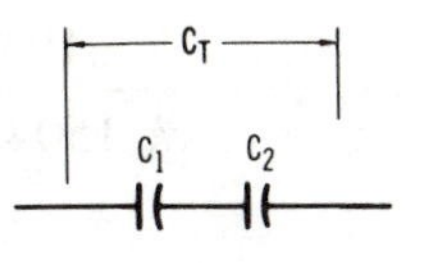

Fig. 1-13

The parallel-resistance nomograph in Fig. 1-10 can also be used to determine the total capacitance of capacitors in series.

The capacitance of a parallel-plate capacitor is determined by:

$$C = 0.2235 \frac{kA}{d} (N - 1)$$

where

C is the capacitance, in picofarads,
k is the dielectric constant,*
A is the area of one plate, in square inches,
d is the thickness of the dielectric, in inches,
N is the number of plates.

Charge Stored

The charge stored in a capacitor is determined by:

$$Q = CE$$

where

Q is the charge, in coulombs,
C is the capacitance, in farads,
E is the voltage impressed across the capacitor, in volts.

Energy Stored

The energy stored in a capacitor can be determined by:

$$W = \frac{CE^2}{2}$$

where

W is the energy, in joules (watt-seconds),
C is the capacitance, in farads,
E is the applied voltage, in volts.

Voltage Across Series Capacitors

When an AC voltage is applied across a group of capacitors connected in series (Fig. 1-14), the voltage drop across the combination is, of course, equal to the applied voltage. The drop across each individual capacitor is inversely proportional to its capacitance. The drop across any capacitor in a group of series capacitors is calculated by the formula:

$$E_C = \frac{E_A \times C_T}{C}$$

where

E_C is the voltage across the individual capacitor in the series (C_1, C_2, or C_3), in volts,
E_A is the applied voltage, in volts,
C_T is the total capacitance of the series combination, in farads,
C is the capacitance of the individual capacitor under consideration, in farads.

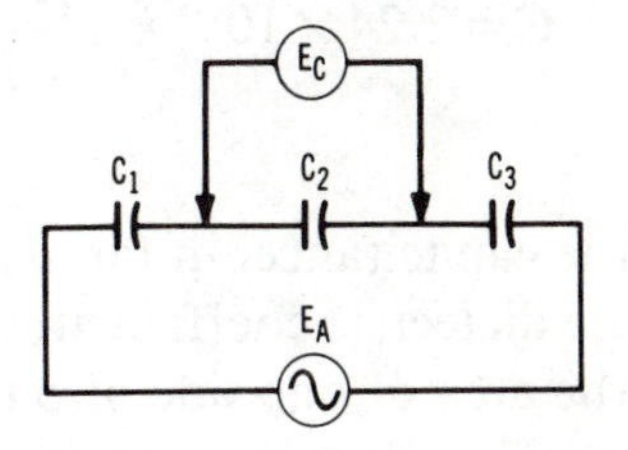

Fig. 1-14

Since a capacitor is composed of a pair of metal plates separated by an insulator, such as air, a unit capacitor could be a pair of metal plates separated by 0.001 in, with

*For a list of dielectric constants of materials, see section entitled Constants and Standards.

375 w

an area of 4.46×10^9 in^2. This unit capacitor will have a capacitance of 1 F. Voltage is potential energy per unit charge. In turn, if this capacitor is charged to a potential-energy difference of 1 V (potential difference of 1 V), the plates will attract each other with a force of approximately 4400 lb, or about two long tons. This force is exerted through a distance of 0.001 in. In other words, the potential difference gives the plates potential energy (energy of position).

As an example of voltage generation (potential-energy generation) by charge separation, suppose that the capacitor described above has been charged to a potential-energy difference of 1 V. Then, if the separation between the plates is increased from 0.001 in to 0.002 in, the potential-energy difference increases to 2 V. In other words, $Q = CE$, and E is inversely proportional to the separation between the plates. Q remains constant (1 C), E is doubled (2 V), and C is halved (0.5 F). The separation between unit charges has been increased through unit distance, with the result that a potential-energy difference of 1 V has been generated.

The formula for calculating the capacitance is:

$$C = 2.24 \times 10^{-13}\, kA \frac{N-1}{d}$$

where

C is the capacitance, in farads,
k is the dielectric coefficient,
A is the area of one side of one plate, in square inches,
d is the separation between the plates, in inches,
N is the number of plates.

The formula for calculating the force of attraction between the two plates is:

$$F = \frac{AV^2}{k(1504S)^2}$$

where

F is the attractive force, in dynes,
A is the area of one plate, in square centimeters,
F is the potential-energy difference, in volts,
k is the dielectric coefficient,
S is the separation between the plates, in centimeters.

A dyne is about 1/980 g; there are 454 g in 1 lb. When the separation between plates is doubled, the voltage (potential-energy difference) between the plates is doubled, but the charge and the force of attraction between the plates remain the same. Because the initial unit separation has been doubled, twice as much work has been done (the initial voltage has been doubled). Initial unit separation was assigned as 0.001 in in the foregoing example. The initial potential-energy difference will, in turn, be assigned as 1 mV when calculating basic relations.

INDUCTANCE

The following formulas can be used for calculating the total inductance in a circuit.

Inductors in series with no mutual inductance (Fig. 1-15):

$$L_T = L_1 + L_2 + L_3 + \cdots$$

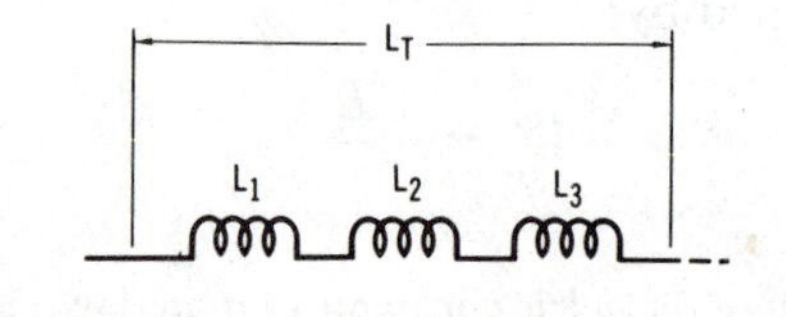

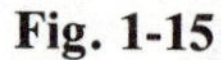

Fig. 1-15

Inductors in parallel with no mutual inductance (Fig. 1-16):

$$L_T = \frac{1}{\frac{1}{L_1} + \frac{1}{L_2} + \frac{1}{L_3} + \cdots}$$

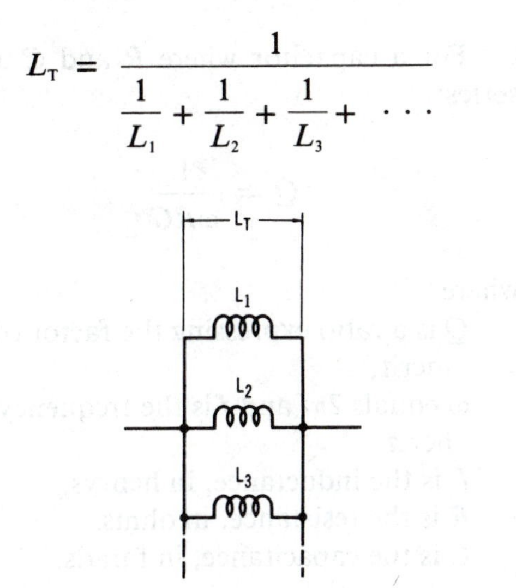

Fig. 1-16

Two inductors in parallel with no mutual inductance (Fig. 1-17):

$$L_T = \frac{L_1 \times L_2}{L_1 + L_2}$$

where

L_T is the total inductance of the circuit, in henrys,

L_1 and L_2 are the inductances of the individual inductors (coils).

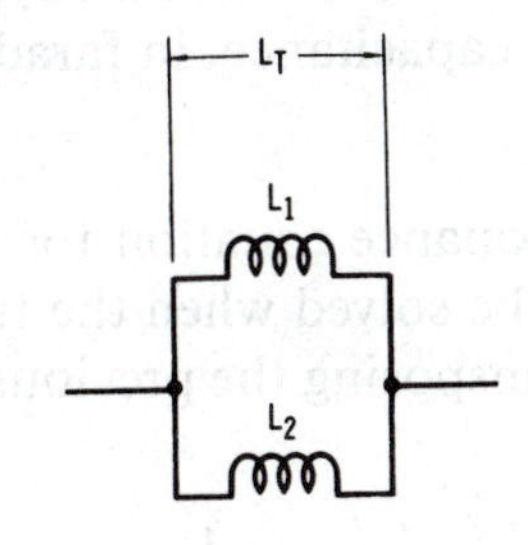

Fig. 1-17

The parallel-resistance nomograph in Fig. 1-10 can also be used to determine the total inductance of inductors in parallel.

Mutual Inductance

The mutual inductance of two coils with fields interacting can be determined by:

$$M = \frac{L_A - L_B}{4}$$

where

M is the mutual inductance of L_A and L_B, in henrys,

L_A is the total inductance of coils L_1 and L_2 with fields aiding, in henrys,

L_B is the total inductance of coils L_1 and L_2 with fields opposing, in henrys.

Coupled Inductance

The coupled inductance can be determined by the following formulas.

In parallel with fields aiding:

$$L_T = \frac{1}{\frac{1}{L_1 + M} + \frac{1}{L_2 + M}}$$

In parallel with fields opposing:

$$L_T = \frac{1}{\frac{1}{L_1 - M} - \frac{1}{L_2 - M}}$$

In series with fields aiding:

$$L_T = L_1 + L_2 + 2M$$

In series with fields opposing:

$$L_T = L_1 + L_2 - 2M$$

where

L_T is the total inductance, in henrys,

L_1 and L_2 are the inductances of the

individual coils, in henrys,
M is the mutual inductance, in henrys.

Coupling Coefficient

When two coils are inductively coupled to give transformer action, the coupling coefficient is determined by:

$$K = \frac{M}{\sqrt{L_1 L_2}}$$

where
K is the coupling coefficient,
M is the mutual inductance, in henrys,
L_1 and L_2 are the inductances of the two coils, in henrys.

Note. An inductor in a circuit has a reactance of $j2\pi fL\ \Omega$. Mutual inductance in a circuit also has a reactance equal to $j2\pi fM\ \Omega$. The operator j denotes that the reactance dissipates no energy, although the reactance opposes current flow.

Energy Stored

The energy stored in an inductor can be determined by:

$$W = \frac{LI^2}{2}$$

where
W is the energy, in joules (watt-seconds),
L is the inductance, in henrys,
I is the current, in amperes.

Q FACTOR

The ratio of reactance to resistance is known as the Q factor. It can be determined by the following formulas.

For a coil where R and L are in series:

$$Q = \frac{\omega L}{R}$$

For a capacitor where R and C are in series:

$$Q = \frac{1}{\omega RC}$$

where
Q is a ratio expressing the factor of merit,
ω equals $2\pi f$ and f is the frequency, in hertz,
L is the inductance, in henrys,
R is the resistance, in ohms,
C is the capacitance, in farads.

RESONANCE

The resonant frequency, or the frequency at which the reactances of the circuit add up to zero ($X_L = X_C$), is determined by:

$$f_R = \frac{1}{2\pi\sqrt{LC}}$$

where
f_R is the resonant frequency, in hertz,
L is the inductance, in henrys,
C is the capacitance, in farads.

The resonance equation for either L or C can also be solved when the frequency is known. Transposing the previous formula:

$$L = \frac{1}{4\pi^2 f_R^2 C}$$

$$C = \frac{1}{4\pi^2 f_R^2 L}$$

The resonant frequency of various combinations of inductance and capacitance can also be obtained from the reactance charts in Fig. 1-18. Simply lay a straightedge across the values of inductance and capacitance, and read the resonant frequency from the frequency scale of the chart.

ADMITTANCE

The measure of the ease with which alternating current flows in a circuit is the admittance of the circuit.

Admittance of a series circuit is given by:

$$Y = \frac{1}{\sqrt{R^2 + X^2}}$$

Admittance is also expressed as the reciprocal of impedance; thus:

$$Y = \frac{1}{Z}$$

where

Y is the admittance, in siemens,
R is the resistance, in ohms,
X is the reactance, in ohms,
Z is the impedance, in ohms.

Admittance is equal to conductance plus susceptance. Conductance is the reciprocal of resistance. The unit of conductance is the siemens (formerly the mho). Inductive reactance is positive, and capacitive reactance is negative. Inductive susceptance is negative, and capacitive susceptance is positive. If an impedance has a positive phase angle, its corresponding admittance will have a negative phase angle, and the values of the two phase angles will be the same.

SUSCEPTANCE

The susceptance of a series circuit is given by:

$$B = \frac{X}{R^2 + X^2}$$

When the resistance is zero, susceptance becomes the reciprocal of reactance; thus:

$$B = \frac{1}{X}$$

where

B is the susceptance, in siemens,
X is the reactance, in ohms,
R is the resistance, in ohms.

CONDUCTANCE

Conductance is the measure of the ability of a component to conduct electricity. Conductance for DC circuits is expressed as the reciprocal of resistance; therefore:

$$G = \frac{1}{R}$$

where

G is the conductance, in siemens,
R is the resistance, in ohms.

Ohm's law formulas when conductance is considered are:

$$I = EG$$

$$G = \frac{I}{E}$$

$$E = \frac{I}{G}$$

where

I is the current, in amperes,
E is the voltage, in volts,
G is the conductance, in siemens.

ENERGY UNITS

Energy is the capacity or ability to do work. The joule is a unit of energy. One joule is the amount of energy required to maintain a current of 1 A for 1 s through a resistance of 1 Ω. It is equivalent to a watt-second. The watt-hour is the practical unit of energy; 3600 Ws equals 1 Wh. The number of watt-hours is calculated:

$$\text{Watt-hours} = P \times T$$

where

P is the power, in watts,
T is the time, in hours, the power is dissipated.

See the section entitled Capacitance to determine the energy stored in a capacitor and the section entitled Inductance to determine the energy stored in an inductor.

REACTANCE

The opposition to the flow of alternating current by the inductance or capacitance of a component or circuit is called the reactance.

Capacitive Reactance

The reactance of a capacitor may be calculated by the formula:

$$X_C = \frac{1}{2\pi fC}$$

where

X_C is the reactance, in ohms,
f is the frequency, in hertz,
C is the capacitance, in farads.

Inductive Reactance

The reactance of an inductor may be calculated by the formula:

$$X_L = 2\pi fL$$

where

X_L is the reactance, in ohms,
f is the frequency, in hertz,
L is the inductance, in henrys.

Reactance Charts

Charts for determining unknown values of reactance, inductance, capacitance, and frequency are shown in Figs. 1-18A through 1-18C. The chart in Fig. 1-18A covers 1–1000 Hz, Fig. 1-18B covers 1–1000 kHz, and Fig. 1-18C covers 1–1000 MHz.

To find the amount of reactance of a capacitor at a given frequency, lay the straightedge across the values for the capacitor and the frequency. Then read the reactance from the reactance scale. By extending the line, the value of an inductance, which will give the same reactance, can be obtained.

Since $X_C = X_L$ at resonance, by laying the straightedge across the capacitance and inductance values, the resonant frequency of the combination can be determined.

Example. If the frequency is 10 Hz and the capacitance is 50 μF, what is the reactance of the capacitor? What value of inductance will give this same reactance?

Answer. 310 Ω. The inductance needed to produce this same reactance is 5 H. Thus, it follows that a 50-μF capacitor and a 5-H choke are resonant at 10 Hz. (Place the straightedge, on the proper chart [Fig. 1-18A], across 10 Hz and 50 μF. Read the values indicated on the reactance and inductance scales.)

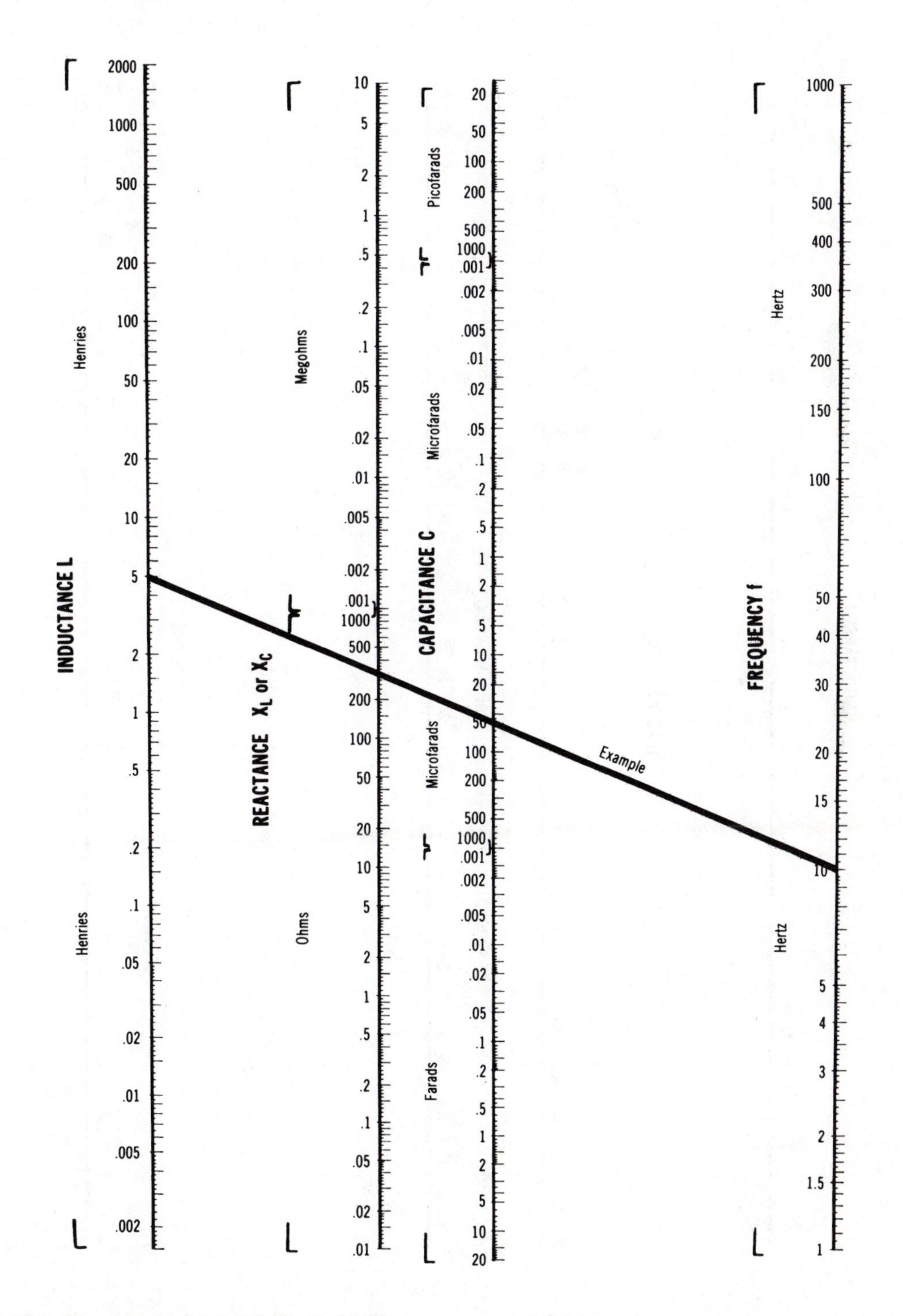

Fig. 1-18A. Reactance chart—1 Hz to 1 kHz.

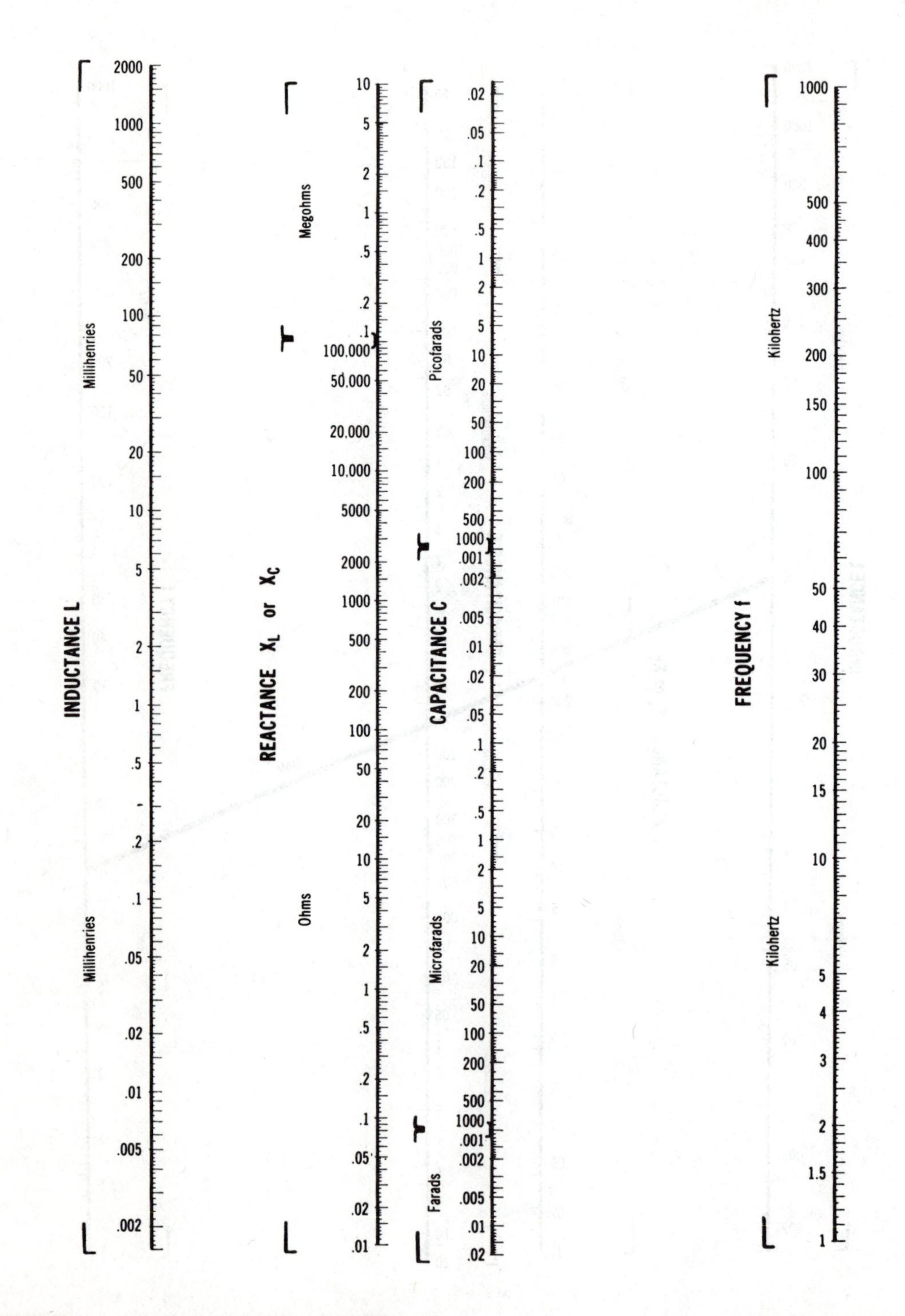

Fig. 1-18B. Reactance chart—1 kHz to 1 MHz.

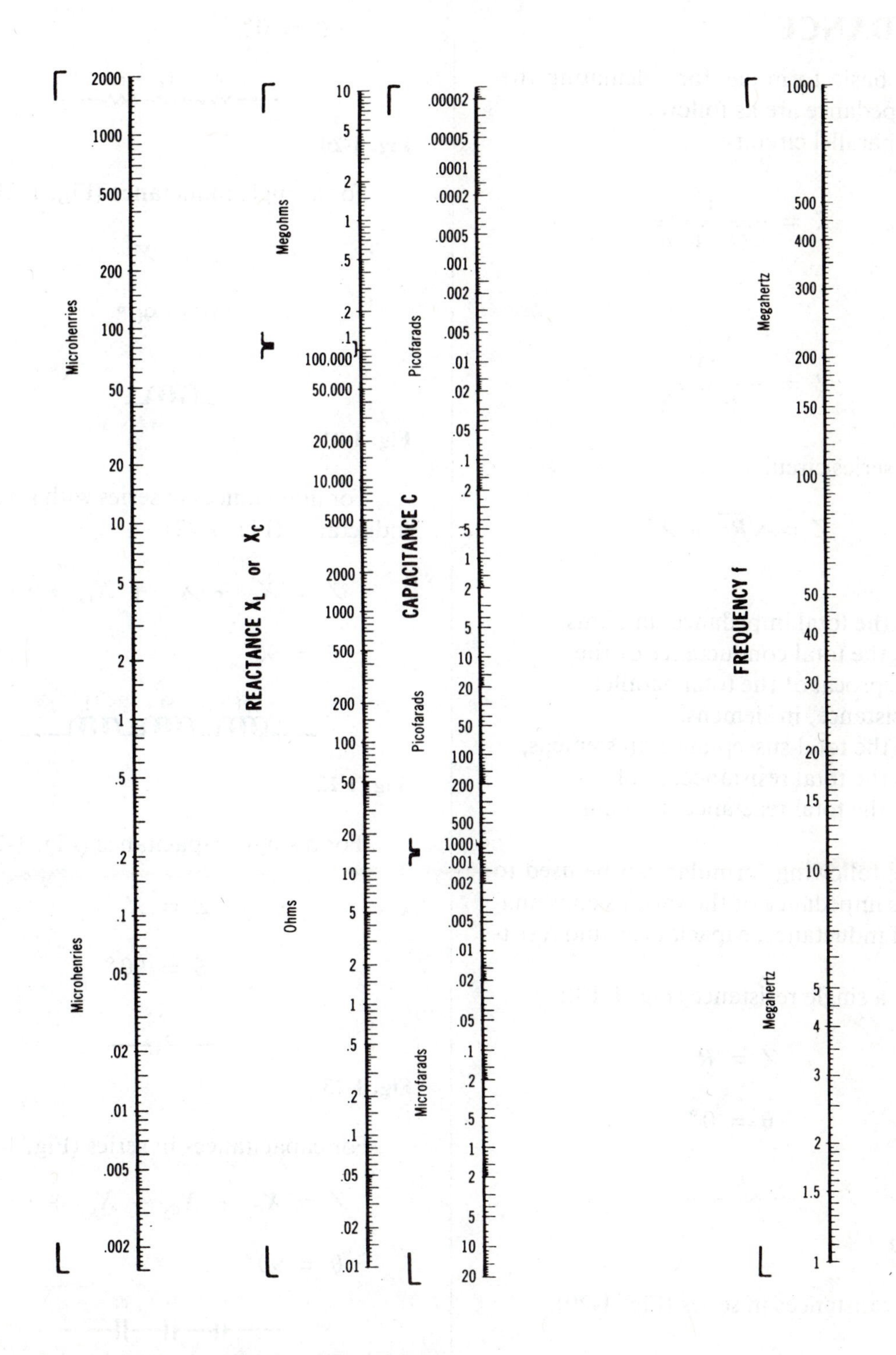

Fig. 1-18C. Reactance chart—1 MHz to 1000 MHz.

IMPEDANCE

The basic formulas for calculating the total impedance are as follows:

For parallel circuits:

$$Z = \frac{1}{\sqrt{G^2 + B^2}}$$

or

$$Z = \frac{RX}{\sqrt{R^2 + X^2}}$$

For series circuits:

$$Z = \sqrt{R^2 + X^2}$$

where

Z is the total impedance, in ohms,
G is the total conductance or the reciprocal of the total parallel resistance, in siemens,
B is the total susceptance, in siemens,
R is the total resistance, in ohms,
X is the total reactance, in ohms.

The following formulas can be used to find the impedance of the various combinations of inductance, capacitance, and resistance.

For a single resistance (Fig. 1-19):

$$Z = R$$

$$\theta = 0°$$

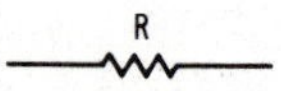

R

Fig. 1-19

For resistances in series (Fig. 1-20):

$$Z = R_1 + R_2 + R_3 + \cdots$$

$$\theta = 0°$$

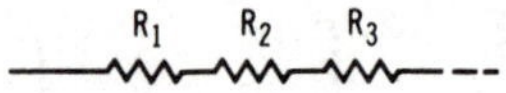

R_1 R_2 R_3

Fig. 1-20

For a single inductance (Fig. 1-21):

$$Z = X_L$$

$$\theta = 90°$$

L

Fig. 1-21

For inductances in series with no mutual inductance (Fig. 1-22):

$$Z = X_{L_1} + X_{L_2} + X_{L_3} + \cdots$$

$$\theta = 90°$$

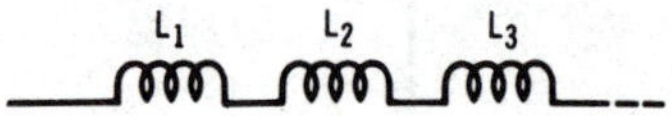

L_1 L_2 L_3

Fig. 1-22

For a single capacitance (Fig. 1-23):

$$Z = X_C$$

$$\theta = 90°$$

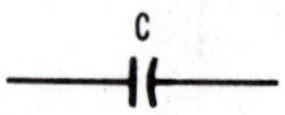

C

Fig. 1-23

For capacitances in series (Fig. 1-24):

$$Z = X_{C_1} + X_{C_2} + X_{C_3} + \cdots$$

$$\theta = 90°$$

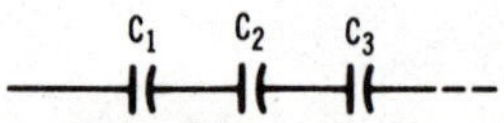

C_1 C_2 C_3

Fig. 1-24

For resistance and inductance in series (Fig. 1-25):

$$Z = \sqrt{R^2 + X_L^2}$$

$$\theta = \arc\tan \frac{X_L}{R}$$

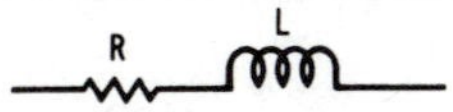

Fig. 1-25

For resistance and capacitance in series (Fig. 1-26):

$$Z = \sqrt{R^2 + X_C^2}$$

$$\theta = \text{arc tan} \frac{X_C}{R}$$

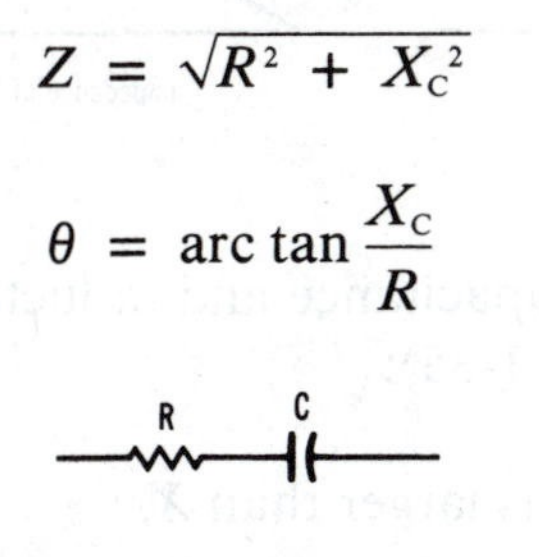

Fig. 1-26

For inductance and capacitance in series (Fig. 1-27):

When X_L is larger than X_C:

$$Z = X_L - X_C$$

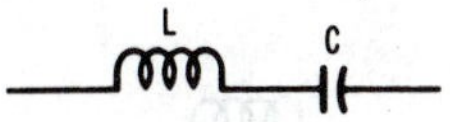

Fig. 1-27

When X_C is larger than X_L:

$$Z = X_C - X_L$$

Note. $\theta = 0°$ when $X_L = X_C$.

For resistance, inductance, and capacitance in series (Fig. 1-28):

$$Z = \sqrt{R^2 + (X_L - X_C)^2}$$

$$\theta = \text{arc tan} \frac{X_L - X_C}{R}$$

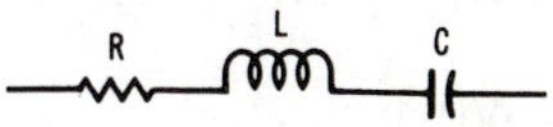

Fig. 1-28

For resistances in parallel (Fig. 1-29):

$$Z = \frac{1}{\frac{1}{R_1} + \frac{1}{R_2} + \frac{1}{R_3} + \cdots}$$

$$\theta = 0°$$

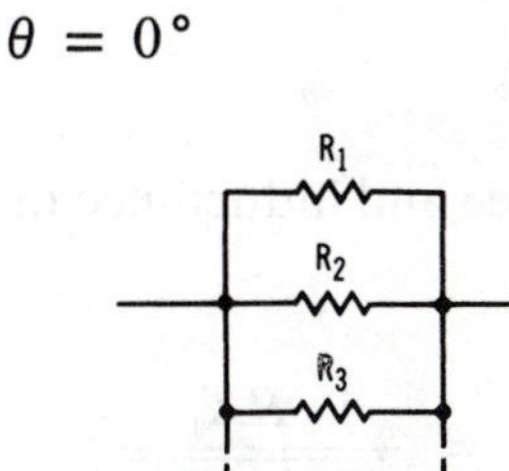

Fig. 1-29

For inductances in parallel with no mutual inductance (Fig. 1-30):

$$Z = \frac{1}{\frac{1}{X_{L_1}} + \frac{1}{X_{L_2}} + \frac{1}{X_{L_3}} + \cdots}$$

$$\theta = 90°$$

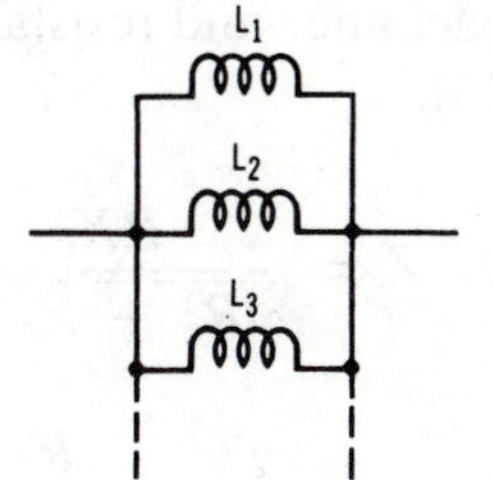

Fig. 1-30

For capacitances in parallel (Fig. 1-31):

$$Z = \frac{1}{\frac{1}{X_{C_1}} + \frac{1}{X_{C_2}} + \frac{1}{X_{C_3}} + \cdots}$$

$$\theta = 90°$$

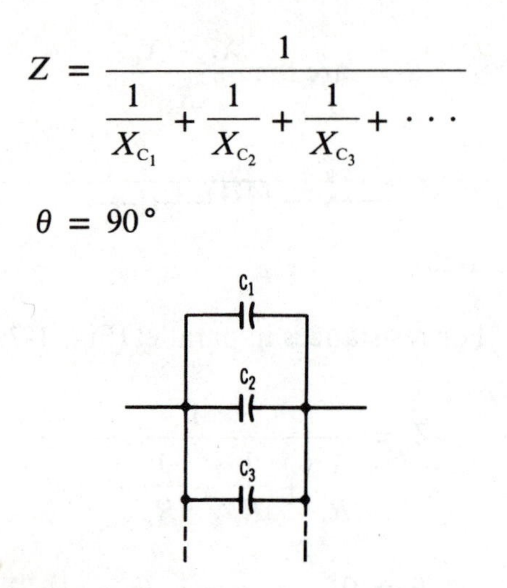

Fig. 1-31

For resistance and inductance in parallel (Fig. 1-32):

$$Z = \frac{RX_L}{\sqrt{R^2 + X_L^2}}$$

$$\theta = \arctan \frac{R}{X_L}$$

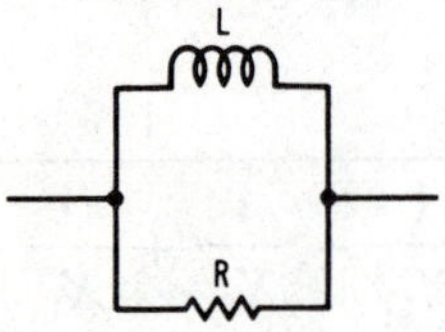

Fig. 1-32

For capacitance and resistance in parallel (Fig. 1-33):

$$Z = \frac{RX_C}{\sqrt{R^2 + X_C^2}}$$

$$\theta = \arctan \frac{R}{X_C}$$

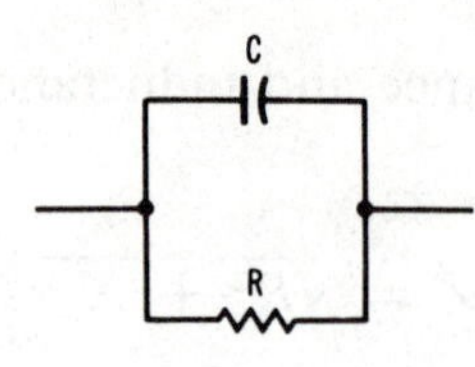

Fig. 1-33

The graphical solution for capacitance and resistance in series or in parallel (Fig. 1-34):

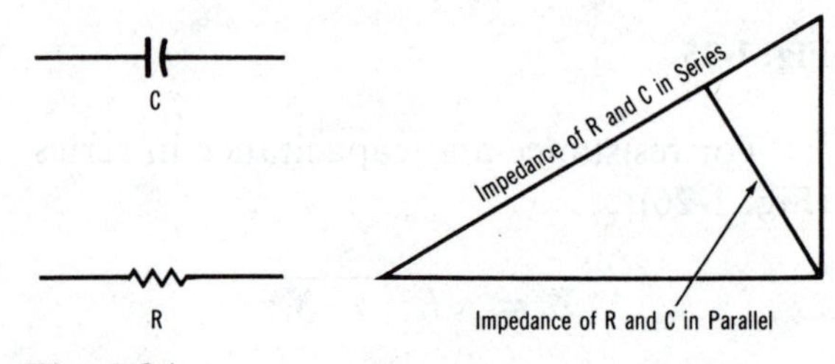

Fig. 1-34

For capacitance and inductance in parallel (Fig. 1-35):

When X_L is larger than X_C:

$$Z = \frac{X_L X_C}{X_L - X_C}$$

When X_C is larger than X_L:

$$Z = \frac{X_C X_L}{X_C - X_L}$$

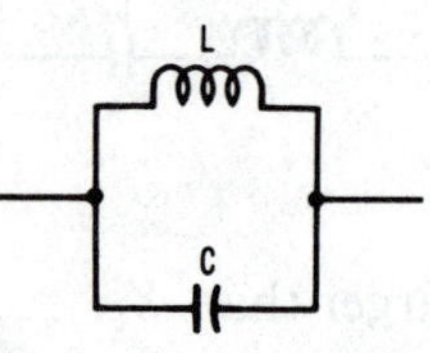

Fig. 1-35

Note. $\theta = 0°$ when $X_L = X_C$.

The graphical solution for resultant reactance of parallel inductive and capacitive reactances (Figs. 1-36A and 1-36B):

When X_L is larger than X_C:

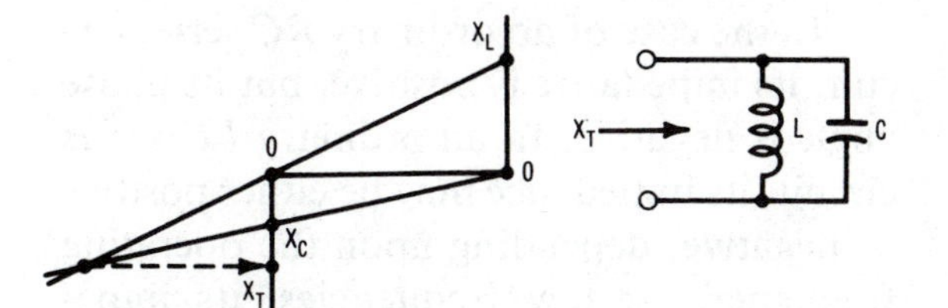

Fig. 1-36A

When X_C is larger than X_L:

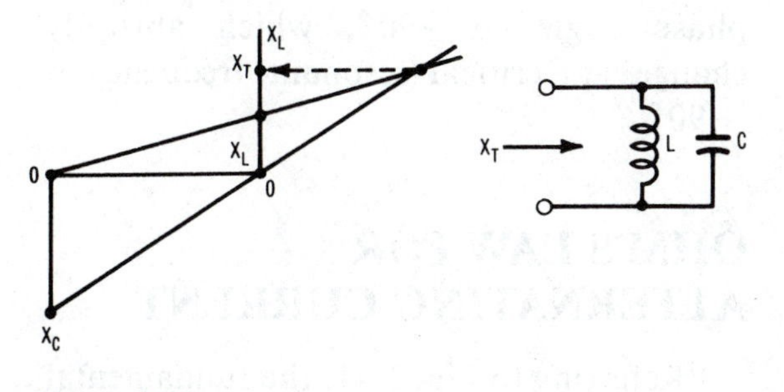

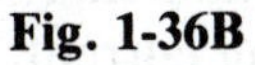

Fig. 1-36B

Note. In Figs. 1-36A and 1-36B, the base line 0–0 may have any finite length. The input impedance of any network can be represented at a given frequency by *R* and *C* connected in series or by *R* and *L* connected in series. Or the input impedance can be represented at a given frequency by *R* and *C* connected in parallel or by *R* and *L* connected in parallel. Conversely, the output impedance of any network can be similarly represented at a given frequency.

For inductance, capacitance, and resistance in parallel (Fig. 1-37):

$$Z = \frac{RX_LX_C}{\sqrt{X_L^2X_C^2 + R^2(X_L - X_C)^2}}$$

$$\theta = \arctan \frac{R(X_L - X_C)}{X_LX_C}$$

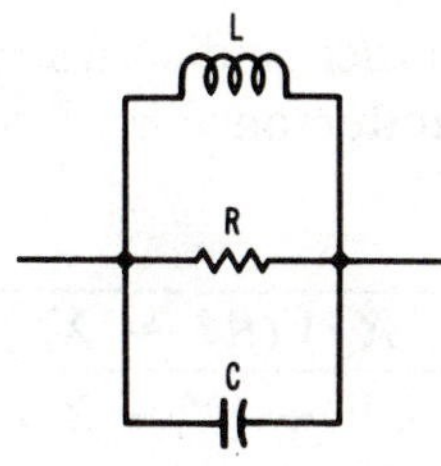

Fig. 1-37

For inductance and series resistance in parallel with resistance (Fig. 1-38):

$$Z = R_2\sqrt{\frac{R_1^2 + X_L^2}{(R_1 + R_2)^2 + X_L^2}}$$

$$\theta = \arctan \frac{X_LR_2}{R_1^2 + X_L^2 + R_1R_2}$$

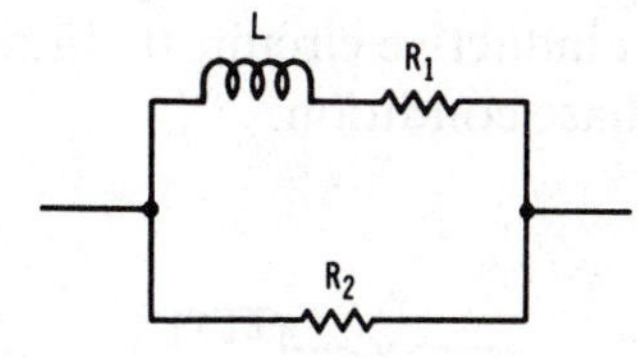

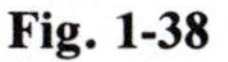

Fig. 1-38

For inductance and series resistance in parallel with capacitance (Fig. 1-39):

$$Z = X_C\sqrt{\frac{R^2 + X_L^2}{R^2 + (X_L - X_C)^2}}$$

$$\theta = \arctan \frac{X_L(X_C - X_L) - R^2}{RX_C}$$

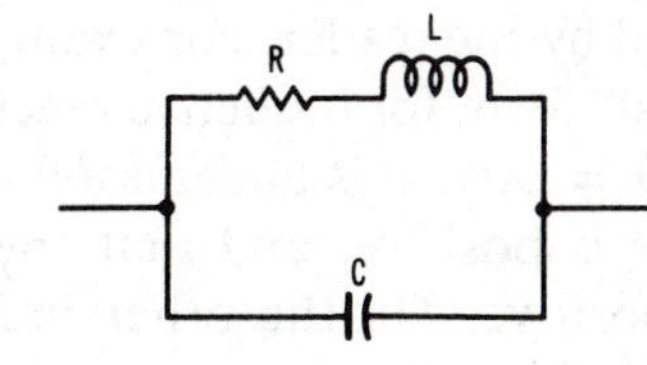

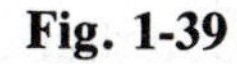

Fig. 1-39

For capacitance and series resistance in parallel with inductance and series resistance (Fig. 1-40):

$$Z = \sqrt{\frac{(R_1^2 + X_L^2)(R_2^2 + X_C^2)}{(R_1 + R_2)^2 + (X_L - X_C)^2}}$$

$$\theta = \text{arc tan} \frac{X_L(R_2^2 + X_C^2) - X_C(R_1^2 + X_L^2)}{R_1(R_2^2 + X_C^2) + R_2(R_1^2 + X_L^2)}$$

where

Z is the impedance, in ohms,
R is the resistance, in ohms,
L is the inductance, in henrys,
X_L is the inductive reactance, in ohms,
X_C is the capacitive reactance, in ohms,
θ is the phase angle, in degrees, by which the current leads the voltage in a capacitive circuit or lags the voltage in an inductive circuit. 0° indicates an in-phase condition.

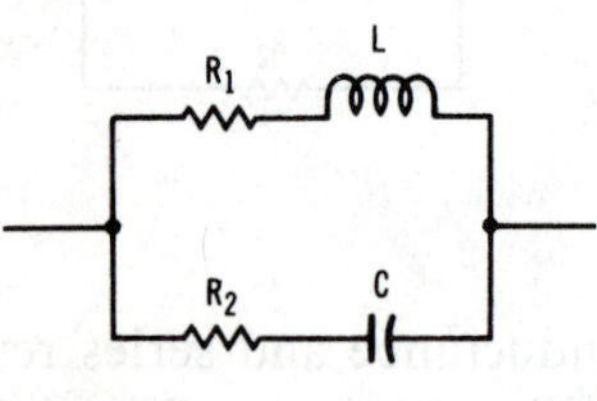

Fig. 1-40

The formulas in this section are written in "shorthand" form, wherein the signs of quantities and absolute values of quantities are implied rather than expressed. In turn, when the formulas are applied to a circuit-action problem, the appropriate signs must be supplied by the reader. For example, the "shorthand" form for inductive reactance is $Z = X_L, \theta = 90°$. It is understood that the impedance is positive, and that the phase angle is positive. On the other hand, the "shorthand" form for capacitive reactance is $Z = X_C, \theta = 90°$, and it is understood that the impedance is negative and that the phase angle is negative. In other words, these formulas state absolute values wherein signs are disregarded.

In the case of an ordinary *RC* series circuit, its impedance is positive, but its phase angle is negative. In an ordinary *LC* series circuit, its impedance may be either positive or negative, depending upon the operating frequency. At low frequencies, its impedance is negative (the circuit is capacitive). However, at high frequencies, its impedance is positive (the circuit is inductive). In turn, at low frequencies, an *LC* series circuit has a phase angle of −90°, which abruptly changes at a critical (resonant) frequency to +90°.

OHM'S LAW FOR ALTERNATING CURRENT

Referring to Fig. 1-41, the fundamental Ohm's law formulas for alternating current are given by:

$$E = IZ$$

$$I = \frac{E}{Z}$$

$$Z = \frac{E}{I}$$

where

E is the voltage, in volts,
I is the current, in amperes,
Z is the impedance, in ohms.

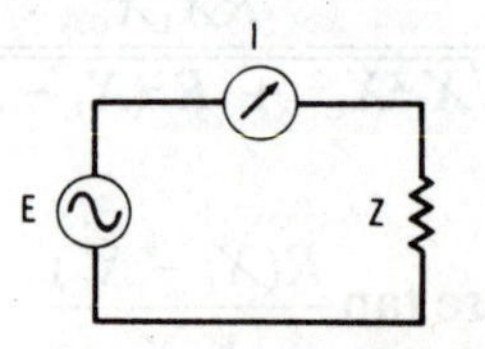

Fig. 1-41

The power expended in an AC circuit is calculated by the formula:

$$P = EI \cos \theta$$

where
- P is the power, in watts,
- E is the voltage, in volts,
- I is the current, in amperes,
- θ is the phase angle, in degrees.

The phase angle is the difference in degrees by which the current leads or lags the voltage in a reactive circuit. In a series circuit, the phase angle is determined by the formula:

$$\theta = \arctan \frac{X}{R}$$

where
- θ is the phase angle, in degrees,
- X is the inductive or capacitive reactance, in ohms,
- R is the nonreactive resistance, in ohms.

Therefore:

For a purely resistive circuit:

$$\theta = 0°$$

$$\cos \theta = 1$$

$$P = EI$$

For a resonant circuit:

$$\theta = 0°$$

$$\cos \theta = 1$$

$$P = EI$$

For a purely reactive circuit:

$$\theta = 90°$$

$$\cos \theta = 0$$

$$P = 0$$

where
- P is the power, in watts,
- E is the voltage, in volts,
- I is the current, in amperes,
- θ is the phase angle, in degrees.

AVERAGE, RMS, PEAK, AND PEAK-TO-PEAK VOLTAGE AND CURRENT

Table 1-1 can be used to convert sinusoidal voltage (or current) values from one method of measurement to another. To use the table, first find the given type of reading in the left-hand column, then find the desired type of reading across the top of the table. To convert the given value to the desired value, multiply the given value by the factor listed under the desired value.

Example. What factor must peak voltage be multiplied by to obtain root mean square (rms) voltage?

Answer. 0.707.

TABLE 1-1
Average, RMS, Peak, and Peak-to-Peak Values

	Multiplying factor to get			
Given value	*Average*	*RMS*	*Peak*	*Peak-to-peak*
average	—	1.11	1.57	3.14
rms	0.9	—	1.414	2.828
peak	0.637	0.707	—	2.0
peak-to-peak	0.32	0.3535	0.5	—

POWER FACTOR

Power factor is the ratio of true power to apparent power in an AC circuit. Thus, using Fig. 1-42:

$$pf = \frac{P_T}{P_A} = \frac{EI \cos\theta}{EI}$$
$$= \cos\theta$$

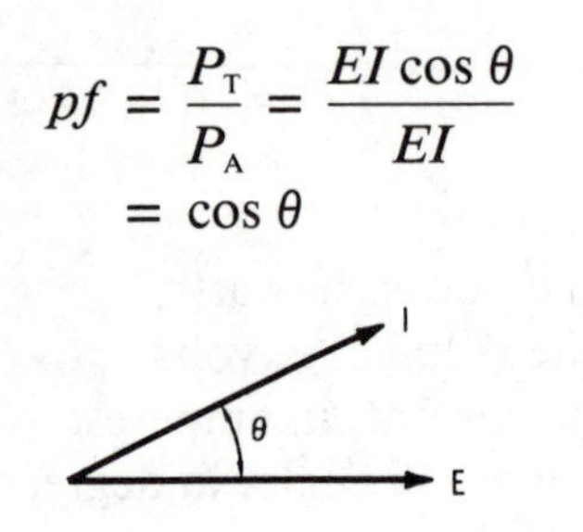

Fig. 1-42

where

pf is the power factor,
P_T is the true power, in watts,
P_A is the apparent power, in volt-amperes,
$EI \cos\theta$ (Fig. 1-42) is the true power, in watts,
EI is the apparent power, in volt-amperes.

Therefore:

For a purely resistive circuit:

$$\theta = 0°$$

$$pf = 1$$

For a resonant circuit:

$$\theta = 0°$$

$$pf = 1$$

For a purely reactive circuit:

$$\theta = 90°$$

$$pf = 0$$

POWER

As shown in Fig. 1-43, the following relationships exist:

True power ($EI \cos\theta$) does work

Apparent power ($EI \sin\theta$) does no work

Reactive power is measured in vars (volt-amperes-reactive)

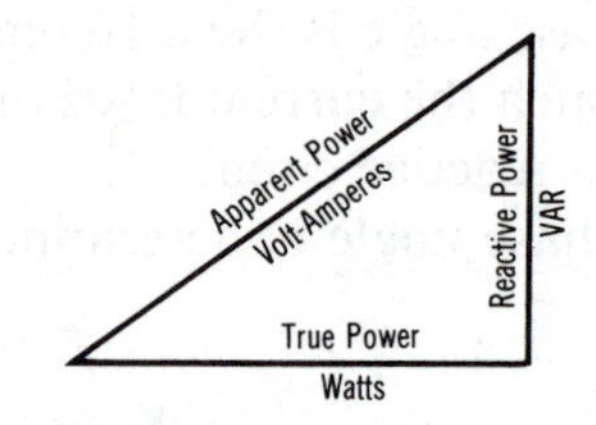

Fig. 1-43. Power triangle.

Power is the rate of doing work (kilowatt, watt, erg). A kilowatt is equal to 1000 W. A watt equals 1 J/s. One joule is equal to the movement of 1 C through a potential difference of 1 V. One coulomb is the quantity of electricity that passes a point in 1 s when the current is 1 A and is equal to the passage of 6.281×10^{18} electrons. Energy is equal to work; thus, energy is measured in such units as kilowatt-hours and watt-hours.

Work is the product of force times the distance through which the force has been applied. For example, a horsepower is equal to 746 W and represents the application of 550 lbf through 1 ft each second (or, raising 33,000 lb a distance of 1 ft in 1 min).

Potential energy is energy of position, as when a weight has been raised above ground level. Kinetic energy is the energy of motion, as when a weight falls to the ground. Voltage is basically a unit of work per unit charge; it is basically potential energy.

Book Mark

Sams books cover a wide range of technical topics. We are always interested in hearing from our readers regarding their informational needs. Please complete this questionnaire and return it to us with your suggestions. We appreciate your comments.

1. Which brand and model of computer do you use?

☐ Apple ________
☐ Commodore ________
☐ IBM ________
☐ Other (please specify) ________

2. Where do you use your computer?

☐ Home ☐ Work

3. Are you planning to buy a new compüter?

☐ Yes ☐ No

If yes, what brand are you planning to buy? ________

4. Please specify the brand/ type of software, operating systems or languages you use.

☐ Word Processing ________
☐ Spreadsheets ________
☐ Data Base Management ________
☐ Integrated Software ________
☐ Operating Systems ________
☐ Computer Languages ________

5. Are you interested in any of the following electronics or technical topics?

☐ Amateur radio
☐ Antennas and propagation
☐ Artificial intelligence/ expert systems
☐ Audio
☐ Data communications/ telecommunications
☐ Electronic projects
☐ Instrumentation and measurements
☐ Lasers
☐ Power engineering
☐ Robotics
☐ Satellite receivers

6. Are you interested in servicing and repair of any of the following (please specify)?

☐ VCRs ________
☐ Compact disc players ________
☐ Microwave ovens ________
☐ Television ________
☐ Computers ________
☐ Automotive electronics ________
☐ Mobile telephones ________
☐ Other ________

7. How many computer or electronics books did you buy in the last year?

☐ One or two ☐ Three or four
☐ Five or six ☐ More than six

8. What is the average price you paid per book?

☐ Less than $10 ☐ $10-$15
☐ $16-$20 ☐ $21-$25 ☐ $26+

9. What is your occupation?

☐ Manager
☐ Engineer
☐ Technician
☐ Programmer/analyst
☐ Student
☐ Other ________

10. Please specify your educational level.

☐ High school
☐ Technical school
☐ College graduate
☐ Postgraduate

11. Are there specific books you would like to see us publish? ________

Comments ________

Name ________
Address ________
City ________
State/Zip ________

22469

SAMS
Book Mark
Book Mark

SAMS

Generally, we speak of potential difference instead of potential-energy difference. When a potential difference is applied across a load resistor, potential energy is transformed into heat energy. Energy is transformed only when the power factor is unity; if the power factor is zero, potential energy merely surges back and forth and does no real work, as when an AC voltage is applied across an ideal capacitor.

Electromotive force (emf) is source voltage and is measured in volts. The term "electromotive force" is somewhat of a misnomer inasmuch as voltage and, consequently, emf are potential energy.

TIME CONSTANTS

A certain amount of time is required after a DC voltage has been applied to an *RC* or *RL* circuit and before the capacitor can charge or the current can build up to a portion of the full value. This time is called the time constant of the circuit. However, the time constant is not the time required for the voltage or current to reach the full value; instead, it is the time required to reach 63.2% of full value. During the next time constant, the capacitor is charged or the current builds up to 63.2% of the remaining difference, which is 86.5% of the full value. Table 1-2 gives the percent of full charge on a capacitor (or current buildup in an inductance after each time constant). Theoretically, neither the charge on the capacitor nor the current through the coil can ever reach 100%. However, it is usually considered to be 100% after five time constants.

TABLE 1-2
Time Constants Versus Percent of Voltage or Current

No. of time constants	Percent charge or buildup	Percent discharge or decay
1	63.2	36.8
2	86.5	13.5
3	95.0	5.0
4	98.2	1.8
5	99.3	0.7

Likewise, when the voltage source is removed, the capacitor will discharge or the current will decay 63.2%, which is 36.8% of full value during the first time constant. Table 1-2 also gives the percent of full voltage after each time constant for discharge of a capacitor or decay of the current through a coil.

The time per time constant is calculated as follows.

For an *RC* circuit (Fig. 1-44):

$$T = RC$$

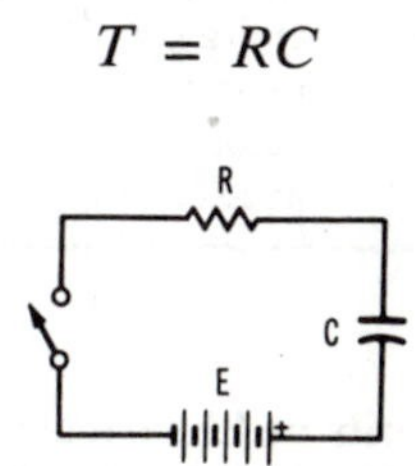

Fig. 1-44

For an *RL* circuit (Fig. 1-45):

$$T = \frac{L}{R}$$

where

T is the time, in seconds,
R is the resistance, in ohms,
C is the capacitance, in farads,
L is the inductance, in henrys.

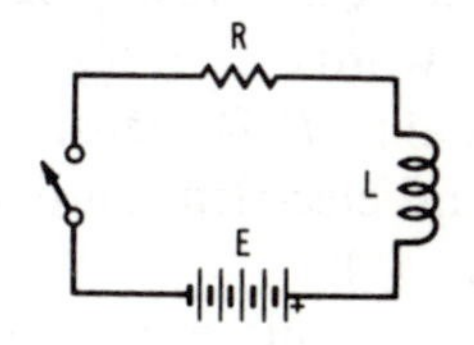

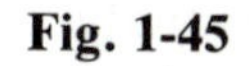
Fig. 1-45

In addition, the values can also be expressed by the following relationships:

T	*R*	*C* or *L*
seconds	megohms	microfarads
seconds	megohms	microhenrys
microseconds	ohms	microfarads
microseconds	megohms	picofarads
microseconds	ohms	microhenrys

TABLE 1-3
Dimensional Units of Mechanical Quantities

Symbol	Physical unit	Dimensional unit
F	force	F
L	length	L
t	time	T
M	mass	FT^2L^{-1}
W	energy, work	FL
P	power	FLT^{-1}
v	velocity	LT^{-1}
a	acceleration	LT^{-2}

TABLE 1-4
Dimensional Units of Electrical Quantities

Symbol	Physical unit	Dimensional unit
Q	charge	Q
I	current	QT^{-1}
V	voltage	FLQ^{-1}
R	resistance	$FLTQ^{-2}$
L	inductance	FLT^2Q^{-2}
C	capacitance	$F^{-1}L^{-1}Q^2$
XL	inductive reactance	$jFLTQ^{-2}$
XC	capacitive reactance	$-jFLTQ^{-2}$

Dimensional units show why the product of capacitance and resistance is equal to time. In other words, $F^{-1}L^{-1}Q^2$ multiplied by $FLTQ^{-2}$ is equal to T. Dimensional units for mechanical and electrical units are listed in Tables 1-3 and 1-4.

Dimensional units are used extensively in calculating with formulas and in analyzing circuit action. As a basic example, dimensional units provide a quick check concerning whether an algebraic error has been made. In other words, no matter how the terms in a formula may be transposed or substituted back and forth, the dimensional units must always be the same on either side of the equals sign. A dimensional check of a derived electrical formula is comparable to a check of an addition problem by first adding the columns up and then adding the columns down.

Example. If we write $I = E/R$, then $QT^{-1} = FLQ^{-1}/FLTQ^{-2} = QT^{-1}$. Again, if we write $P = EI = E^2/R$, then $FLT^{-1} = FLQ^{-1}QT^{-1} = F^2L^2Q^{-2}/FLTQ^{-2} = FLT^{-1}$.

Formulas are customarily simplified insofar as possible. In turn, the terms of a formula and the answer that is obtained may require interpretation. For example, an ideal coaxial cable has a certain capacitance per unit length and a certain inductance per unit length. In turn, the formula $R_0 = \sqrt{L/C}$ is used to calculate the characteristic resistance of the cable. This formula provides a resistance in ohms, when L is in henrys and C is in farads. However, the characteristic resistance R_0 is a representational resistance and not a simple physical resistance. A representational resistance dissipates no power, whereas a simple physical resistance dissipates power.

Resistance has the dimensions $FLTQ^{-2}$, inductance has the dimensions FLT^2Q^{-2}, capacitance has the dimensions $F^{-1}L^{-1}Q^2$. Accordingly, $R_0 = \sqrt{(FLT^2Q^{-2})/(F^{-1}L^{-1}Q^2)}$ or $R_0 = \sqrt{F^2L^2T^2Q^{-4}}$, so that $R_0 = FLTQ^{-2}$. Thus, the resistance term is dimensionally correct, and the correct numerical value will be obtained for R_0 when the square root is taken of the L/C ratio. On the other hand, the R_0 value cannot be assumed to be a simple physical resistance; it is a representational resistance (since it cannot dissipate power).

The foregoing interpretation is based on the circumstance that the L/C ratio has the dimensions $F^2L^2T^2Q^{-4}$, which are the dimensions of R^2. It is a fundamental principle of circuit action that whenever two electrical units are multiplied or divided (or squared or rooted), a new electrical unit is obtained. In this practical example, the new electrical unit of representational resistance is obtained. As previously noted, the circuit action of representational resistance is not the same as the circuit action of simple physical resistance, although some of its aspects are similar.

This and related principles of circuit action are summarized by the basic principle that although $Y = 2X = \sqrt{4X^2}$ is a mathematically correct series of relations, each term has a particular interpretation insofar as circuit action is concerned.

TRANSFORMER FORMULAS

In a transformer, the relationships between the number of turns in the primary and secondary, the voltage across each winding, and the current through the windings are expressed by the following equations:

$$\frac{E_p}{E_s} = \frac{N_p}{N_s}$$

and

$$\frac{E_p}{E_s} = \frac{I_s}{I_p}$$

By rearranging these equations and by referring to Fig. 1-46, any unknown can be determined from the following formulas:

$$E_p = \frac{E_s N_p}{N_s} = \frac{E_s I_s}{I_p}$$

$$E_s = \frac{E_p N_s}{N_p} = \frac{E_p I_p}{I_s}$$

$$N_p = \frac{E_p N_s}{E_s} = \frac{N_s I_s}{I_p}$$

$$N_s = \frac{E_s N_p}{E_p} = \frac{N_p I_p}{I_s}$$

$$I_p = \frac{E_s I_s}{E_p} = \frac{N_s I_s}{N_p}$$

$$I_s = \frac{E_p I_p}{E_s} = \frac{N_p I_p}{N_s}$$

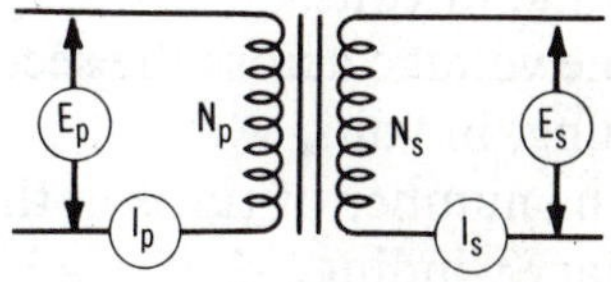

Fig. 1-46

The turns ratio of a transformer is determined by the following formulas:

For a step-up transformer:

$$T = \frac{N_s}{N_p}$$

For a step-down transformer:

$$T = \frac{N_p}{N_s}$$

The impedance ratio of a transformer is determined by:

$$Z = T^2$$

The impedance of an unknown winding is determined by the following:

For a step-up transformer:

$$Z_p = \frac{Z_s}{Z}$$

$$Z_s = Z \times Z_p$$

For a step-down transformer:

$$Z_p = Z \times Z_s$$

$$Z_s = \frac{Z_p}{Z}$$

where
- E_p is the voltage across the primary winding, in volts,
- E_s is the voltage across the secondary winding, in volts,
- N_p is the number of turns in the primary winding,
- N_s is the number of turns in the secondary winding,
- I_p is the current through the primary winding, in amperes,
- I_s is the current through the secondary winding, in amperes,
- T is the turns ratio,
- Z is the impedance ratio,
- Z_p is the impedance of the primary winding, in ohms,
- Z_s is the impedance of the secondary winding, in ohms.

VOLTAGE REGULATION

When a load is connected to a power supply, the output voltage drops because more current flows through the resistive elements of the power supply. Voltage regulation is a measure of how much the voltage drops and is usually expressed as a percentage. It is determined by the following formula:

$$\%R = \frac{E_1 - E_2}{E_2} \times 100$$

where
- $\%R$ is the voltage regulation, in percent,
- E_1 is the no-load voltage, in volts,
- E_2 is the voltage under load, in volts.

DC-METER FORMULAS

The basic instrument for testing current and voltage is the moving-coil meter. The meter can be either a DC milliammeter or a DC microammeter. A series resistor converts the meter to a DC voltmeter, and a parallel resistor converts the meter to a DC ammeter. The resistance of the meter movement is determined first, as follows. Connects a suitable variable resistor R_a and a battery as shown in Fig. 1-47. Adjust resistor R_a until full-scale deflection is obtained. Then connect a variable resistor R_b in parallel with the meter, and adjust R_b until half-scale deflection is obtained. Disconnect R_b and measure its resistance. The measured value is the resistance of the meter movement.

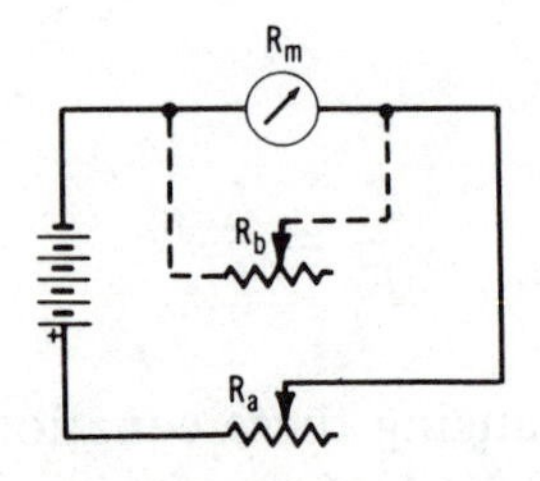

Fig. 1-47

Voltage Multipliers (Fig. 1-48)

$$R = \frac{E_s}{I_s} - R_m$$

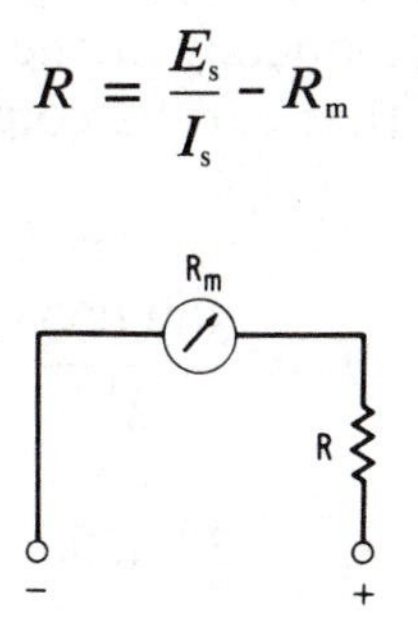

Fig. 1-48

where
R is the multiplier resistance, in ohms,
E_s is the full-scale reading, in volts,
I_s is the full-scale reading, in amperes,
R_m is the meter resistance, in ohms.

Shunt-Type Ohmmeter for Low Resistance (Fig. 1-49)

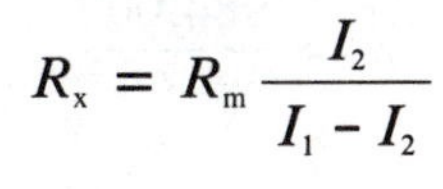

$$R_x = R_m \frac{I_2}{I_1 - I_2}$$

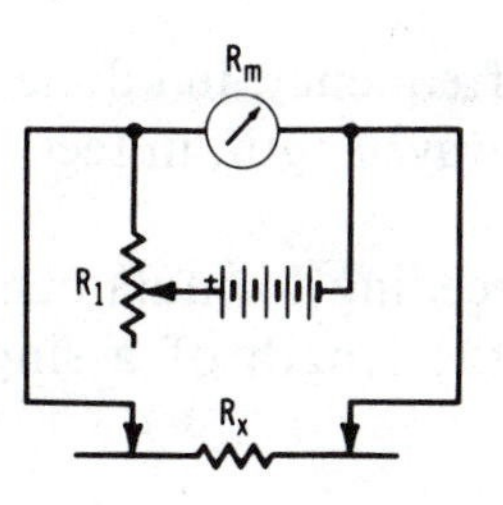

Fig. 1-49

where
R_x is the unknown resistance, in ohms,
R_m is the meter resistance, in ohms,
I_1 is the current reading with probes open, in amperes,
I_2 is the current reading with probes connected across unknown resistor, in amperes.

R_1 in Fig. 1-49 is a variable resistance for current limiting to keep meter adjusted for full-scale reading with probes open.

Series-Type Ohmmeter for High Resistance (Fig. 1-50)

$$R_x = (R_1 + R_m)\frac{I_1 - I_2}{I_2}$$

where
R_x is the unknown resistance, in ohms,
R_1 is a variable resistance adjusted for full-scale reading with probes shorted together, in ohms,
R_m is the meter resistance, in ohms,
I_1 is the current reading with probes shorted, in amperes,
I_2 is the current reading with unknown resistor connected, in amperes.

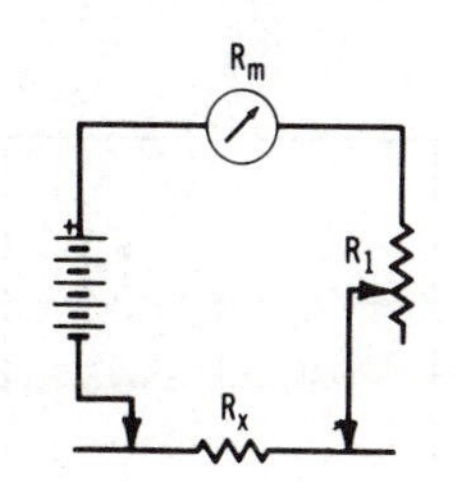

Fig. 1-50

Ammeter Shunts (Fig. 1-51)

$$R = \frac{R_m}{N - 1} = \frac{I_m R_m}{I_s}$$

where
R is the resistance of the shunt, in ohms,
R_m is the meter resistance, in ohms,
N is the scale multiplication factor,
I_m is the meter current, in amperes,
I_s is the shunt current, in amperes.

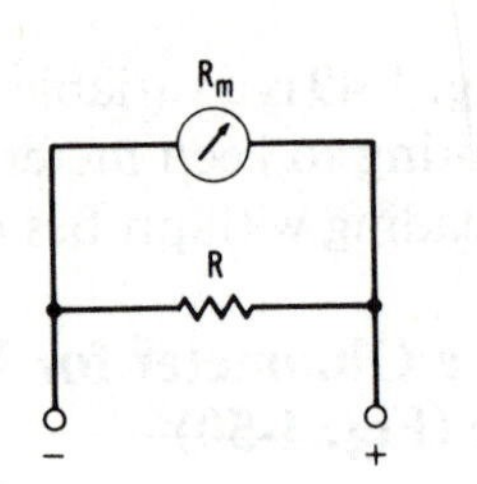

Fig. 1-51

Ammeter With Multirange Shunt (Fig. 1-52)

$$R_2 = \frac{(R_1 + R_2) + R_m}{N}$$

where

R_2 is the intermediate value, in ohms,
$R_1 + R_2$ is the total shunt resistance for lowest full-scale reading, in ohms,
R_m is the meter resistance, in ohms,
N is the scale multiplication factor.

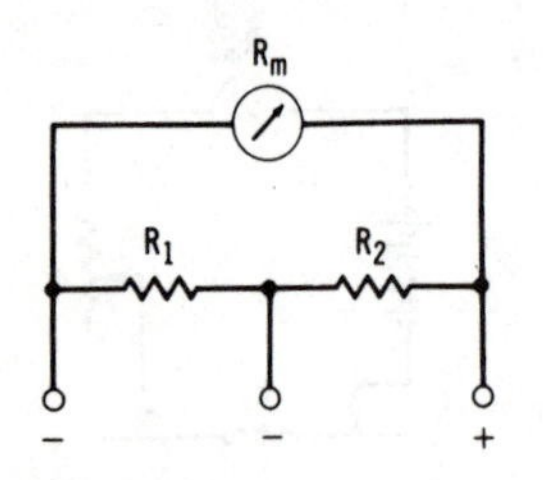

Fig. 1-52

FREQUENCY AND WAVELENGTH

Formulas

Since frequency is defined as the number of complete hertz and since all radio waves travel at a constant speed, it follows that a complete cycle occupies a given distance in space. The distance between two corresponding parts of two waves (the two positive or negative crests or the points where the two waves cross the zero axis in a given direction) constitutes the wavelength. If either the frequency or the wavelength is known, the other can be computed as follows:

$$f = \frac{300{,}000}{\lambda}$$

$$\lambda = \frac{300{,}000}{f}$$

where

f is the frequency, in kilohertz,
λ is the wavelength, in meters.

To calculate wavelength in feet, the following formulas should be used:

$$f = \frac{984{,}000}{\lambda}$$

$$\lambda = \frac{984{,}000}{f}$$

where

f is the frequency, in kilohertz,
λ is the wavelength, in feet.

The preceding formula can be used to determine the length of a single-wire antenna.

For a half-wave antenna:

$$L = \frac{492}{f}$$

For a quarter-wave antenna:

$$L = \frac{246}{f}$$

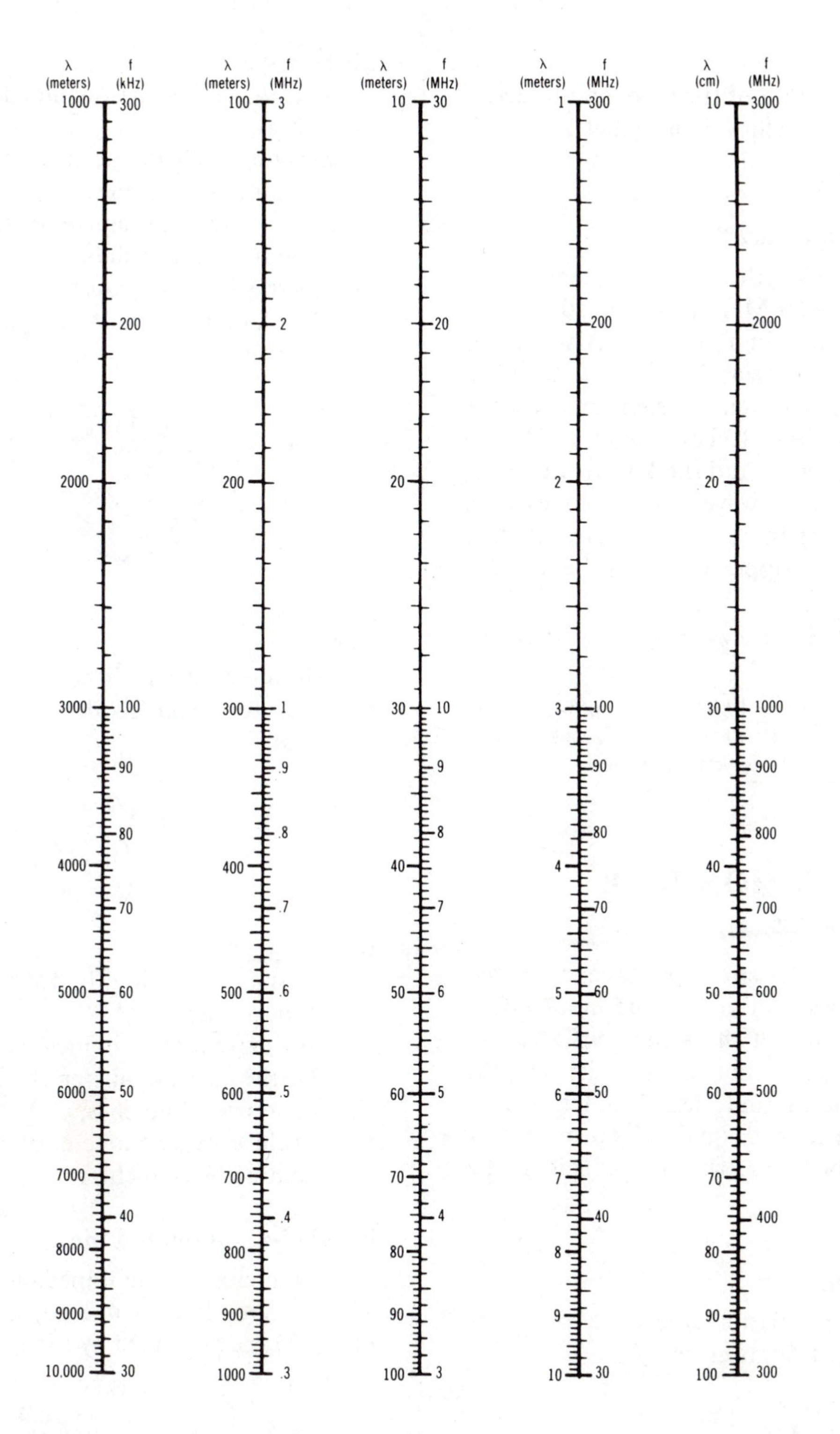

Fig. 1-53. Frequency-wavelength conversion chart.

where
L is the length of the antenna, in feet,
f is the frequency, in megahertz.

Conversion Chart

The wavelength of any frequency from 30 kHz to 3000 MHz can be read directly from the chart in Fig. 1-53. Also, if the wavelength is known, the corresponding frequency can be obtained from the chart for wavelengths from 10 cm to 1000 m. To use the chart, merely find the known value (either frequency or wavelength) on one of the scales, and then read the corresponding value from the opposite side of the scale.

Example. What is the wavelength of a 4-MHz signal?

Answer. 75 m. (Find 4 MHz on the third scale from the left. Opposite 4 MHz on the frequency scale find 75 m on the wavelength scale.)

TRANSMISSION-LINE FORMULAS

The characteristic impedance of a transmission line is defined as the input impedance of a line of the same configuration and dimensions but of infinite length. When a line of finite length is terminated with an impedance equal to its own characteristic impedance, the line is said to be matched.

Coaxial Line

The characteristic impedance of a coaxial line (Fig. 1-54) is given by:

$$Z_0 = \frac{138}{\sqrt{k}} \log \frac{D}{d}$$

where
Z_0 is the characteristic impedance, in ohms,
D is the inside diameter of the outer conductor, in inches,
d is the outside diameter of the inner conductor, in inches,
k is the dielectric constant of the insulating material* (k equals 1 for dry air).

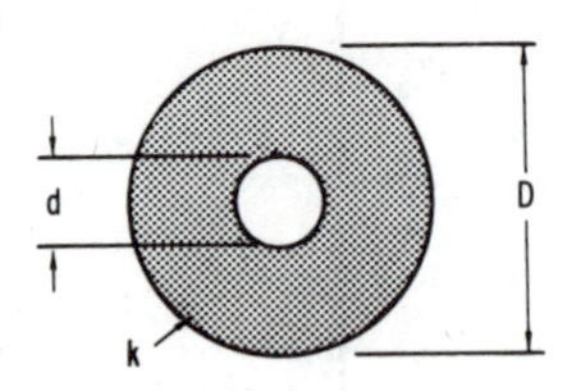

Fig. 1-54

The attenuation of a coaxial line in decibels per foot can be determined by the formula:

$$a = \frac{4.6\sqrt{f}\,(D + d)}{D \times d \left(\log \frac{D}{d}\right)} \times 10^{-6}$$

where
a is the attenuation, in decibels per foot of line,
f is the frequency, in megahertz,
D is the inside diameter of the outer conductor, in inches,
d is the outside diameter of the inner conductor, in inches.

Parallel-Conductor Line

The characteristic impedance of parallel-conductor line (twin-lead) as shown in Fig. 1-55 is determined by the formula:

$$Z_0 = \frac{276}{\sqrt{k}} \log \frac{2D}{d}$$

where

Z_0 is the characteristic impedance, in ohms,
D is the center-to-center distance between conductors, in inches,
d is the diameter of the conductors, in inches,
k is the dielectric constant of the insulating material between conductors* (k equals 1 for dry air).

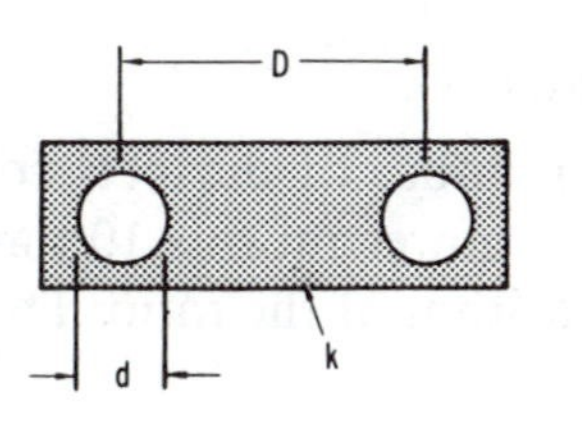

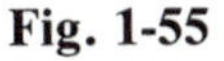
Fig. 1-55

MODULATION FORMULAS

Amplitude Modulation

The amount of modulation of an amplitude-modulated carrier shown in Fig. 1-56 is referred to as the percentage of modulation. It can be determined by the following formulas:

$$\%M = \frac{E_C - E_T}{2E_{av}} \times 100$$

or

$$\%M = \frac{E_C - E_T}{E_C + E_T} \times 100$$

where

$\%M$ is the percentage of modulation,
E_C is the amplitude of the crest of the modulated carrier,
E_T is the amplitude of the trough of the modulated carrier,
E_{av} is the average amplitude of the modulated carrier.

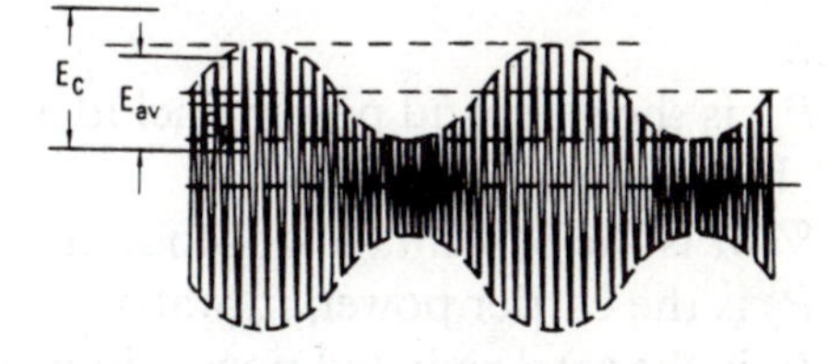

Fig. 1-56

Also, the percentage of modulation can be determined by applying the modulated carrier wave to the vertical plates and the modulating voltage wave to the horizontal plates of an oscilloscope. This produces a trapezoidal wave, as shown in Fig. 1-57. The dimensions A and B are proportional to the crest and trough amplitudes, respectively. The percentage of modulation can be determined by measuring the height of A and B and using the formula:

$$\%M = \frac{A - B}{A + B} \times 100$$

where

$\%M$ is the percentage of modulation,
A and B are the dimensions measured in Fig. 1-57.

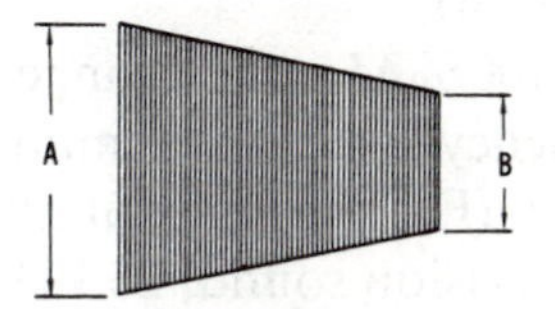

Fig. 1-57

The sideband power of an amplitude-modulated carrier is determined by:

$$P_{SB} = \frac{\%M^2}{2} \times P_C$$

*For a list of dielectric constants of materials, see Table 2-1.

The total radiated power is the sum of the carrier and the radiated powers:

$$P_T = P_{SB} + P_C$$

where
P_{SB} is the sideband power (includes both sidebands), in watts,
$\%M$ is the percentage of modulation,
P_C is the carrier power, in watts,
P_T is the total radiated power, in watts.

Note. The carrier power does not change with modulation.

Frequency Modulation

In a frequency-modulated carrier, the amount the carrier frequency changes is determined by the amplitude of the modulating signal, and the number of times the changes occur per second is determined by the frequency of the modulating signal.

The percentage of modulation of a frequency-modulated carrier can be computed from:

$$\%M = \frac{\Delta f}{\Delta f \text{ for } 100\%M} \times 100$$

where
$\%M$ is the percentage of modulation,
Δf is the change in frequency (or the deviation),
Δf for $100\%M$ is the change in frequency for a 100% modulated carrier. (For commercial fm, 75 kHz; for television sound, 25 kHz; for two-way radio, 15 kHz.)

The modulation index of a frequency-modulated carrier is determined by:

$$M = \frac{f_d}{f_a}$$

where
M is the modulation index,
f_d is the deviation in frequency, in hertz,
f_a is the modulating audio frequency, in the same units as f_d.

DECIBELS AND VOLUME UNITS

Equations

The number of decibels corresponding to a given power ratio is 10 times the common logarithm of the ratio. Thus:

$$\text{dB} = 10 \log \frac{P_2}{P_1}$$

where
P_1 and P_2 are the individual power readings, in watts.

The number of decibels corresponding to a given voltage or current ratio is 20 times the common logarithm of the ratio. Thus, when the impedances across which the signals are being measured are equal, the equations are:

$$\text{dB} = 20 \log \frac{E_2}{E_1}$$

$$\text{dB} = 20 \log \frac{I_2}{I_1}$$

where
E_1 and E_2 are the individual voltage readings, in volts,
I_1 and I_2 are the individual current readings, in amperes.

If the impedances across which the signals are measured are not equal, the equations become:

$$dB = 20 \log \frac{E_2\sqrt{Z_1}}{E_1\sqrt{Z_2}}$$

$$dB = 20 \log \frac{I_2\sqrt{Z_2}}{I_1\sqrt{Z_1}}$$

where

E_1 and E_2 are the individual voltage readings, in volts,
I_1 and I_2 are the individual current readings, in amperes,
Z_1 and Z_2 are the individual impedances across which the signals were read, in ohms.

Reference Levels

The decibel is not an absolute value; it is a means of stating the ratio of a level to a certain reference level. Usually, when no reference level is given, it is 6 mV across a 500-Ω impedance. However, the reference level should be stated whenever a value in decibels is given. Other units, which do have specific reference levels, have been established. Some of the more common are:

dBj 1 mV

dBk 1 kW

dBm mW, 600 Ω

dBs Japanese designation for dBm system

dBv 1 V (no longer in use)

dBw 1 W

dBvg voltage gain

dBrap decibels above a reference acoustical power of 10^{-16} W

VU 1 mW, 600 Ω (complex waveforms varying in both amplitude and frequency)

Decibel Table

Tables 1-5 and 1-6 are decibel tables that list most of the power, current, and voltage ratios commonly encountered, with their decibel values. Figure 1-58 shows the relationship between power and voltage or current. If a decibel value is not listed in Tables 1-5 and 1-6, first subtract one of the given values from the unlisted value (select a value so the remainder will also be listed). Then multiply the ratios given in the chart for each value. To covert a ratio not given in the tables to a decibel value, first factor the ratio so that each factor will be a listed value; then find the decibel equivalents for each factor and add them.

TABLE 1-5
Decibel Table (0–1.9 dB)*

	Power ratio		Current or voltage ratio	
dB	*Gain*	*Loss*	*Gain*	*Loss*
0	1.000	1.0000	1.000	1.0000
0.1	1.023	0.9772	1.012	0.9886
0.2	1.047	0.9550	1.023	0.9772
0.3	1.072	0.9333	1.035	0.9661
0.4	1.096	0.9120	1.047	0.9550
0.5	1.122	0.8913	1.059	0.9441
0.6	1.148	0.8710	1.072	0.9333
0.7	1.175	0.8511	1.084	0.9226
0.8	1.202	0.8318	1.096	0.9120
0.9	1.230	0.8128	1.109	0.9016
1.0	1.259	0.7943	1.122	0.8913
1.1	1.288	0.7762	1.135	0.8810
1.2	1.318	0.7586	1.148	0.8710
1.3	1.349	0.7413	1.161	0.8610
1.4	1.380	0.7244	1.175	0.8511
1.5	1.413	0.7079	1.189	0.8414
1.6	1.445	0.6918	1.202	0.8318
1.7	1.479	0.6761	1.216	0.8222
1.8	1.514	0.6607	1.230	0.8128
1.9	1.549	0.6457	1.245	0.8035

TABLE 1-5 Cont.
Decibel Table (2.0-10.9)

	Power ratio		Current or voltage ratio	
dB	*Gain*	*Loss*	*Gain*	*Loss*
2.0	1.585	0.6310	1.259	0.7943
2.1	1.622	0.6166	1.274	0.7852
2.2	1.660	0.6026	1.288	0.7762
2.3	1.698	0.5888	1.303	0.7674
2.4	1.738	0.5754	1.318	0.7586
2.5	1.778	0.5623	1.334	0.7499
2.6	1.820	0.5495	1.349	0.7413
2.7	1.862	0.5370	1.365	0.7328
2.8	1.905	0.5248	1.380	0.7244
2.9	1.950	0.5129	1.396	0.7161
3.0	1.995	0.5012	1.413	0.7079
3.1	2.042	0.4898	1.429	0.6998
3.2	2.089	0.4786	1.445	0.6918
3.3	2.138	0.4677	1.462	0.6839
3.4	2.188	0.4571	1.479	0.6761
3.5	2.239	0.4467	1.496	0.6683
3.6	2.291	0.4365	1.514	0.6607
3.7	2.344	0.4266	1.531	0.6531
3.8	2.399	0.4169	1.549	0.6457
3.9	2.455	0.4074	1.567	0.6383
4.0	2.512	0.3981	1.585	0.6310
4.1	2.570	0.3890	1.603	0.6237
4.2	2.630	0.3802	1.622	0.6166
4.3	2.692	0.3715	1.641	0.6095
4.4	2.754	0.3631	1.660	0.6026
4.5	2.818	0.3548	1.679	0.5957
4.6	2.884	0.3467	1.698	0.5888
4.7	2.951	0.3388	1.718	0.5821
4.8	3.020	0.3311	1.738	0.5754
4.9	3.090	0.3236	1.758	0.5689
5.0	3.162	0.3162	1.778	0.5623
5.1	3.236	0.3090	1.799	0.5559
5.2	3.311	0.3020	1.820	0.5495
5.3	3.388	0.2951	1.841	0.5433
5.4	3.467	0.2884	1.862	0.5370
5.5	3.548	0.2818	1.884	0.5309
5.6	3.631	0.2754	1.905	0.5248
5.7	3.715	0.2692	1.928	0.5188
5.8	3.802	0.2630	1.950	0.5129
5.9	3.890	0.2570	1.972	0.5070
6.0	3.981	0.2512	1.995	0.5012
6.1	4.074	0.2455	2.018	0.4955
6.2	4.169	0.2399	2.042	0.4898
6.3	4.266	0.2344	2.065	0.4842
6.4	4.365	0.2291	2.089	0.4786
6.5	4.467	0.2239	2.113	0.4732
6.6	4.571	0.2188	2.138	0.4677
6.7	4.677	0.2138	2.163	0.4624
6.8	4.786	0.2089	2.188	0.4571
6.9	4.898	0.2042	2.213	0.4519
7.0	5.012	0.1995	2.239	0.4467
7.1	5.129	0.1950	2.265	0.4416
7.2	5.248	0.1905	2.291	0.4365
7.3	5.370	0.1862	2.317	0.4315
7.4	5.495	0.1820	2.344	0.4266
7.5	5.623	0.1778	2.371	0.4217
7.6	5.754	0.1738	2.399	0.4169
7.7	5.888	0.1698	2.427	0.4121
7.8	6.026	0.1660	2.455	0.4074
7.9	6.166	0.1622	2.483	0.4027
8.0	6.310	0.1585	2.512	0.3981
8.1	6.457	0.1549	2.541	0.3936
8.2	6.607	0.1514	2.570	0.3890
8.3	6.761	0.1479	2.600	0.3846
8.4	6.918	0.1445	2.630	0.3802
8.5	7.079	0.1413	2.661	0.3758
8.6	7.244	0.1380	2.692	0.3715
8.7	7.413	0.1349	2.723	0.3673
8.8	7.586	0.1318	2.754	0.3631
8.9	7.762	0.1288	2.786	0.3589
9.0	7.943	0.1259	2.818	0.3548
9.1	8.128	0.1230	2.851	0.3508
9.2	8.318	0.1202	2.884	0.3467
9.3	8.511	0.1175	2.917	0.3428
9.4	8.710	0.1148	2.951	0.3388
9.5	8.913	0.1122	2.985	0.3350
9.6	9.120	0.1096	3.020	0.3311
9.7	9.333	0.1072	3.055	.3273
9.8	9.550	0.1047	3.090	0.3236
9.9	9.772	0.1023	3.126	0.3199
10.0	10.000	0.1000	3.162	0.3162
10.1	10.23	0.09772	3.199	0.3126
10.2	10.47	0.09550	3.236	0.3090
10.3	10.72	0.09333	3.273	0.3055
10.4	10.96	0.09120	3.311	0.3020
10.5	11.22	0.08913	3.350	0.2985
10.6	11.48	0.08710	3.388	0.2951
10.7	11.75	0.08511	3.428	0.2917
10.8	12.02	0.08318	3.467	0.2884
10.9	12.30	0.08128	3.508	0.2851

TABLE 1-5 Cont.
Decibel Table (11.0-19.9)

dB	Power ratio		Current or voltage ratio	
	Gain	*Loss*	*Gain*	*Loss*
11.0	12.59	0.07943	3.548	0.2818
11.1	12.88	0.07762	3.589	0.2786
11.2	13.18	0.07586	3.631	0.2754
11.3	13.49	0.07413	3.673	0.2723
11.4	13.80	0.07244	3.715	0.2692
11.5	14.13	0.07079	3.758	0.2661
11.6	14.45	0.06918	3.802	0.2630
11.7	14.79	0.06761	3.846	0.2600
11.8	15.14	0.06607	3.890	0.2570
11.9	15.49	0.06457	3.936	0.2541
12.0	15.85	0.06310	3.981	0.2512
12.1	16.22	0.06166	4.027	0.2483
12.2	16.60	0.06026	4.074	0.2455
12.3	16.98	0.05888	4.121	0.2427
12.4	17.38	0.05754	4.169	0.2399
12.5	17.78	0.05623	4.217	0.2371
12.6	18.20	0.05495	4.266	0.2344
12.7	18.62	0.05370	4.315	0.2317
12.8	19.05	0.05248	4.365	0.2291
12.9	19.50	0.05129	4.416	0.2265
13.0	19.95	0.05012	4.467	0.2239
13.1	20.42	0.04898	4.519	0.2213
13.2	20.89	0.04786	4.571	0.2188
13.3	21.38	0.04677	4.624	0.2163
13.4	21.88	0.04571	4.677	0.2138
13.5	22.39	0.04467	4.732	0.2113
13.6	22.91	0.04365	4.786	0.2089
13.7	23.44	0.04266	4.842	0.2065
13.8	23.99	0.04169	4.898	0.2042
13.9	24.55	0.04074	4.955	0.2018
14.0	25.12	0.03981	5.012	0.1995
14.1	25.70	0.03890	5.070	0.1972
14.2	26.30	0.03802	5.129	0.1950
14.3	26.92	0.03715	5.188	0.1928
14.4	27.54	0.03631	5.248	0.1905
14.5	28.18	0.03548	5.309	0.1884
14.6	28.84	0.03467	5.370	0.1862
14.7	29.51	0.03388	5.433	0.1841
14.8	30.20	0.03311	5.495	0.1820
14.9	30.90	0.03236	5.559	0.1799
15.0	31.62	0.03162	5.623	0.1778
15.1	32.36	0.03090	5.689	0.1758
15.2	35.11	0.03020	5.754	0.1738
15.3	33.88	0.02951	5.821	0.1718
15.4	34.67	0.02884	5.888	0.1698
15.5	35.48	0.02818	5.957	0.1679
15.6	36.31	0.02754	6.026	0.1660
15.7	37.15	0.02692	6.095	0.1641
15.8	38.02	0.02630	6.166	0.1622
15.9	38.90	0.02570	6.237	0.1603
16.0	39.81	0.02512	6.310	0.1585
16.1	40.74	0.02455	6.383	0.1567
16.2	41.69	0.02399	6.457	0.1549
16.3	42.66	0.02344	6.531	0.1531
16.4	43.65	0.02291	6.607	0.1514
16.5	44.67	0.02239	6.683	0.1496
16.6	45.71	0.02188	6.761	0.1479
16.7	46.77	0.02138	6.839	0.1462
16.8	47.86	0.02089	6.918	0.1445
16.9	48.98	0.02042	6.998	0.1429
17.0	50.12	0.01995	7.079	0.1413
17.1	51.29	0.01950	7.161	0.1396
17.2	52.48	0.01905	7.244	0.1380
17.3	53.70	0.01862	7.328	0.1365
17.4	54.95	0.01820	7.413	0.1349
17.5	56.23	0.01778	7.499	0.1334
17.6	57.54	0.01738	7.586	0.1318
17.7	58.88	0.01698	7.674	0.1303
17.8	60.26	0.01660	7.762	0.1288
17.9	61.66	0.01622	7.852	0.1274
18.0	63.10	0.01585	7.943	0.1259
18.1	64.57	0.01549	8.035	0.1245
18.2	66.07	0.01514	8.128	0.1230
18.3	67.61	0.01479	8.222	0.1216
18.4	69.18	0.01445	8.318	0.1202
18.5	70.79	0.01413	8.414	0.1189
18.6	72.44	0.01380	8.511	0.1175
18.7	74.13	0.01349	8.610	0.1161
18.8	75.86	0.01318	8.710	0.1148
18.9	77.62	0.01288	8.810	0.1135
19.0	79.43	0.01259	8.913	0.1122
19.1	81.28	0.01230	9.016	0.1109
19.2	83.18	0.01202	9.120	0.1096
19.3	85.11	0.01175	9.226	0.1084
19.4	87.10	0.01148	9.333	0.1072
19.5	89.13	0.01122	9.441	0.1059
19.6	91.20	0.01096	9.550	0.1047
19.7	93.33	0.01072	9.661	0.1035
19.8	95.50	0.01047	9.772	0.1023
19.9	97.72	0.01023	9.886	0.1012

*For values from 20 to 100 dB, see Table 1-6.

TABLE 1-6
Decibel Table (20–100 dB)*

<table>
<tr><th rowspan="2">dB</th><th colspan="2">Power ratio</th><th colspan="2">Current or voltage ratio</th></tr>
<tr><th>Gain</th><th>Loss</th><th>Gain</th><th>Loss</th></tr>
<tr><td>20</td><td>10^{2}
Use the same numbers as 0–10 dB, but shift point one step to the right.</td><td>10^{-2}
Use the same numbers as 0–10 dB, but shift point one step to the left.</td><td rowspan="2">10.00
Use the same numbers as 0–20 dB, but shift point one step to the right.</td><td rowspan="2">0.1000
Use the same numbers as 0–20 dB, but shift point one step to the left.</td></tr>
<tr><td>30</td><td>10^{3}
Use the same numbers as 0–10 dB, but shift point one step to the right.</td><td>10^{-3}
Use the same numbers as 0–10 dB, but shift point one step to the left.</td></tr>
<tr><td>40</td><td>10^{4}
Use the same numbers as 0–10 dB, but shift point two steps to the right.</td><td>10^{-4}
Use the same numbers as 0–10 dB, but shift point two steps to the left.</td><td rowspan="2">100
Use the same numbers as 0–20 dB, but shift point two steps to the right.</td><td rowspan="2">0.01
Use the same numbers as 0–20 dB, but shift point two steps to the left.</td></tr>
<tr><td>50</td><td>10^{5}
Use the same numbers as 0–10 dB, but shift point two steps to the right.</td><td>10^{-5}
Use the same numbers as 0–10 dB, but shift point two steps to the left.</td></tr>
<tr><td>60</td><td>10^{6}
Use the same numbers as 0–10 dB, but shift point three steps to the right.</td><td>10^{-6}
Use the same numbers as 0–10 dB, but shift point three steps to the left.</td><td rowspan="2">1000
Use the same numbers as 0–20 dB, but shift point three steps to the right.</td><td rowspan="2">0.001
Use the same numbers as 0–20 dB, but shift point three steps to the left.</td></tr>
<tr><td>70</td><td>10^{7}
Use the same numbers as 0–10 dB, but shift point three steps to the right.</td><td>10^{-7}
Use the same numbers as 0–10 dB, but shift point three steps to the left.</td></tr>
<tr><td>80</td><td>10^{8}
Use the same numbers as 0–10 dB, but shift point four steps to the right.</td><td>10^{-8}
Use the same numbers as 0–10 dB, but shift point four steps to the left.</td><td rowspan="2">10,000
Use the same numbers as 0–20 dB, but shift point four steps to the right.</td><td rowspan="2">0.0001
Use the same numbers as 0–20 dB, but shift point four steps to the left.</td></tr>
<tr><td>90</td><td>10^{9}
Use the same numbers as 0–10 dB, but shift point four steps to the right.</td><td>10^{-9}
Use the same numbers as 0–10 dB, but shift point four steps to the left.</td></tr>
<tr><td>100</td><td>10^{10}
Use the same numbers as 0–10 dB, but shift point five steps to the right.</td><td>10^{-10}
Use the same numbers as 0–10 dB, but shift point five steps to the left.</td><td>100,000
Use the same numbers as 0–20 dB, but shift point five steps to the right.</td><td>0.00001
Use the same numbers as 0–20 dB, but shift point five steps to the left.</td></tr>
</table>

*For values from 0 to 19.9 dB, see Table 1-5.

Example 1. Find the decibel equivalent of a power ratio of 0.631.

Answer. 2-dB loss.

Example 2. Find the current ratio corresponding to a gain of 43 dB.

Answer. 141. (First, find the current ratio for 40 dB [100]; then find the current ratio for 3 dB [1.41]. Multiply 100 × 1.41 = 141.)

Example 3. Find the decibel value corresponding to a voltage ratio of 150.

Answer. 43.5 (First, factor 150 into 1.5 × 100. The decibel value for a voltage ratio of 100 is 40; the decibel value for a voltage ratio of 1.5 is 3.5 [approximately]. Therefore, the decibel value for a voltage ratio is 40 + 3.5 or 43.5 dB.)

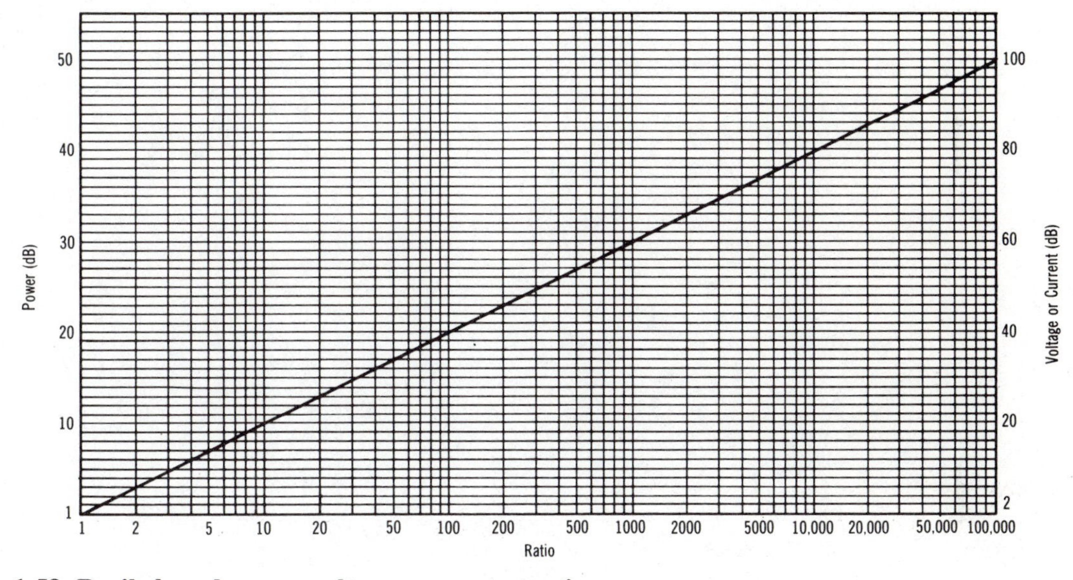

Fig. 1-58. Decibels and power, voltage, or current ratios.

Chapter 2

CONSTANTS AND STANDARDS

DIELECTRIC CONSTANTS OF MATERIALS

The dielectric constants of most materials vary for different temperatures and frequencies. Likewise, small differences in the composition of materials cause differences in the dielectric constants. A list of materials and the approximate range (where available) of their dielectric constants is given in Table 2-1. The values shown are accurate enough for most applications. The dielectric constants of some materials (such as quartz, Styrofoam, and Teflon) do not change appreciably with frequency. Figure 2-1 shows the relationship between temperature and change in capacitance.

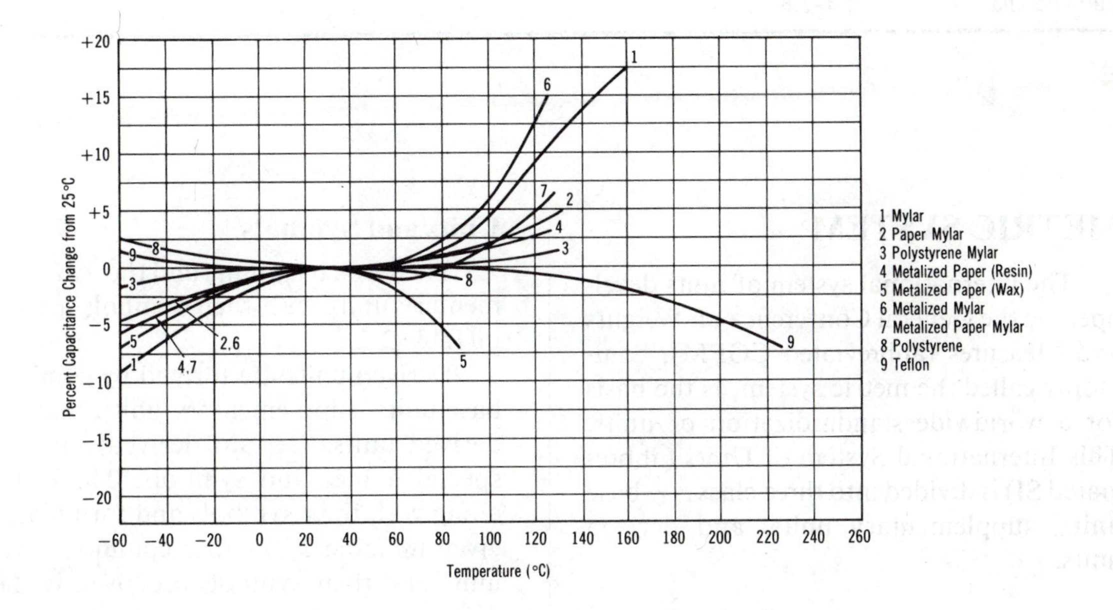

Fig. 2-1. Capacitance variation versus temperature for typical commercial capacitors.

TABLE 2-1
Dielectric Constants of Materials

Material	Dielectric constant (approx.)	Material	Dielectric constant (approx.)	Material	Dielectric constant (approx.)
air	1.0	Isolantite	6.1	porcelain (wet process)	5.8–6.5
amber	2.6–2.7	Lucite	2.5	quartz	5.0
asbestos fiber	3.1–4.8	mica (electrical)	4.0–9.0	quartz (fused)	3.78
Bakelite (asbestos base)	5.0–22	mica (clear India)	7.5	rubber (hard)	2.0–4.0
Bakelite (mica filled)	4.5–4.8	mica (filled phenolic)	4.2–5.2	ruby mica	5.4
barium titanate	100–1250	Micaglass (titanium dioxide)	9.0–9.3	selenium (amorphous)	6.0
beeswax	2.4–2.8			shellac (natural)	2.9–3.9
cambric (varnished)	4.0	Micarta	3.2–5.5	silicone (glass) (molding)	3.2–4.7
carbon tetrachloride	2.17	Mycalex	7.3–9.3	silicone (glass) (laminate)	3.7–4.3
Celluloid	4.0	Mylar	4.7	slate	7.0
cellulose acetate	2.9–4.5	neoprene	4.0–6.7	soil (dry)	2.4–2.9
Durite	4.7–5.1	nylon	3.4–22.4	steatite (ceramic)	5.2–6.3
ebonite	2.7	paper (dry)	1.5–3.0	steatite (low loss)	4.4
epoxy resin	3.4–3.7	paper (paraffin coated)	2.5–4.0	Styrofoam	1.03
ethyl alcohol (absolute)	6.5–25	paraffin (solid)	2.0–3.0	Teflon	2.1
fiber	5.0	Plexiglas	2.6–3.5	titanium dioxide	100
Formica	3.6–6.0	polycarbonate	2.9–3.2	Vaseline	2.16
glass (electrical)	3.8–14.5	polyethylene	2.5	vinylite	2.7–7.5
glass (photographic)	7.5	polyimide	3.4–3.5	water (distilled)	34–78
glass (Pyrex)	4.6–5.0	polystyrene	2.4–3.0	waxes, mineral	2.2–2.3
glass (window)	7.6	porcelain (dry process)	5.0–6.5	wood (dry)	1.4–2.9
gutta percha	2.4–2.6				

METRIC SYSTEM

The international system of units developed by the General Conference on Weights and Measures (abbreviated CGPM), commonly called the metric system, is the basis for a worldwide standardization of units. This International System of Units (abbreviated SI) is divided into three classes—base units, supplementary units, and derived units.

Units and Symbols

The seven base units and the two supplementary units with their symbols are given in Table 2-2.

Derived units are formed by combining base units, supplementary units, and other derived units. Certain derived units have special names and symbols. These units, along with their symbols and formulas, are given in Table 2-3. Other common derived units and their symbols are given in Table 2-4.

TABLE 2-2
SI Base and Supplementary Units

Quantity	Unit	Symbol
length	meter	m
mass	kilogram	kg
time	second	s
electric current	ampere	A
thermodynamic temperature	kelvin*	K
amount of substance	mole	mol
luminous intensity	candela	cd
plane angle	radian†	rad
solid angle	steradian†	sr

*The degree Celsius is also used for expressing temperature.
†Supplementary units.

TABLE 2-3
SI Derived Units with Special Names

Quantity	Unit	Symbol	Formula
frequency (of a periodic phenomenon)	hertz	Hz	1/s
force	newton	N	$kg \cdot m/s^2$
pressure, stress	pascal	Pa	N/m^2
energy, work, quantity of heat	joule	J	$N \cdot m$
power, radiant flux	watt	W	J/s
quantity of electricity, electric charge	coulomb	C	$A \cdot s$
electric potential, potential difference, electromotive force	volt	V	W/A
capacitance	farad	F	C/V
electric resistance	ohm	Ω	V/A
conductance	siemens	S	A/V
magnetic flux	weber	Wb	$V \cdot s$
magnetic flux density	tesla	T	Wb/m^2
inductance	henry	H	Wb/A
luminous flux	lumen	lm	$cd \cdot sr$
illuminance	lux	lx	lm/m^2
activity (of radionuclides)	becquerel	Bq	1/s
absorbed dose	gray	Gy	J/kg

TABLE 2-4
Common SI Derived Units

Quantity	Unit	Symbol
acceleration	meter per second squared	m/s^2
angular acceleration	radian per second squared	rad/s^2
angular velocity	radian per second	rad/s
area	square meter	m^2
concentration (of amount of substance)	mole per cubic meter	mol/m^3
current density	ampere per square meter	A/m^2
density, mass	kilogram per cubic meter	kg/m^3
electric charge density	coulomb per cubic meter	C/m^3
electric field strength	volt per meter	V/m
electric flux density	coulomb per square meter	C/m^2
energy density	joule per cubic meter	J/m^3
entropy	joule per kelvin	J/K
heat capacity	joule per kelvin	J/K
heat flux density, irradiance	watt per square meter	W/m^2
luminance	candela per square meter	cd/m^2
magnetic field strength	ampere per meter	A/m
molar energy	joule per mole	J/mol
molar entropy	joule per mole kelvin	$J/(mol \cdot K)$
molar heat capacity	joule per mole kelvin	$J/(mol \cdot K)$
moment of force	newton meter	$N \cdot m$
permeability	henry per meter	H/m
permittivity	farad per meter	F/m
radiance	watt per square meter steradian	$W/(m^2 \cdot sr)$
radiant intensity	watt per steradian	W/sr
specific heat capacity	joule per kilogram kelvin	$J/(kg \cdot K)$
specific energy	joule per kilogram	J/kg
special entropy	joule per kilogram kelvin	$J/(kg \cdot K)$
specific volume	cubic meter per kilogram	m^3/kg
surface tension	newton per meter	N/m
thermal conductivity	watt per meter kelvin	$W/(m \cdot K)$

TABLE 2-4 Cont.
Common SI Derived Units

Quantity	Unit	Symbol
velocity	meter per second	m/s
viscosity, dynamic	pascal second	Pa·s
viscosity, kinematic	square meter per second	m^2/s
volume	cubic meter	m^3
wavenumber	1 per meter	1/m

Some units, not part of SI, are so widely used they are impractical to abandon. These units (listed in Table 2-5) are acceptable for continued uses in the United States.

TABLE 2-5
Units in Use with SI

Quantity	Unit	Symbol	Value
time	minute	min	1 min = 60 s
	hour	h	1 h = 60 min = 3600 s
	day	d	1 d = 24 h = 86,400 s
	week		
	month		
	year		
plane angle	degree	°	1° = (π/180) rad
	minute	′	1′ = (1/60)° = (π 10,800) rad
	second	″	1″ = (1/60)′ = (π/648,000) rad
volume	liter	L*	1 L = 1 dm^3 = $10^{-3}m^3$
mass	metric ton	t	1 t = 10^3 kg
area (land)	hectare	ha	1 ha = 10^4 m^2

*The international symbol for liter is the lowercase "l," which can be easily confused with the number "1." Therefore, the symbol "L" or spelling out the term liter is preferred for United States use.

Prefixes

The sixteen prefixes in Table 2-6 are used to form multiples and submultiples of the SI units. The use of more than one prefix is to be avoided (e.g., use pico instead of micromicro and giga instead of kilomega). The preferred U.S. pronunciation of the terms is also included in the table. The accent is on the first syllable of each prefix.

TABLE 2-6
Metric Prefixes

Multiplication factor	Prefix	Abbreviation	Pronunciation (U.S.)
10^{18}	exa	E	ex′a (a as in about)
10^{15}	peta	P	as in petal
10^{12}	tera	T	as in terrace
10^9	giga	G	jig′a (a as in about)
10^6	mega	M	as in megaphone
10^3	kilo	k	as in kilowatt
10^2	hecto	h*	heck′ toe
10^1	deka	da*	deck′a (a as in about)
10^{-1}	deci	d*	as in decimal
10^{-2}	centi	c*	as in sentiment
10^{-3}	milli	m	as in military
10^{-6}	micro	μ	as in microphone
10^{-9}	nano	n	nan′oh (an as in ant)
10^{-12}	pico	p	peek′oh
10^{-15}	femto	f	fem′toe (fem as in feminine)
10^{-18}	atto	a	as in anatomy

*The use of hecto, deka, deci, and centi should be avoided for SI unit multiples except for area and volume, and the nontechnical use of centimeter for body and clothing measurements.

Conversion Table

Table 2-7 gives the number of places and the direction the decimal point must be moved to convert from one metric notation to another. The value labeled "Units" is the basic unit of measurement (e.g., ohms and farads). To use the table, find the desired

TABLE 2-7
Metric Conversion Table*

Desired value	Original value																	
	Exa-	*Peta-*	*Tera-*	*Giga-*	*Mega-*	*Myria-**	*Kilo-*	*Hecto-*	*Deka-*	*Units*	*Deci-*	*Centi-*	*Milli-*	*Micro-*	*Nano-*	*Pico-*	*Femto-*	*Atto-*
exa-		← 3	← 6	← 8	← 9	←10	←11	←12	←13	←14	←15	←18	←21	←24	←27	←30	←33	←36
peta-	3→		← 3	← 5	← 6	← 7	← 8	← 9	←10	←11	←12	←15	←18	←21	←24	←27	←30	←33
tera-	6→	3→		← 3	← 6	← 8	← 9	←10	←11	←12	←13	←14	←15	←18	←21	←24	←27	←30
giga-	8→	5→	3→		← 3	← 5	← 6	← 7	← 8	← 9	←10	←11	←12	←15	←18	←21	←24	←27
mega-	9→	6→	6→	3→		← 2	← 3	← 4	← 5	← 6	← 7	← 8	← 9	←12	←15	←18	←21	←24
myria-†	10→	7→	8→	5→	2→		← 1	← 2	← 3	← 4	← 5	← 6	← 7	←10	←13	←16	←19	←22
kilo-	11→	8→	9→	6→	3→	1→		← 1	← 2	← 3	← 4	← 5	← 6	← 9	←12	←15	←18	←21
hecto-	12→	9→	10→	7→	4→	2→	1→		← 1	← 2	← 3	← 4	← 5	← 8	←11	←14	←17	←20
deka-	13→	10→	11→	8→	5→	3→	2→	1→		← 1	← 2	← 3	← 4	← 7	←10	←13	←16	←19
units	14→	11→	12→	9→	6→	4→	3→	2→	1→		← 1	← 2	← 3	← 6	← 9	←12	←15	←18
deci-	15→	12→	13→	10→	7→	5→	4→	3→	2→	1→		← 1	← 2	← 5	← 8	←11	←14	←17
centi-	18→	15→	14→	11→	8→	6→	5→	4→	3→	2→	1→		← 1	← 4	← 7	←10	←13	←16
milli-	21→	18→	15→	12→	9→	7→	6→	5→	4→	3→	2→	1→		← 3	← 6	← 9	←12	←15
micro-	24→	21→	18→	15→	12→	10→	9→	8→	7→	6→	5→	4→	3→		← 3	← 6	← 9	←12
nano-	27→	24→	21→	18→	15→	13→	12→	11→	10→	9→	8→	7→	6→	3→		← 3	← 6	← 9
pico-	30→	27→	24→	21→	18→	16→	15→	14→	13→	12→	11→	10→	9→	6→	3→		← 3	← 6
femto-	33→	30→	27→	24→	21→	19→	18→	17→	16→	15→	14→	13→	12→	9→	6→	3→		← 3
atto-	36→	33→	30→	27→	24→	22→	21→	20→	19→	18→	17→	16→	15→	12→	9→	6→	3→	

*Arrow indicates direction decimal moves.
†The prefix "myria" is not normally used.

value in the left-hand column; then follow the horizontal line across to the column with the prefix in which the original value is stated. The number and arrow at this point indicate the number of places and the direction the decimal point must be moved to change the original value to the desired value.

Miscellaneous

The terms "liter" and "meter" are also spelled "litre" and "metre." However, the most widely accepted U.S. practice is with the "er," and this spelling is recommended by the U.S. Department of Commerce.

While the SI unit for temperature is the kelvin (K), wide use is also made of the degree celsius (°C) in expressing temperature and temperature intervals in SI. The Celsius scale (formerly called centigrade) is directly related to thermodynamic temperature (kelvin) as follows:

$$t = T - T_0$$

where

t is the temperature, in degrees Celsius,
T is the thermodynamic temperature, in kelvins,
T_0 equals 273.15 K.

Note. A temperature interval of 1 °C is exactly equal to 1 K. Thus 0 °C = 273.15 K.

The special name "liter" is used for a cubic decimeter, but use of this term should be restricted to measurements of liquids and gases. Do not use any prefix other than "milli" with "liter."

Note that "kilogram" is the only base unit employing a prefix. However, to form multiples, the prefix is added to "gram," not to "kilogram." The megagram may be used for large masses. However, the term "metric ton" is also used in commercial applications. No prefixes should be used with "metric ton."

Other units have been used through the years as part of the metric (cgs) system. Avoid using them in SI.

CONVERSION FACTORS

Table 2-8 lists the multiplying factors necessary to convert from one unit of measure to another. To use this table, locate either the unit of measure you are converting from or the one you are converting to in the left-hand column. Opposite this listing are the multiplying factors for converting either unit of measure to the other unit of measure.

Note. Conversions from one metric prefix to another (e.g., from kilo to mega) are not included in Table 2-8. For these conversions, see the preceding section.

TABLE 2-8
Conversion Factors

To convert	Into	Multiply by	Conversely, multiply by
acres	square feet	4.356×10^{4}	2.296×10^{-5}
acres	square meters	4047	2.471×10^{-4}
acres	square miles	1.5625×10^{-3}	640
ampere-hours	coulombs	3600	2.778×10^{-4}
ampere-turns	gilberts	1.257	0.7958
ampere-turns per centimeter	ampere-turns per inch	2.54	0.3937
angstrom units	inches	3.937×10^{-9}	2.54×10^{8}
angstrom units	meters	10^{-10}	10^{10}
ares	square meters	10^{2}	10^{-2}
atmospheres	feet of water	33.90	0.02950
atmospheres	inch of mercury at 0 °C	29.92	3.342×10^{-2}
atmospheres	kilogram per square meter	1.033×10^{4}	9.678×10^{-5}
atmospheres	millimeter of mercury at 0 °C	760	1.316×10^{-3}
atmospheres	pascals	1.0133×10^{5}	0.9869×10^{-5}
atmospheres	pounds per square inch	14.70	0.06804
barns	square centimeters	10^{-24}	10^{24}
bars	atmospheres	9.870×10^{-7}	1.0133
bars	dynes per square centimeter	10^{6}	10^{-6}
bars	pascals	10^{5}	10^{-5}
bars	pounds per square inch	14.504	6.8947×10^{-2}
board feet	cubic meters	2.3597×10^{-3}	4.238×10^{2}
Btu	ergs	1.0548×10^{10}	9.486×10^{-11}
Btu	foot-pounds	778.3	1.285×10^{-3}
Btu	joules	1054.8	9.480×10^{-4}
Btu	kilogram-calories	0.252	3.969
Btu per hour	horsepower-hours	3.929×10^{-4}	2545
bushels	cubic feet	1.2445	0.8036
bushels	cubic meters	3.5239×10^{-2}	28.38
calories, gram	joules	4.185	0.2389
carats (metric)	grams	0.2	5
Celsius	Fahrenheit	$(°C \times 9/5) + 32 = °F$	$(°F - 32) \times 5/9 = °C$

TABLE 2-8 Cont.
Conversion Factors

To convert	Into	Multiply by	Conversely, multiply by
Celsius	kelvin	°C + 273.1 = K	K - 273.1 = °C
chains (surveyor's)	feet	66	1.515×10^{-2}
circular mils	square centimeters	5.067×10^{-6}	1.973×10^{5}
circular mils	square mils	0.7854	1.273
cords	cubic meters	3.625	0.2758
cubic feet	cords	7.8125×10^{-3}	128
cubic feet	gallons (liquid U.S.)	7.481	0.1337
cubic feet	liters	28.32	3.531×10^{-2}
cubic inches	cubic centimeters	16.39	6.102×10^{-2}
cubic inches	cubic feet	5.787×10^{-4}	1728
cubic inches	cubic meters	1.639×10^{-5}	6.102×10^{4}
cubic inches	gallons (liquid U.S.)	4.329×10^{-3}	231
cubic meters	cubic feet	35.31	2.832×10^{-2}
cubic meters	cubic yards	1.308	0.7646
cups	cubic centimeter	2.366×10^{2}	4.227
curies	becquerels	3.7×10^{10}	2.7×10^{-11}
cycles per second	hertz	1	1
degrees (angle)	mils	17.45	5.73×10^{-2}
degrees (angle)	radians	1.745×10^{-2}	57.3
dynes	pounds	2.248×10^{-6}	4.448×10^{5}
electron volts	joules	1.602×10^{-19}	0.624×10^{18}
ergs	foot-pounds	7.376×10^{-8}	1.356×10^{7}
ergs	joules	10^{-7}	10^{7}
ergs per second	watts	10^{-7}	10^{7}
ergs per square centimeter	watts per square centimeter	10^{-3}	10^{3}
Fahrenheit	kelvin	(°F + 459.67)/1.8	1.8K - 459.67
Fahrenheit	Rankine	°F + 459.67 = °R	°R - 459.67 = °F
faradays	ampere-hours	26.8	3.731×10^{-2}
fathoms	feet	6	0.16667
fathoms	meters	1.8288	0.5467
feet	centimeters	30.48	3.281×10^{-2}
feet	meters	0.3048	3.281
feet	mils	1.2×10^{4}	8.333×10^{-5}
feet of water at 4 °C	inches of mercury at 0 °C	0.8826	1.133
feet of water at 4 °C	kilogram per square meter	304.8	3.281×10^{-3}
feet of water at 4 °C	pascals	2.989×10^{3}	3.346×10^{-4}
fermis	meters	10^{-15}	10^{15}
foot candles	lux	10.764	0.0929
foot lamberts	candelas per square meter	3.4263	0.2918
foot-pounds	gram-centimeters	1.383×10^{4}	1.235×10^{-5}
foot-pounds	horsepower-hours	5.05×10^{-7}	1.98×10^{6}
foot-pounds	kilogram-meters	0.1383	7.233
foot-pounds	kilowatt-hours	3.766×10^{-7}	2.655×10^{6}
foot-pounds	ounce-inches	192	5.208×10^{-3}
gallons	meters per second	9.807	0.102
gallons (liquid U.S.)	cubic meters	3.785×10^{-3}	264.2
gallons (liquid U.S.)	gallons (liquid British Imperial)	0.8327	1.201

TABLE 2-8 Cont.
Conversion Factors

To convert	Into	Multiply by	Conversely, multiply by
gammas	teslas	10^{-9}	10^{9}
gausses	lines per square centimeter	1.0	1.0
gausses	lines per square inch	6.452	0.155
gausses	teslas	10^{-4}	10^{4}
gausses	webers per square inch	6.452×10^{-8}	1.55×10^{7}
gilberts	amperes	0.7958	1.257
grads	radians	1.571×10^{-2}	63.65
grams	dynes	980.7	1.02×10^{-3}
grams	grains	15.43	6.481×10^{-2}
grams	ounces (avdp)	3.527×10^{-2}	28.35
grams	poundals	7.093×10^{-2}	14.1
grams per centimeter	pounds per inch	5.6×10^{-3}	178.6
grams per cubic centimeter	pounds per cubic inch	3.613×10^{-2}	27.68
grams per square centimeter	pounds per square foot	2.0481	0.4883
hectares	acres	2.471	0.4047
horsepower	Btu per minute	42.418	2.357×10^{-2}
horsepower	foot-pounds per minute	3.3×10^{4}	3.03×10^{-5}
horsepower	foot-pounds per second	550	1.182×10^{-3}
horsepower	horsepower (metric)	1.014	0.9863
horsepower	kilowatts	0.746	1.341
horsepower (metric)	Btu per minute	41.83	2.390×10^{-2}
horsepower (metric)	kilogram-calories per minute	10.54	9.485×10^{-2}
horsepower (metric)	watts	7.355×10^{2}	745.7
inches	centimeters	2.54	0.3937
inches	feet	8.333×10^{-2}	12
inches	meters	2.54×10^{-2}	39.37
inches	miles	1.578×10^{-5}	6.336×10^{4}
inches	mils	10^{3}	10^{-3}
inches	yards	2.778×10^{-2}	36
inches of mercury at 0 °C	pascals	3.386×10^{3}	2.953×10^{-4}
inches of mercury at 0 °C	pounds per square inch	0.4912	2.036
inches of water at 4 °C	inches of mercury	7.355×10^{-2}	13.60
inches of water at 4 °C	kilograms per square meter	25.40	3.937×10^{-2}
inches of water at 15.6 °C	pascals	2.488×10^{2}	4.02×10^{-3}
joules	foot-pounds	0.7376	1.356
joules	watt-hours	2.778×10^{-4}	3600
kilogram-calories	kilogram-meters	426.9	2.343×10^{-3}
kilograms	tonnes	10^{3}	10^{-3}
kilograms	tons (long)	9.842×10^{-4}	1016
kilograms	tons (short)	1.102×10^{-3}	907.2
kilograms	pounds (avdp)	2.205	0.4536
kilograms per square meter	pounds per square foot	0.2048	4.882

TABLE 2-8 Cont.
Conversion Factors

To convert	Into	Multiply by	Conversely, multiply by
kilometers	feet	3281	3.408×10^{-4}
kilometers	inches	3.937×10^{4}	2.54×10^{-5}
kilometers	light years	1.0567×10^{-13}	9.4637×10^{12}
kilometers per hour	feet per minute	54.68	1.829×10^{-2}
kilometers per hour	knots	0.5396	1.8532
kilowatt-hours	Btu	3413	2.93×10^{-4}
kilowatt-hours	foot-pounds	2.655×10^{6}	3.766×10^{-7}
kilowatt-hours	horsepower-hours	1.341	0.7457
kilowatt-hours	joules	3.6×10^{6}	2.778×10^{-7}
kilowatt-hours	kilogram-calories	860	1.163×10^{-3}
kilowatt-hours	kilogram-meters	3.671×10^{5}	2.724×10^{-6}
kilowatt-hours	pounds water evaporated from and at 212°F	3.53	0.284
kilowatt-hours	watt-hours	10^{3}	10^{-3}
knots	feet per second	1.688	0.5925
knots	meters per minute	30.87	0.0324
knots	miles per hour	1.1508	0.869
lamberts	candles per square centimeter	0.3183	3.142
lamberts	candles per square inch	2.054	0.4869
leagues	miles	3	0.33
links	chains	0.01	100
links (surveyor's)	inches	7.92	0.1263
liters	bushels (dry U.S.)	2.838×10^{-2}	35.24
liters	cubic centimeters	10^{3}	10^{-3}
liters	cubic inches	61.02	1.639×10^{-2}
liters	cubic meters	10^{-3}	10^{3}
liters	gallons (liquid U.S.)	0.2642	3.785
liters	pints (liquid U.S.)	2.113	0.4732
$\log_e N$	$\log_{10} N$	0.4343	2.303
lumens per square foot	foot-candles	1	1
lumens per square meter	foot-candles	0.0929	10.764
lux	foot-candles	0.0929	10.764
maxwells	kilolines	10^{-3}	10^{3}
maxwells	megalines	10^{-6}	10^{6}
maxwells	webers	10^{-8}	10^{8}
meters	feet	3.28	30.48×10^{-2}
meters	inches	39.37	2.54×10^{-2}
meters	miles	6.214×10^{-4}	1609.35
meters	yards	1.094	0.9144
meters per minute	feet per minute	3.281	0.3048
meters per minute	kilometers per hour	0.06	16.67
Mhos	siemens	1	1
miles (nautical)	feet	6076.1	1.646×10^{-4}
miles (nautical)	meters	1852	5.4×10^{-4}
miles (statute)	feet	5280	1.894×10^{-4}
miles (statute)	kilometers	1.609	0.6214
miles (statute)	light years	1.691×10^{-13}	5.88×10^{12}
miles (statute)	miles (nautical)	0.869	1.1508
miles (statute)	yards	1760	5.6818×10^{-4}
miles per hour	feet per minute	88	1.136×10^{-2}

TABLE 2-8 Cont.
Conversion Factors

To convert	Into	Multiply by	Conversely, multiply by
miles per hour	feet per second	1.467	0.6818
miles per hour	kilometers per hour	1.609	0.6214
miles per hour	kilometers per minute	2.682×10^{-2}	37.28
miles per hour	knots	0.8684	1.152
millimeters	inches	3.937×10^{-2}	25.4
millimeters	microns	10^{3}	10^{-3}
mils	meters	2.54×10^{-5}	3.94×10^{4}
mils	minutes	3.438	0.2909
minutes (angle)	degrees	1.666×10^{-2}	60
minutes (angle)	radians	2.909×10^{-4}	3484
nepers	decibels	8.686	0.1151
newtons	dynes	10^{5}	10^{-5}
newtons	kilograms	0.1020	9.807
newtons per square meter	pascals	1	1
newtons	pounds (avdp)	0.2248	4.448
oersteds	amperes per meter	7.9577×10	1.257×10^{-2}
ohms	ohms (international)	0.99948	1.00052
ohms circular-mil per foot	ohms per square millimeter per meter	1.66×10^{-3}	6.024×10^{2}
ohms per foot	ohms per meter	0.3048	3.281
ounces (fluid)	quarts	3.125×10^{-2}	32
ounces (avdp)	pounds	6.25×10^{-2}	16
pints	quarts (liquid U.S.)	0.50	2
poundals	dynes	1.383×10^{4}	7.233×10^{-5}
poundals	pounds (avdp)	3.108×10^{-2}	32.17
pounds	grams	453.6	2.205×10^{-3}
pounds (force)	newtons	4.4482	0.2288
pounds carbon oxidized	Btu	14,544	6.88×10^{-5}
pounds carbon oxidized	horsepower-hours	5.705	0.175
pounds carbon oxidized	kilowatt-hours	4.254	0.235
pounds of water (dist)	cubic feet	1.603×10^{-2}	62.38
pounds of water (dist)	gallons	0.1198	8.347
pounds per foot	kilograms per meter	1.488	0.6720
pounds per square inch	dynes per square centimeter	6.8946×10^{4}	1.450×10^{-5}
pounds per square inch	pascals	6.895×10^{3}	1.45×10^{-4}
quadrants	degrees	90	11.111×10^{-2}
quadrants	radians	1.5708	0.637
quarts (U.S. dry)	cubic centimeters	1101.4	9.9079×10^{-4}
quarts (U.S. liquid)	cubic centimeters	946.4	1.057×10^{-3}
radians	mils	10^{3}	10^{-3}
radians	minutes	3.438×10^{3}	2.909×10^{-4}
radians	seconds	2.06265×10^{5}	4.848×10^{-6}
revolutions per minute	degrees per second	6.0	0.1667
revolutions per minute	radians per second	0.1047	9.549
revolutions per minute	revolutions per second	1.667×10^{-2}	60
rods	feet	16.5	6.061×10^{-2}
rods	miles	3.125×10^{-3}	320
rods	yards	5.5	0.1818
roentgens	coulombs per kilogram	2.58×10^{-4}	3.876×10^{3}

TABLE 2-8 Cont.
Conversion Factors

To convert	Into	Multiply by	Conversely, multiply by
slugs	kilograms	1.459	0.6854
slugs	pounds (avdp)	32.174	3.108×10^{-2}
square feet	square centimeters	929.034	1.076×10^{-3}
square feet	square inches	144	6.944×10^{-3}
square feet	square meters	9.29×10^{-2}	10.764
square feet	square miles	3.587×10^{-8}	27.88×10^{6}
square feet	square yards	11.11×10^{-2}	9
square inches	circular mils	1.273×10^{6}	7.854×10^{-7}
square inches	square centimeters	6.452	0.155
square inches	square mils	10^{6}	10^{-6}
square inches	square millimeters	645.2	1.55×10^{-3}
square kilometers	square miles	0.3861	2.59
square meters	square yards	1.196	0.8361
square miles	acres	640	1.562×10^{-3}
square miles	square yards	3.098×10^{6}	3.228×10^{-7}
square millimeters	circular mils	1973	5.067×10^{-4}
square mils	circular mils	1.273	0.7854
steres	cubic meters	1	1
stokes	square meter per second	10^{-4}	10^{-4}
tablespoons	cubic centimeters	14.79	6.761×10^{-2}
teaspoons	cubic centimeters	4.929	0.203
tonnes	kilograms	10^{3}	10^{-3}
tonnes	pounds	2204.63	4.536×10^{-4}
tons (long)	pounds (avdp)	2240	4.464×10^{-4}
tons (metric)	kilograms	10^{3}	10^{-3}
tons (short)	pounds	2000	5×10^{-4}
torrs	newtons per square meter	133.32	7.5×10^{-3}
varas	feet	2.7777	0.36
watts	Btu per hour	3.413	0.293
watts	Btu per minute	6.589×10^{-2}	17.58
watts	foot-pounds per minute	44.26	2.26×10^{-2}
watts	foot-pounds per second	0.7378	1.356
watts	horsepower	1.341×10^{-3}	746
watts	kilogram-calories per minute	1.433×10^{-2}	69.77
watt-seconds	gram-calories (mean)	0.2389	4.186
watt-seconds	joules	1	1
webers	maxwells	10^{8}	10^{-8}
webers per square meter	gausses	10^{4}	10^{-4}
yards	feet	3	0.3333
yards	varas	1.08	0.9259

STANDARD FREQUENCIES AND TIME SIGNALS

WWV, WWVH, and WWVB

Time signals and audiofrequencies are broadcast continuously day and night from WWV, operated by the National Bureau of Standards at Fort Collins, Colorado 80521. The WWV broadcast frequencies are 2.5, 5, 10, and 15 MHz and the 1-s marker tone consists of a 5-ms pulse at 1000 Hz. A similar station, WWVH, is located at Kekaha, Kauai, Hawaii. It broadcasts on frequencies 2.5, 5, 10, and 15 MHz and the 1-s marker tone consists of a 5-ms pulse at 1200 Hz.

The broadcasts of WWV may also be heard by the use of the telephone by dialing (303) 499-7111, Boulder, Colorado. WWVH may be heard by dialing (808) 335-4363. Neither is a toll free number. The telephone user will hear the live broadcasts from the station called. With the instabilities and variable delays of propagation, the accuracy of the telephone time signals will not be better than 30 ms. This service is automatically limited to 3 minutes per call.

Station WWVB broadcasts on a frequency of 60 kHz. The station broadcasts a time code continuously. No voice announcements are broadcast from WWVB.

WWV and WWVH broadcast frequencies are consistent with the internationally agreed time scale Coordinated Universal Time (UTC). These changes became effective January 1, 1972. This coordination provides a more uniform system of time and frequency transmission throughout the world. It also aids in the solution of many scientific and technical problems such as radio communications, geodesy, navigation, and artificial-satellite tracking. At WWV and WWVH, the carrier and modulation frequencies are derived from cesium-controlled oscillators. These broadcasts are in conformity with the international Radio Consultation Committee. The frequency offset of UTC was made permanently zero, effective 0000 hours UTC January 1, 1972.

The hourly broadcast of WWV and WWVH is shown in Fig. 2-2. Standard audiofrequencies of 440, 500, and 600 Hz are broadcast on each radio carrier frequency by WWV and WWVH. The 600-Hz tone is broadcast during the odd minutes by WWV and during even minutes by WWVH. The 500-Hz tone is broadcast during alternate minutes unless voice announcements or silent periods are scheduled. The 440-Hz tone is broadcast beginning 2 minutes after the hour at WWV and 1 minute after the hour at WWVH. The 440-Hz tone period is omitted during the first hour of the UTC day. The duration of each transmitted tone is approximately 45 seconds, as indicated in Fig. 2-2.

The frequencies transmitted from WWV and WWVH are accurate to within one part of 10^{12}.

There are no audiotones or special announcements during the semisilent period from either station. The period for WWV is from 45 to 50 minutes after the hour and from 15 to 20 minutes after the hour from WWVH.

A voice announcement of Coordinated Universal Time is given during the last 7.5 seconds of every minute. The announcement is as follows: "At the tone ____ hours ____ minutes Coordinated Universal Time." A voice announcement at WWVH occurs during the period 45 seconds to 52.5 seconds after the minute. The voice announcement at WWVH precedes that of WWV by 7.5 seconds. However, the tone markers referred to in both announcements occur simultaneously, but they may not be received at the same time due to propagation effects.

Before January 1, 1972, time signals broadcast from WWV and WWVH were kept in close agreement with UT2 (astronomical time) by making step adjustments of 100 ms when necessary. On December 1, 1971, at 23 hours 59 minutes 60.107600 seconds UTC (i.e., GMT), UTC (NBS) "was retarded 0.107600 second" to give the new UTC scale an initial difference of 10 seconds late with respect to International Atomic Time (IAT) as maintained by the

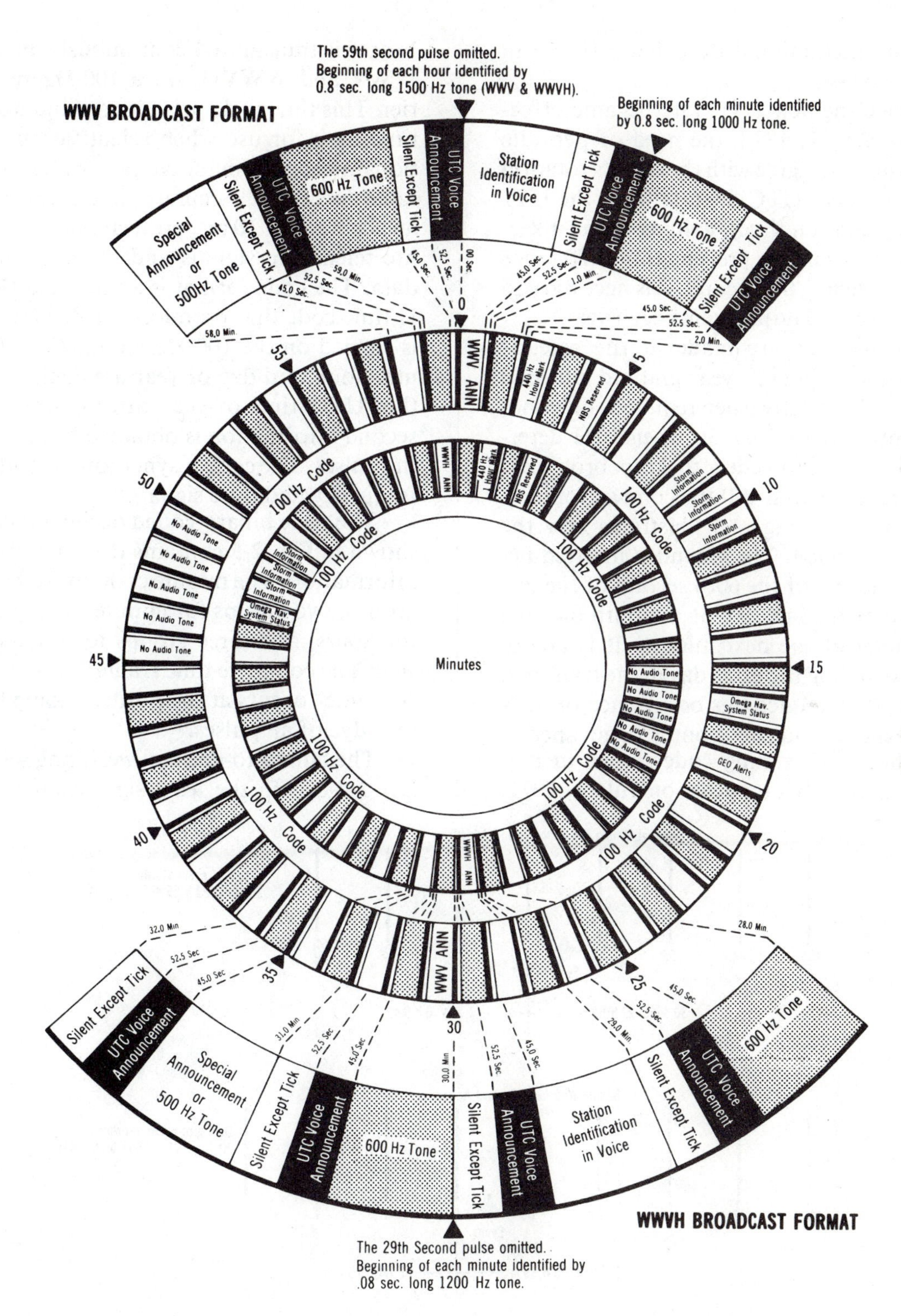

WWV BROADCAST FORMAT
The 59th second pulse omitted.
Beginning of each hour identified by
0.8 sec. long 1500 Hz tone (WWV & WWVH).
Beginning of each minute identified
by 0.8 sec. long 1000 Hz tone.
Special Announcement or 500Hz Tone
Silent Except Tick
UTC Voice Announcement
600 Hz Tone
Station Identification in Voice
WWV ANN
440 Hz 1 Hour Mark
NBS Reserved
Storm Information
Omega Nav. System Status
GEO Alerts
No Audio Tone
100 Hz Code
WWVH ANN
Minutes
WWVH BROADCAST FORMAT
The 29th Second pulse omitted.
Beginning of each minute identified by
.08 sec. long 1200 Hz tone.

Fig. 2-2

Bureau International de l'Heure (BIH) in Paris, France.

Since the new UTC rate became effective January 1, 1972, the need of periodic adjustment to agree with the earth's rotation is not needed. UTC departs from the UT1 (earth's rotation time), gaining about 1 second each year. To prevent this difference from exceeding 0.7 second, it is necessary to make 1-second adjustments each year.

Corrections are made at the rate of about 1 second each year and are adjusted by 1 second exactly when required, on either December 31 or June 30 when BIH determines they are needed to keep broadcast time signals within ± 0.9 second astronomical time UT1. Fig. 2-3 illustrates how the second is added. The second is inserted between the end of the 60th second of the last minute of the last day of a month and the beginning of the next minute. It is analogous to adding the extra day in the leap year. BIH will announce this occurrence of adding to the second two months in advance.

The WWV timing code shown in Fig. 2-4 was initially broadcast on July 1, 1971. It now is transmitted continuously, both by WWV and WWVH, on a 100-Hz subcarrier. This time code provides a standardized time base for use when scientific observations are being made at two widely separated locations. Accurate time markers, to an accuracy of 10 ms, are available for satellite telemetric signals and other scientific data. The code format is a modified IRIG-H time code that produces a 1-pps rate and is carried on the 100-Hz modulation. Minute, hour, and day of year are contained in this UTC time-of-year information. The second information is obtained by counting the pulses. The code is synchronous with the frequency and time signals.

The code binary-coded decimal (bcd) as shown in Fig. 2-5 contains the time-of-year information. The minute contains seven bcd groups, two groups for minutes, two groups for hours, and three groups for the day of year. The complete time frame is 1 min. The "on time" occurs at the positive-going leading edge of all pulses.

The binary-to-decimal weighting scheme is 1-2-4-8 with the least significant binary

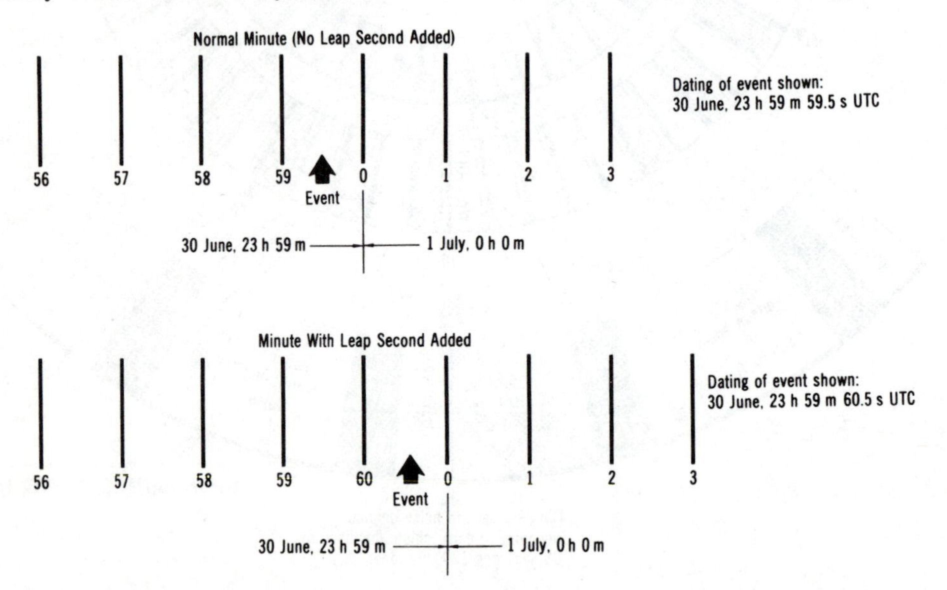

Fig. 2-3

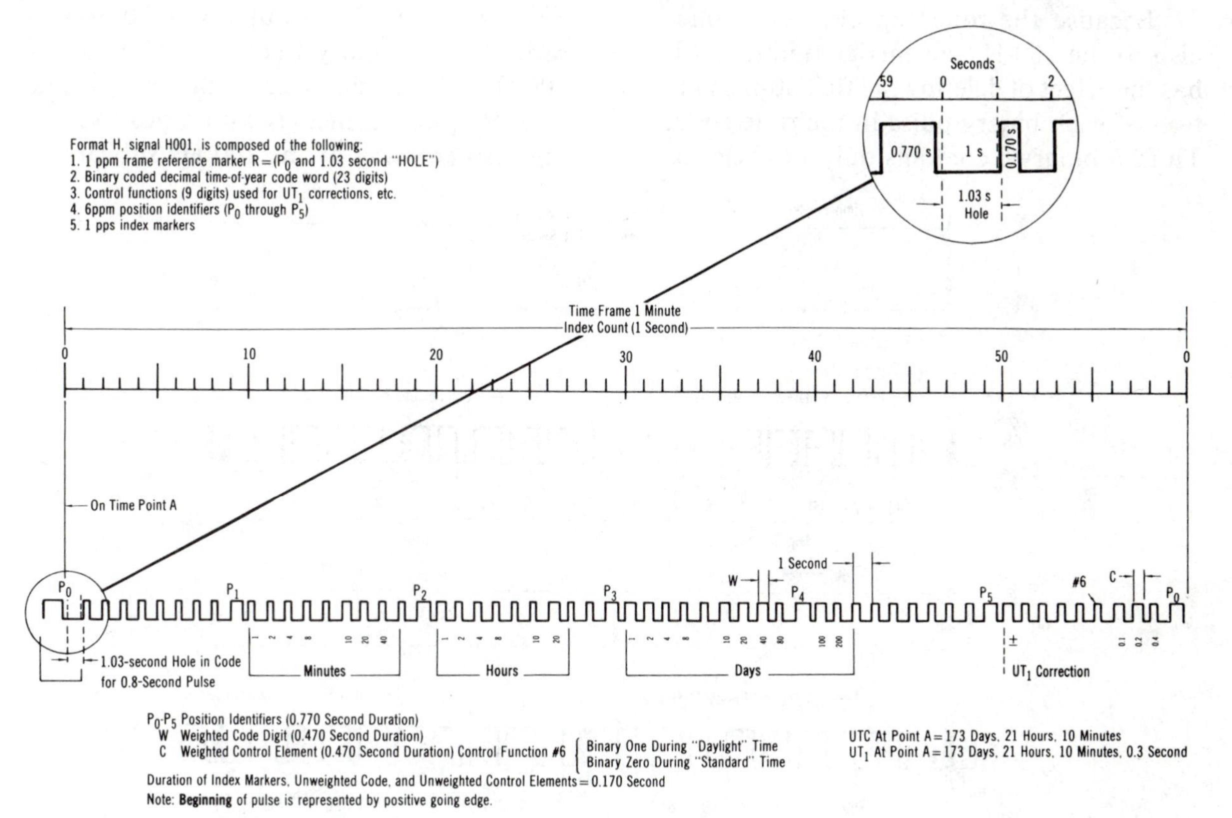

Fig. 2-4

digit always transmitted first. The binary groups and their basic decimal equivalents are shown in Table 2-9. The decimal equivalent of a bcd group is derived by multiplying each binary digit times the weight factor of its respective column and adding the four products together.

Example. The binary sequence 1010 in the 1-2-4-8 scheme means $(1 \times 1) + (0 \times 2) + (1 \times 4) + (0 \times 8) = 1 + 0 + 4 + 0 = 5$. If fewer than nine decimal digits are needed, one or more of the binary columns may be omitted.

In the standard IRIG-H code, a binary 0 pulse consists of exactly 20 cycles of 100-Hz amplitude modulation (200-ms duration), whereas a binary 1 consists of 50 cycles of 100 Hz (500-ms duration). In the WWV/WWVH broadcast format, however, all tones are suppressed briefly while the seconds pulses are transmitted.

TABLE 2-9
Binary and Decimal Equivalents

Binary group				Decimal
1	*2*	*4*	*8*	equivalent
0	0	0	0	0
1	0	0	0	1
0	1	0	0	2
1	1	0	0	3
0	0	1	0	4
1	0	1	0	5
0	1	1	0	6
1	1	1	0	7
0	0	0	1	8
1	0	0	1	9

Because the tone suppression applies also to the 100-Hz subcarrier frequency, it has the effect of deleting the first 30-ms portion of each binary pulse in the time code. Thus, a binary 0 contains only 17 cycles of 100-Hz amplitude modulation (170-ms duration) and a binary 1 contains 47 cycles of 100 Hz (470-ms duration). The leading edge of every pulse coincides with a positive-going zero crossing of the 100-Hz subcarrier,

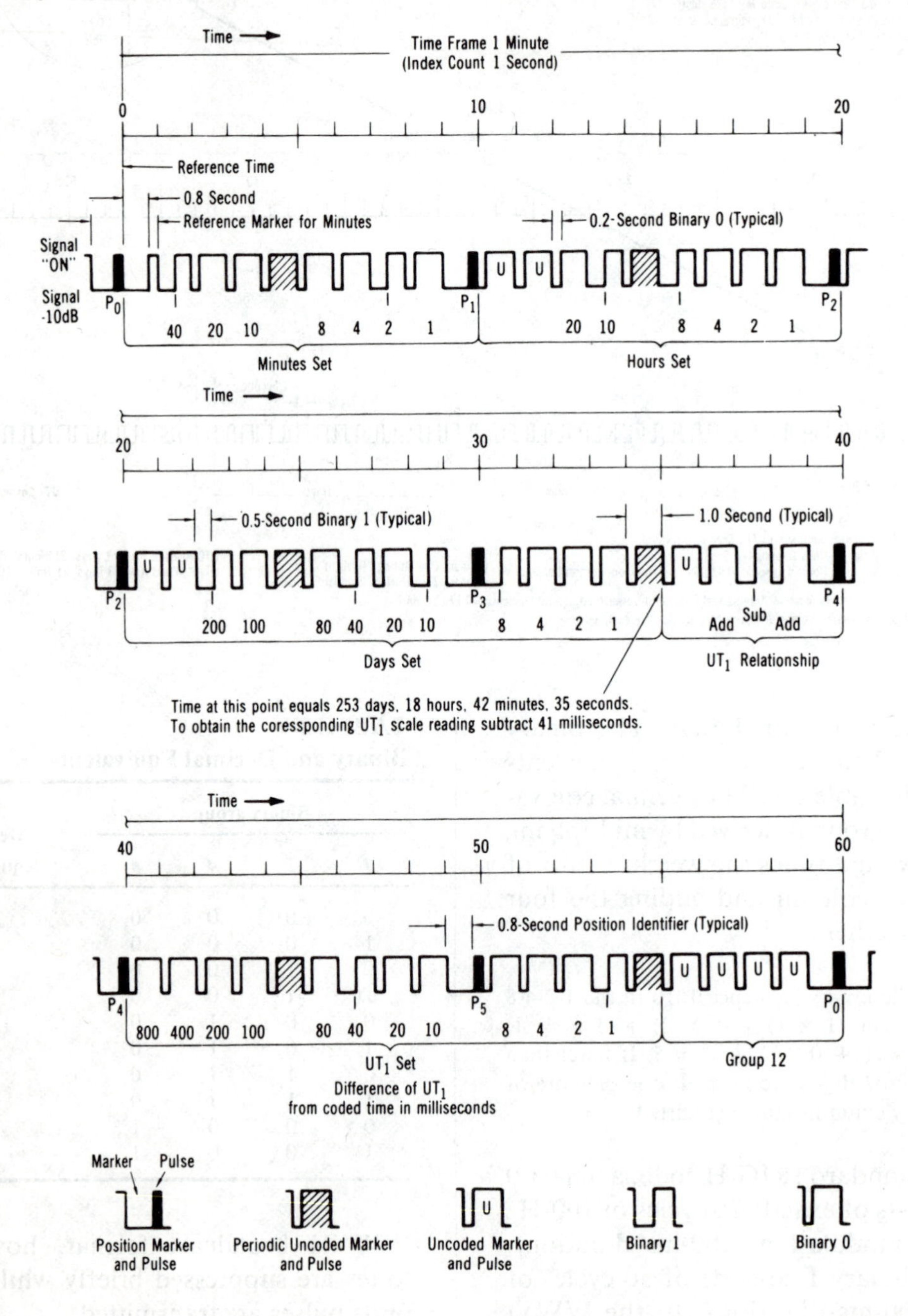

Fig. 2-5

but it occurs 30 ms after the beginning of the second.

Within a time frame of 1 min, enough pulses are transmitted to convey in bcd language the current minute, hour, and day of year. Two bcd groups are needed to express the hour (00 through 23); three groups are needed to express the day of year (001 through 366). When representing units, tens, or hundreds, the basic 1-2-4-8 weights are simply multiplied by 1, 10, or 100 as appropriate. The coded information always refers to time at the beginning of the 1-min frame. Seconds may be determined by counting pulses within the frame.

Each frame starts with a unique spacing of pulses to mark the beginning of a new minute. No pulse is transmitted during the first second of the minute. Instead, a 1-s space or hole occurs in the pulse train at that time. Because all pulses in the time code are 30 ms late with respect to UTC, each minute actually begins 1030 ms (or 1.03 s) prior to the leading edge of the first pulse in the new frame.

For synchronization purposes, every 10 s a so-called position identifier pulse is transmitted. Unlike the bcd data pulses, the position identifiers consist of 77 cycles of 100 Hz (770-ms duration).

UT1 corrections to the nearest 0.1 s are broadcast via bcd pulses during the final 10 s of each frame. The coded pulses that occur between the 50th and 59th seconds of each frame are called control functions. Control function No. 1, which occurs at 50 s, tells whether the UT1 correction is negative or positive. If control function No. 1 is a binary 0, the correction is negative; if it is a binary 1, the correction is positive. Control functions No. 7, 8, and 9, which occur, respectively, at 56, 57, and 58 s, specify the amount of UT1 correction. Because the UT1 corrections are expressed in tenths of a second, the basic binary-to-decimal weights are multiplied by 0.1 when applied to these control functions.

Control function No. 6, which occurs at 55 s, is programmed as a binary 1 throughout those weeks when Daylight Saving Time is in effect and as a binary 0 when Standard Time is in effect. The setting of this function is changed at 0000 UTC on the date of change. Throughout the U.S. mainland, this schedule allows several hours for the function to be received before the change becomes effective locally (i.e., at 2:00 AM local time). Thus, control function No. 6 allows clocks or digital recorders operating on local time to be programmed to make an automatic 1-hr adjustment in changing from Daylight Saving Time to Standard Time and vice versa.

Figure 2-4 shows one frame of the time code as it might appear after being rectified, filtered, and recorded. In this example, the leading edge of each pulse is considered to be the positive-going excursion. The pulse train in the figure is annotated to show the characteristic features of the time code format. The six position identifiers are denoted by symbols P_1, P_2, P_3, P_4, P_5, and P_0. The minutes, hours, days, and UT1 sets are marked by brackets, and the applicable weighting factors are printed beneath the coded pulses in each bcd group. With the exception of the position identifiers, all uncoded pulses are set permanently to binary 0.

The first 10 s of every frame always include the 1.03-s hole followed by eight uncoded pulses and the position identifier P_1. The minutes set follows P_1 and consists of two bcd groups separated by an uncoded pulse. Similarly, the hours set follows P_2. The days set follows P_3 and extends for two pulses beyond P_4 to allow enough elements to represent three decimal digits. The UT1 set follows P_5, and the last pulse in the frame is always P_0.

In Fig. 2-4, the least significant digit of the minutes set is $(0 \times 1) + (0 \times 2) + (0 \times 4) + (0 \times 8) = 0$; the most significant digit of that set is $(1 \times 10) + (0 \times 20) + (0 \times 40) = 10$. Hence, at the beginning of the 1.03-s hole in that frame, the time was exactly 10 min past the hour. By decoding the hours set and the days set, the time of day is in the 21st hour on the 173rd day of the year. The UT1 correction is +0.3 s. Therefore, at point A, the correct time on the UT1 scale is 173 days, 21 hours, 10 minutes, 0.3 second.

Weather information about major storms in the Atlantic and Pacific areas is broadcast from WWV and WWVH. Times of broadcast are 0500, 1100, 1700, and 2300 UTC by WWV and 0000, 0600, 1200, and 1800 UTC by WWVH. These broadcasts are given in voice during the 8th, 9th, and 10th minute from WWV and during the 48th, 49th, and 50th minute from WWVH.

Omega Navigation System status reports are broadcast in voice from WWV at 16 minutes after the hour and from WWVH at 47 minutes after the hour. The International Omega Navigation System is a very low frequency (VLF) radio navigation aid operating in the 10- to 14-kHz frequency band. Eight stations operate around the world. Omega, like other radio navigation systems, is subject to signal degradation caused by ionospheric disturbances at high latitudes. The Omega announcements on WWV and WWVH are given to provide users with immediate notification of such events and other information on the status of the Omega system.

Station identifications are made by voice every 30 min by WWV and WWVH.

Station WWVB broadcasts a continuous binary coded decimal (bcd) signal, which is synchronized with the 60-kHz carrier signal. WWVB uses a level-shift carrier time code. The signal consists of 60 markers each minute, as shown in Fig. 2-5, with one marker occurring each second.

When a marker is generated, the carrier power is reduced by 10 dB at the beginning of the corresponding second and restored 0.2 s later for an uncoded marker or binary 0, 0.5 s later for a binary 1, and 0.8 s later for a 10-s position marker or for a minute reference marker.

The bcd are set up in groups. The 1st and 2nd bcd groups specify the minute of the hour; the 3rd and 4th bcd groups specify the hour of the day; the 5th, 6th, and 7th bcd groups specify the day of the year; and the 9th, 10th, and 11th bcd groups specify the number of milliseconds to be added or subtracted from the code time in order to obtain UT1 (astronomical time). The 8th bcd group specifies if the UT1 is fast or slow with the respect to the code time.

If UT1 is slow, a binary 1 labeled SUB (subtract) will be broadcast during the 38th second of the minute. If UT1 is fast, binary 1's labeled ADD will be broadcast during the 37th and 39th seconds of the minute. The 12th bcd group is not used to convey information.

CHU

The National Research Council of Ottawa, Ontario, Canada, broadcasts time signals that can be heard throughout North America and many other parts of the world. The frequencies are 3330, 7335, and 14,670 kHz, and the transmission is continuous on all frequencies. The transmitter has a power output of 3 kW at frequencies of 3330 and 14,670 kHz and a power output of 10 kW at 7335 kHz.

The frequencies and time signals are derived from a cesium atomic clock that is accurate to within a few microseconds per year.

A chart of the broadcast signal is shown in Fig. 2-6. The seconds pips consist of 300 cycles of a 1000-Hz tone. The seconds pips are broadcast continuously except for the 29th and the 51st to the 59th pips, which are omitted each minute. In addition, the 1st to 10th pips are omitted during the first minute of the hour. The beginning of the pip marks the exact second. The zero pip of each minute has a duration of 0.5 s, and the zero pip of each hour has a duration of 1 s. The remaining seconds pips have a duration of 0.3 s. An FSK time code is inserted after 10 cycles on the 31st to 39th seconds.

A voice announcement of the time is given each minute during the 10-s interval between the 50th and 60th second when the pips are omitted. The announcement is as follows: "CHU, Dominion Observatory Canada, Eastern Standard Time, ____ hours, ____ minutes." The time given refers to the beginning of the minute pip that follows and is on the 24-hr system.

Other Standards Stations

Throughout the world, there are many other stations that broadcast similar data. Table 2-10 lists some of them as well as some other data about stations operating on the standards frequencies. It also lists some other stations in the low frequency (LF) and very low frequency (VLF) bands, which broadcast similar data, but not on the frequencies assigned for standard-frequency operation.

WORLD TIME CONVERSION CHART

The standard time in any time zone can be converted to Greenwich Mean Time (GMT) (i.e., UTC) or to any time zone in other parts of the world by using the chart in Fig. 2-7. To use this chart, visualize the horizontal line as making a complete circle. From one time zone, trace horizontally to the right (counterclockwise); it will be tomorrow when passing through midnight and yesterday when passing the international date line. Moving to the left (clockwise), it will be yesterday when passing midnight and tomorrow when passing the international date line. There is no date change when passing both the international date line and midnight, moving in one direction. Always trace in the shortest direction between time zones.

Example. At 9 PM in New York Eastern Standard Time, it is 4 AM tomorrow in Moscow, Russia (moving left, clockwise).

At 10 AM in the Philippines, it will be 4 PM yesterday in Hawaii (moving right, counterclockwise).

At 7 AM Chicago Central Standard Time, it is 10 PM in Tokyo, Japan the same day (moving left, clockwise).

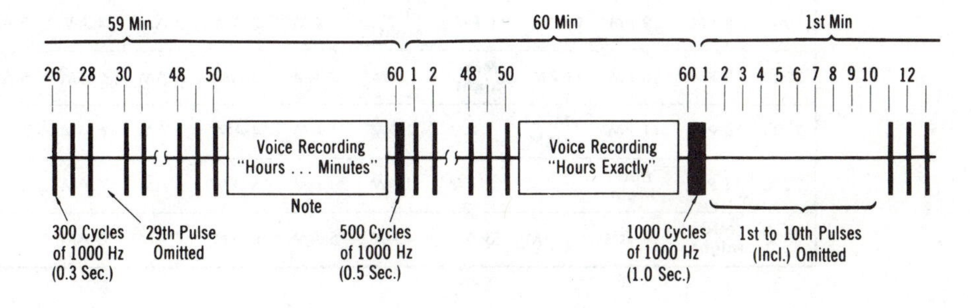

Fig. 2-6

Greenwich Meridian Time. London, England.	Central Europe. Berlin, Geneva, Stockholm, Vienna.	Eastern Europe, Athens, Cape Town, Cairo, Moscow.	Arabia, Armenia, Ethiopia, Madagascar.	Mauritius, Iran, Reunion Island.	Central Russia, Bombay, India.	Calcutta, Novosibirsk, Russia, Tibet.	Sumatra. Thailand, Laos.	Philippines. Perth, Australia.	Central Australia. Tokyo, Japan.	Eastern Australia. Melbourne, Sydney.	New Caledonia. New Zealand.
0000	1 AM	2 AM	3 AM	4 AM	5 AM	6 AM	7 AM	8 AM	9 AM	10 AM	11 AM
0100	2 AM	3 AM	4 AM	5 AM	6 AM	7 AM	8 AM	9 AM	10 AM	11 AM	Noon
0200	3 AM	4 AM	5 AM	6 AM	7 AM	8 AM	9 AM	10 AM	11 AM	Noon	**1 PM**
0300	4 AM	5 AM	6 AM	7 AM	8 AM	9 AM	10 AM	11 AM	Noon	**1 PM**	**2 PM**
0400	5 AM	6 AM	7 AM	8 AM	9 AM	10 AM	11 AM	Noon	**1 PM**	**2 PM**	**3 PM**
0500	6 AM	7 AM	8 AM	9 AM	10 AM	11 AM	Noon	**1 PM**	**2 PM**	**3 PM**	**4 PM**
0600	7 AM	8 AM	9 AM	10 AM	11 AM	Noon	**1 PM**	**2 PM**	**3 PM**	**4 PM**	**5 PM**
0700	8 AM	9 AM	10 AM	11 AM	Noon	**1 PM**	**2 PM**	**3 PM**	**4 PM**	**5 PM**	**6 PM**
0800	9 AM	10 AM	11 AM	Noon	**1 PM**	**2 PM**	**3 PM**	**4 PM**	**5 PM**	**6 PM**	**7 PM**
0900	10 AM	11 AM	Noon	**1 PM**	**2 PM**	**3 PM**	**4 PM**	**5 PM**	**6 PM**	**7 PM**	**8 PM**
1000	11 AM	Noon	**1 PM**	**2 PM**	**3 PM**	**4 PM**	**5 PM**	**6 PM**	**7 PM**	**8 PM**	**9 PM**
1100	Noon	**1 PM**	**2 PM**	**3 PM**	**4 PM**	**5 PM**	**6 PM**	**7 PM**	**8 PM**	**9 PM**	**10 PM**
1200	**1 PM**	**2 PM**	**3 PM**	**4 PM**	**5 PM**	**6 PM**	**7 PM**	**8 PM**	**9 PM**	**10 PM**	**11 PM**
1300	**2 PM**	**3 PM**	**4 PM**	**5 PM**	**6 PM**	**7 PM**	**8 PM**	**9 PM**	**10 PM**	**11 PM**	**Midnight**
1400	**3 PM**	**4 PM**	**5 PM**	**6 PM**	**7 PM**	**8 PM**	**9 PM**	**10 PM**	**11 PM**	**Midnight**	1 AM
1500	**4 PM**	**5 PM**	**6 PM**	**7 PM**	**8 PM**	**9 PM**	**10 PM**	**11 PM**	**Midnight**	1 AM	2 AM
1600	**5 PM**	**6 PM**	**7 PM**	**8 PM**	**9 PM**	**10 PM**	**11 PM**	**Midnight**	1 AM	2 AM	3 AM
1700	**6 PM**	**7 PM**	**8 PM**	**9 PM**	**10 PM**	**11 PM**	**Midnight**	1 AM	2 AM	3 AM	4 AM
1800	**7 PM**	**8 PM**	**9 PM**	**10 PM**	**11 PM**	**Midnight**	1 AM	2 AM	3 AM	4 AM	5 AM
1900	**8 PM**	**9 PM**	**10 PM**	**11 PM**	**Midnight**	1 AM	2 AM	3 AM	4 AM	5 AM	6 AM
2000	**9 PM**	**10 PM**	**11 PM**	**Midnight**	1 AM	2 AM	3 AM	4 AM	5 AM	6 AM	7 AM
2100	**10 PM**	**11 PM**	**Midnight**	1 AM	2 AM	3 AM	4 AM	5 AM	6 AM	7 AM	8 AM
2200	**11 PM**	**Midnight**	1 AM	2 AM	3 AM	4 AM	5 AM	6 AM	7 AM	8 AM	9 AM
2300	**Midnight**	1 AM	2 AM	3 AM	4 AM	5 AM	6 AM	7 AM	8 AM	9 AM	10 AM

Fig. 2-7

International Date Line. Fiji Islands.	Nome, Alaska. Samoa Islands.	Hawaii. Midway Islands.	Eastern Alaska. Dawson.	Pacific Standard Time. Los Angeles, Seattle, Juneau.	Mountain Standard Time. Calgary, Denver, Phoenix.	Central Standard Time. Chicago, Costa Rica.	Eastern Standard Time. Montreal, New York, Peru.	Atlantic Standard Time. Argentina, Nova Scotia.	Greenland. Rio de Janeiro, Brazil.	Azores.	Iceland. Canary Islands.
Noon	1 PM	2 PM	3 PM	4 PM	5 PM	6 PM	7 PM	8 PM	9 PM	10 PM	11 PM
1 PM	2 PM	3 PM	4 PM	5 PM	6 PM	7 PM	8 PM	9 PM	10 PM	11 PM	Mid-night
2 PM	3 PM	4 PM	5 PM	6 PM	7 PM	8 PM	9 PM	10 PM	11 PM	Mid-night	1 AM
3 PM	4 PM	5 PM	6 PM	7 PM	8 PM	9 PM	10 PM	11 PM	Mid-night	1 AM	2 AM
4 PM	5 PM	6 PM	7 PM	8 PM	9 PM	10 PM	11 PM	Mid-night	1 AM	2 AM	3 AM
5 PM	6 PM	7 PM	8 PM	9 PM	10 PM	11 PM	Mid-night	1 AM	2 AM	3 AM	4 AM
6 PM	7 PM	8 PM	9 PM	10 PM	11 PM	Mid-night	1 AM	2 AM	3 AM	4 AM	5 AM
7 PM	8 PM	9 PM	10 PM	11 PM	Mid-night	1 AM	2 AM	3 AM	4 AM	5 AM	6 AM
8 PM	9 PM	10 PM	11 PM	Mid-night	1 AM	2 AM	3 AM	4 AM	5 AM	6 AM	7 AM
9 PM	10 PM	11 PM	Mid-night	1 AM	2 AM	3 AM	4 AM	5 AM	6 AM	7 AM	8 AM
10 PM	11 PM	Mid-night	1 AM	2 AM	3 AM	4 AM	5 AM	6 AM	7 AM	8 AM	9 AM
11 PM	Mid-night	1 AM	2 AM	3 AM	4 AM	5 AM	6 AM	7 AM	8 AM	9 AM	10 AM
Mid-night	1 AM	2 AM	3 AM	4 AM	5 AM	6 AM	7 AM	8 AM	9 AM	10 AM	11 AM
1 AM	2 AM	3 AM	4 AM	5 AM	6 AM	7 AM	8 AM	9 AM	10 AM	11 AM	Noon
2 AM	3 AM	4 AM	5 AM	6 AM	7 AM	8 AM	9 AM	10 AM	11 AM	Noon	1 PM
3 AM	4 AM	5 AM	6 AM	7 AM	8 AM	9 AM	10 AM	11 AM	Noon	1 PM	2 PM
4 AM	5 AM	6 AM	7 AM	8 AM	9 AM	10 AM	11 AM	Noon	1 PM	2 PM	3 PM
5 AM	6 AM	7 AM	8 AM	9 AM	10 AM	11 AM	Noon	1 PM	2 PM	3 PM	4 PM
6 AM	7 AM	8 AM	9 AM	10 AM	11 AM	Noon	1 PM	2 PM	3 PM	4 PM	5 PM
7 AM	8 AM	9 AM	10 AM	11 AM	Noon	1 PM	2 PM	3 PM	4 PM	5 PM	6 PM
8 AM	9 AM	10 AM	11 AM	Noon	1 PM	2 PM	3 PM	4 PM	5 PM	6 PM	7 PM
9 AM	10 AM	11 AM	Noon	1 PM	2 PM	3 PM	4 PM	5 PM	6 PM	7 PM	8 PM
10 AM	11 AM	Noon	1 PM	2 PM	3 PM	4 PM	5 PM	6 PM	7 PM	8 PM	9 PM
11 AM	Noon	1 PM	2 PM	3 PM	4 PM	5 PM	6 PM	7 PM	8 PM	9 PM	10 PM

TABLE 2-10
Other Standards Stations

Station	Location	Frequency (kHz)	Schedule (UT)
ATA	Greater Kailash	5000	12h 30m to 3h 30m
	New Dehli	10,000	continuous
	India	15,000	3h 30m to 12h 30m
BPM	Pucheng	5000	14h to 24h
	China	10,000	continuous
		15,000	0h to 14h
BSF	Chung-Li	5000	continuous (except interruption between 35m
	Taiwan	10,000	and 40m)
	China		
DAM	Elmshorn	8638.5	11h 55m to 12h 06m
	Germany, F.R.	16,980.4	
		4265	23h 55m to 24h 06m from 21 October to 29 March
		8638.5	
		6475.5	23h 55m to 24h 06m from 30 March to 20 October
		12,763.5	
DAN	Osterloog	2614	11h 55m to 12h 06m
	Germany, F.R.		23h 55m to 24h 06m
DAO	Kiel	2775	11h 55m to 12h 06m
	Germany, F.R.		23h 55m to 24h 06m
DCF77	Mainflingen	77.5	continuous
	Germany, F.R.		
DGI	Oranienburg	182	5h 59m 30s to 6h 00m
	Germ. Dem. Rep.		11h 59m 30s to 12h 00m
			17h 59m 30s to 18h 00m
			advanced 1h in summer
EBC	San Fernando	12,008	10h 00m to 10h 25m
	Spain	6840	10h 30m to 10h 55m
FFH	Ste Assise	2500	continuous from 8h to 16h 25m except on Sunday
	France		
FTH42	Ste Assise	7428	at 9h and 21h
FTK77	France	10,775	at 8h and 20h
FTN87		13,873	at 9h 30m, 13h, 22h 30m (may be cancelled)
GBR	Rugby	16	2h 55m to 3h 00m
	United Kingdom		8h 55m to 9h 00m
			14h 55m to 15h 00m
			20h 55m to 21h 00m
HBG	Prangins	75	continuous
	Switzerland		

TABLE 2-10 Cont.
Other Standards Stations

Station	Location	Frequency (kHz)	Schedule (UT)
HLA	Taedok Rep. of Korea	5000	1h to 8h Monday to Friday
IAM	Rome Italy	5000	7h 30m to 8h 30m 10h 30m to 11h 30m except Saturday afternoon, Sunday, and national holidays; advanced 1h in summer
IBF	Torino Italy	5000	during 15m preceding 7h, 9h, 10h, 11h, 12h, 13h, 14h, 15h, 16h, 17h, 18h, advanced by 1 h in summer
JG2AS	Sanwa Ibaraki Japan	40	continuous, except interruptions during communications
JJY	Sanwa Ibaraki Japan	2500 5000 8000 10,000 15,000	continuous, except interruption between 35m and 39m
LOL1	Buenos-Aires Argentina	5000 10,000 15,000	11h to 12h, 14h to 15h, 17h to 18h, 20h to 21h, 23h to 24h
LOL2 LOL3	Buenos-Aires Argentina	4856 8030 17,180	1h, 13h, 21h
MSF	Rugby United Kingdom	60	continuous except for an interruption for maintenance from 10h 0m to 14h 0m on the first Tuesday in each month
MSF	Rugby United Kingdom	2500 5000 10,000	between minutes 0 and 5, 10 and 15, 20 and 25, 30 and 35, 40 and 45, 50 and 55
OLB5	Poděbrady Czechoslovakia	3170	continuous except from 6h to 12h on the first Wednesday of every month
OMA	Liblice Czechoslovakia	50	continuous except from 6h to 12h on the first Wednesday of every month emitted from Poděbrady with reduced power
OMA	Liblice Czechoslovakia	2500	continuous except from 6h to 12h on the first Wednesday of every month
PPE	Rio-de-Janeiro Brazil	8721	0h 30m, 11h 30m, 13h 30m, 19h 30m, 20h 30m, 23h 30m

TABLE 2-10 Cont.
Other Standards Stations

Station	Location	Frequency (kHz)	Schedule (UT)
PPR	Rio-de-Janeiro Brazil	435 4244 8634 13,105 17,194.4 22,603	1h 30m, 14h 30m, 21h 30m
RBU	Moscow U.S.S.R.	66⅔	continuous
RCH	Tashkent U.S.S.R.	2,500 10,000	between minutes 0m and 10m, 30m and 40m 0h to 3h 40m, 5h 30m to 23h 40m 5h 00m to 13h 10m 10h to 13h 10m
RID	Irkutsk U.S.S.R.	5004 10,004 15,004	the station simultaneously operates on three frequencies between minutes 20m and 30m, 50m and 60m
RTA	Novosibirsk U.S.S.R.	10,000 15,000	between 0m and 10m, 30m and 40m 0h to 5h 10m 14h to 23h 40m 6h 30m to 13h 10m
RTZ	Irkutsk U.S.S.R.	50	between 0m and 5m, from 0h to 20h 5m, ending 22h to 23h 5m in winter from 0h to 19h 5m and 21h to 23h 5m in summer
RWM	Moscow U.S.S.R.	4996 9996 14,996	the station simultaneously operates on three frequencies between 10m and 20m, 40m and 50m
UNW3	Molodechno U.S.S.R.	25	from 7h 43m to 7h 52m and 19h 43m to 19h 52m in winter from 7h 43m to 7h 52m and 20h 43m to 20h 52m in summer
UPD8	Arkhangelsk U.S.S.R.	25	from 8h 43m to 8h 52m and 11h 43m to 11h 52m
UQC3	Chabarovsk U.S.S.R.	25	from 0h 43m to 0h 52m, 6h 43m to 6h 52m, and 17h 43m to 17h 52m in winter from 2h 43m to 2h 52m, 6h 43m to 6h 52m, and 18h 43m to 18h 52m in summer
USB2	Frunze U.S.S.R.	25	from 4h 43m to 4h 52m, 9h 43m to 9h 52m, and 21h 43m to 21h 52m in winter from 4h 43m to 4h 52m, 10h 43m to 10h 52m, and 22h 43m to 22h 52m in summer

TABLE 2-10 Cont.
Other Standards Stations

Station	Location	Frequency (kHz)	Schedule (UT)
UTR3	Gorki U.S.S.R.	25	from 5h 43m to 5h 52m, 13h 43m to 13h 52m, and 18h 43m to 18h 52m in winter from 7h 43m to 7h 52m, 14h 43m to 14h 52m, and 19h 43m to 19h 52m in summer
VNG	Lyndhurst Australia	4500 7500 12,000	9h 45m to 21h 30m continuous except 22h 30m to 22h 45m 21h 45m to 9h 30m
Y3S	Nauen Germ. Dem. Rep.	4525	continuous except from 8h 15m to 9h 45m for maintenance if necessary
YVTO	Caracas Venezuela	6100	continuous
ZUO	Olifantsfontein South Africa	2500 5000	18h to 4h continuous
ZUO	Johannesburg South Africa	100,000	continuous

FREQUENCY AND POWER OPERATING TOLERANCES

AM Broadcast

The operating frequency tolerance of each station shall be maintained within ± 20 Hz of the assigned frequency.

The operating power of each AM broadcast station shall be maintained as near as practicable to the licensed power and shall not exceed the limits of 5% above and 10% below the licensed power except in emergencies.

FM Broadcast

Operating frequency tolerance of each station shall be maintained within ± 2000 Hz of the assigned center frequency.

The operating power of each station shall be maintained as near as practicable to the authorized operating power and shall not exceed the limits of 5% above and 10% below the authorized power except in emergencies.

TV Broadcast

The carrier frequency of the visual transmitter shall be maintained within ± 1000 Hz of the authorized carrier frequency.

The center frequency of the aural transmitter shall be maintained 4.5 MHz ± 1000 Hz above the visual carrier frequency.

The peak power shall be monitored by a peak-reading device that reads proportionally to voltages, current, or power in the radiofrequency line. The operating power as so monitored shall be maintained as near as practicable to the authorized operating power and shall not exceed the limits of 10% above and 20% below the authorized power except in emergencies.

The operating power of the aural transmitter shall be maintained as near as practicable to the authorized operating power and shall not exceed the limits of 10% above and 20% below the authorized power except in emergencies.

TABLE 2-11
Power Limits of Personal Radio Services Stations

Class of station	Maximum transmitter output power (W)
general mobile radio service	50
remote control (R/C) service—27.255 MHz	25*
remote control (R/C) service—26.995-27.195 MHz	4
remote control (R/C) service—72-76 MHz	0.75
citizens band (CB) radio service—carrier (where applicable)	4
citizens band (CB) radio service—peak envelope power (where applicable)	12

*A maximum transmitter output of 25 W is permitted on 27.255 MHz only.

TABLE 2-12
Frequency Tolerances of Personal Radio Services Stations

	Frequency tolerance (%)	
Class of station	***Fixed and base***	***Mobile***
general radio service	0.00025	0.0005
remote control (R/C) service	—	0.005*
citizens band (CB) service	—	0.005

*Remote control stations that have a transmitter output of 2.5 W or less, used solely for remote control of objects or devices by radio (other than devices used solely as a means of attracting attention), are permitted a frequency tolerance of 0.01%.

Industrial Radio Service

The carrier frequency of stations operating below 220 MHz in the Industrial Radio Service shall be maintained within ± 0.01% of the authorized power for stations of 3 W or less and within ± 0.005% for stations with an authorized power of more than 3 W. The frequency tolerance of Industrial Radio Service stations operating between 220 and 1000 MHz is specified in the station authorization.

Personal Radio Service (CB)

The maximum power at the transmitter output terminals and delivered to the antenna, antenna transmission line, or other impedance-matched radiofrequency load shall not exceed the values in Table 2-11 under any condition of modulation.

The carrier frequency of a station in this service shall be maintained within the percentages of authorized frequency shown in Table 2-12.

The assigned channel frequencies and upper and lower tolerance limits for citizens band (CB) radio service are listed in Table 2-13.

COMMERCIAL OPERATOR LICENSES

Types of Licenses

Currently, the FCC issues six types of commercial radio licenses and two types of endorsements. They are:

1. *Restricted Radiotelephone Operator Permit.* A Restricted Radiotelephone

TABLE 2-13
Citizens Band Frequencies and Upper and Lower Tolerances

Channel	Assigned frequency (MHz)	Lower limit (MHz)	Upper limit (MHz)
1	26.965000	26.963651	26.966348
2	26.975000	26.973651	26.976348
3	26.985000	26.983650	26.986349
4	27.005000	27.003649	27.006350
5	27.015000	27.013649	27.016450
6	27.025000	27.023648	27.026351
7	27.035000	27.033648	27.036351
8	27.055000	27.053647	27.056352
9	27.065000	27.063646	27.066353
10	27.075000	27.073646	27.076353
11	25.085000	27.083645	27.086354
12	27.105000	27.103644	27.106355
13	27.115000	27.113644	27.116356
14	27.125000	27.123643	27.126356
15	27.135000	27.133643	27.136356
16	27.155000	27.153642	27.156357
17	27.165000	27.163641	27.166358
18	27.175000	27.173641	27.176359
19	27.185000	27.183640	27.186359
20	27.205000	27.203639	27.206360
21	27.215000	27.213639	27.216360
22	27.225000	27.223638	27.226361
23	27.255000	27.253637	27.256363
24	27.235000	27.233638	27.236362
25	27.245000	27.243637	27.246362
26	27.265000	27.263636	27.266364
27	27.275000	27.273636	27.276364
28	27.285000	27.283635	27.286365
29	27.295000	27.293635	27.296365
30	27.305000	27.303634	27.306366
31	27.315000	27.313634	27.316366
32	27.325000	27.323633	27.326366
33	27.335000	27.333633	27.336367
34	27.345000	27.343632	27.346368
35	27.355000	27.353632	27.356368
36	27.365000	27.363631	27.366369
37	27.375000	27.373631	27.376369
38	27.385000	27.383630	27.386369
39	27.395000	27.393639	27.396370
40	27.405000	27.403629	27.406370

Operator Permit allows operation of most aircraft and aeronautical ground stations, maritime radiotelephone stations on pleasure vessels (other than those carrying more than six passengers for hire), and most VHF marine coast and utility stations. It is the only type of license required for transmitter operation, repair, and maintenance (including acting as chief operator) of all types of AM, FM, TV, and international broadcast stations.

There is no examination for this license. To be eligible for it you must:

Be at least 14 years old

Be a legal resident of (eligible for employment in) the U.S. or (if not so eligible) hold an aircraft pilot certificate valid in the U.S. or an FCC radio station license in your name

Be able to speak and hear

Be able to keep at least a rough written log

Be familiar with provisions of applicable treaties, laws, and rules that govern the radio station you will operate

A Restricted Radiotelephone Operator License is normally valid for the lifetime of the holder.

2. *Marine Radio Operator Permit.* A Marine Radio Operator Permit is required to operate radiotelephone stations on board certain vessels sailing the Great Lakes, any tidewater, or the open sea. It is also required to operate certain aviation radiotelephone stations, and certain maritime coast radiotelephone stations. It does not

authorize the operation of AM, FM, or TV broadcast stations.

To be eligible for this license, you must:

Be a legal resident of (eligible for employment in) the U.S.

Be able to receive and transmit spoken messages in English

Pass a written examination covering basic radio law and operating procedures

The Marine Operator Permit is normally valid for a renewable five-year term.

3. *General Radiotelephone Operator License.* A General Radiotelephone Operator License is required for persons responsible for internal repairs, maintenance, and adjustment of FCC licensed radiotelephone transmitters in the Aviation, Maritime, and International Public Fixed radio services. It is also required for operation of maritime land radio transmitters operating with more than 1500 W of peak envelope power and maritime mobile (ship) and aeronautical transmitters with more than 1000 W of peak envelope power.

To be eligible for this license, you must:

Be a legal resident of (eligible for employment in) the U.S.

Be able to receive and transmit spoken messages in English

Pass a written examination covering basic radio law, operating procedures, and basic electronics

The General Radiotelephone Operator License is normally valid for the lifetime of the operator.

4. *Third Class Radiotelegraph Operator Certificate.* A Third Class Radiotelegraph Operator Certificate is required to operate certain coast radiotelegraph stations. It also conveys all the authority of both the Restricted Radiotelephone Operator Permit and the Marine Radio Operator Permit.

To be eligible for this license, you must:

Be a legal resident of (eligible for employment in) the U.S.

Be able to receive and transmit spoken messages in English

Pass Morse code examinations at 16 code groups per minute and 20 words per minute plain language (receive and transmit by hand)

Pass a written examination covering basic radio law, basic operating procedures (telephony), and basic operating procedures (telegraphy)

The Third Class Radiotelegraph Operator Certificate is normally valid for a renewable five-year term.

5. *Second Class Radiotelegraph Operator Certificate.* A Second Class Radiotelegraph Operator Certificate is required to operate ship and coast radiotelegraph stations in the maritime services and to take responsibility for internal repairs, maintenance, and adjustments of any FCC-licensed radiotelegraph transmitter other than an amateur radio transmitter. It also conveys all of the authority of the Third Class Radiotelegraph Operator Certificate.

To be eligible for this license, you must:

Be a legal resident of (eligible for employment in) the U.S.

Be able to receive and transmit spoken messages in English

Pass Morse code examinations at 16 code groups per minute and 20 words per minute plain language (receive and transmit by hand)

Pass a written examination covering basic radio law, basic operating procedures (telephony), basic operating procedures (telegraphy), and electronics technology as applicable to radiotelegraph stations

The Second Class Radiotelegraph Operator Certificate is normally valid for a renewable five-year term.

6. *First Class Radiotelegraph Operator Certificate.* A First Class Radiotelegraph Operator Certificate is required only for those who serve as the chief radio operator on U.S. passenger ships. It also conveys all of the authority of the Second Class Radiotelegraph Operator Certificate.

   To be eligible for this license, you must:

   Be at least 21 years old

   Have at least one year of experience in sending and receiving public correspondence by radiotelegraph at ship stations, coast stations, or both

   Be a legal resident of (eligible for employment in) the U.S.

   Be able to receive and transmit spoken messages in English

   Pass Morse code examinations at 20 code groups per minute and 25 words per minute plain language (receive and transmit by hand)

   Pass a written examination covering basic radio law, basic operating procedures (telephony), basic operating procedures (telegraphy), and electronics technology as applicable to radiotelegraph stations

   The First Class Radiotelegraph Operator Certificate is normally valid for a renewable five-year term.

7. *Ship Radar Endorsement.* The Ship Radar Endorsement is required to service and maintain ship radar equipment.

   To be eligible for this endorsement, you must:

   Hold a valid First or Second Class Radiotelegraph Operator Certificate or a General Radiotelephone Operator License

   Pass a written examination covering the technical fundamentals of radar and radar maintenance techniques

8. *Six-Months Service Endorsement.* The Six-Months Service Endorsement is required to permit the holder to serve as the sole radio operator on board large U.S. cargo ships.

   To be eligible for this endorsement, you must:

   Hold a valid First Class or Second Class Radiotelegraph Operator Certificate

Have at least six months of satisfactory service as a radio officer on board a ship (or ships) of the U.S. equipped with a radiotelegraph station in compliance with Part II of Title III of the Communications Act of 1934

Have held a valid First Class or Second Class Radiotelegraph Operator Certificate while obtaining the six months of service

Have been licensed as a radio officer by the U.S. Coast Guard, in accordance with the Act of May 12, 1948 (46 U.S.C. 229 a-h), while obtaining the six months of service

Discontinued Licenses

The FCC no longer issues the Radiotelephone First or Second Class Operator Licenses, the Radiotelephone Third Class Operator Permit, the Broadcast Endorsement or the Aircraft Radiotelegraph Endorsement. Holders of such licenses should follow the following instructions pertaining to the license held when it is time to renew their license.

1. *Radiotelephone First Class Operator License.* The Radiotelephone First Class Operator License have been abolished and the requirements for holding such licenses to operate and maintain broadcast transmitters have been eliminated. Persons holding such a license will be issued a General Radiotelephone Operator License when they apply at renewal.

2. *Radiotelephone Second Class Operator License.* The Radiotelephone Second Class Operator License has been renamed the General Radiotelephone Operator License. Persons holding the Radiotelephone Second Class Operator License will be issued a General Radiotelephone Operator License when they apply for renewal.

3. *Radiotelephone Third Class Operator Permit.* The Radiotelephone Third Class Operator Permit has been converted to the Marine Radio Operator Pemit. The requirement for its use with a Broadcast Endorsement has been abolished.

   If you are employed as a radio operator aboard vessels or aeronautical stations where its use is required, request issuance of a Marine Radio Operator Permit at time of renewal. (No examination is necessary if your Radiotelephone Third Class Operator Permit expired not more than five years before application.)

   If you hold a Radiotelephone Third Class Operator Permit With Broadcast Endorsement for operating a broadcast station, apply for a Restricted Radiotelephone Operator Permit at time for renewal.

   If you operate stations that require you to hold a Marine Operator Permit and you also operate the transmitter of an AM, FM, or TV broadcast station, you should apply for both a Marine Operator Permit and a Restricted Radiotelephone Operator Permit at time of renewal.

4. *Broadcast Endorsement.* The Broadcast Endorsement to the Radiotelephone Third Class Operator Permit formerly required for operations of some classes of broadcast transmitter has been abolished along with the requirement

for a Radiotelephone Third Class Operator Permit. Holders of this type of license and endorsement who have been using it for broadcast transmitter operation should apply for a Restricted Radiotelephone Operator Permit during the last year of the license term.

5. *Aircraft Radiotelegraph Endorsement.* The use of radiotelegraphy aboard aircraft has been discontinued and the Aircraft Radiotelegraph Endorsement has been abolished. If you hold a license with such an endorsement, the endorsement will be eliminated at renewal.

Examination Elements

Written examinations are composed of questions from various categories called elements. These elements, and the types of questions in each, are:

Element 1. *Basic Marine Radio Law.* Provisions of laws, treaties, and regulations with which every operator in the maritime radio services should be familiar.

Element 2. *Basic Operating Practice.* Radio operating procedures and practices generally followed or required in communicating by radiotelephone in the maritime radio services.

Element 3. *General Radiotelephone.* Provisions of laws, treaties, and regulations with which every radio operator in the maritime radio service should be familiar. Radio operating practices generally followed or required in communicating by radiotelephone in the maritime radio services. Technical matters including fundamentals of electronics technology and maintenance techniques as necessary for repair and maintenance of radio transmitters and receivers.

Element 4. *Radiotelegraph Operating Practice.* Radio-operating procedure and practices generally followed or required in operation of shipboard radiotelegraph stations.

Element 5. *Advanced Radiotelegraph.* Technical, legal, and other matters, including electronics technology and radio maintenance and repair techniques applicable to all classes of radiotelegraph stations.

Element 6. *Ship Radar Techniques.* Specialized theory and practice applicable to the proper installation, servicing, and maintenance of ship radar equipment.

AMATEUR OPERATOR PRIVILEGES

Examination Elements

Examinations for amateur operator privileges are composed of questions from various categories, called elements. The various elements and their requirements are:

Element 1(A). *Beginner's Code Test.* Code test at 5 words per minute.

Element 1(B). *General Code Test.* Code test at 13 words per minute.

Element 1(C). *Expert's Code Test.* Code test at 20 words per minute.

Element 2. *Basic Law.* Rules and regulations essential to beginners' operation, including sufficient elementary radio theory to understand these rules.

Element 3. *General Regulations.* Amateur radio operation and apparatus and provisions of treaties, statutes, and rules and regulations affecting all amateur stations and operators.

Element 4(A). *Intermediate Amateur Practice.* Involving intermediate level for general amateur practice in radio theory and operation as applicable to modern amateur techniques, including—but not limited to—radiotelephony and radiotelegraphy.

Element 4(B). *Advanced Amateur Practice.* Advanced radio theory and operation applicable to modern amateur techniques, including—but not limited to—radiotelephony, radiotelegraphy, and transmission of energy for (1) measurements and observations applied to propagation, (2) radio control of remote objects, and (3) similar experimental purposes.

Examination Requirements

An applicant for an original license must be a U.S. citizen or other U.S. national and will be required to pass examinations as follows:

1. *Amateur Extra Class.* Elements 1(C), 2, 3, 4(A), and 4(B).
2. *Advanced Class.* Elements 1(B), 2, 3, and 4(A).
3. *General Class.* Elements 1(B), 2, and 3.
4. *Technician Class.* Elements 1(A), 2, and 3.
5. *Novice Class.* Elements 1(A) and 2.

Note. Since January 1, 1985 all examinations for amateur radio licenses are given by volunteer amateur examiners. Complete details are given in the FCC rules.

AMATEUR ("HAM") BANDS

The frequency bands for various amateur licenses follow.

1. *Amateur Extra Class.* All amateur bands, including these privileged frequencies:

   3500–3525 kHz
   3775–3800 kHz
   7000–7025 kHz
   14,000–14,025 kHz
   14,150–14,175 kHz
   21,000–21,025 kHz
   21,200–21,225 kHz

2. *Advanced Class.* All amateur bands except those frequencies reserved for Amateur Extra Class, including these privileged frequencies:

   3800–3890 kHz
   7150–7225 kHz
   14,175–14,350 kHz
   21,225–21,300 kHz

3. *General Class.* All amateur bands except those frequencies reserved for Amateur Extra Class and Advanced Class.

4. *Technician Class.* All authorized privileges on amateur frequency bands above 50 MHz and those assigned to the Novice Class.

5. *Novice Class.* The following selected bands, using only Type A1 emission.

   3700–3750 kHz
   7100–7150 kHz
   21.10–21.20 and 28.1-28.2 MHz

   The DC power input to the stage supplying power to the antenna shall not exceed 250 W, and the transmitter shall be crystal controlled.

The various bands of frequencies used by amateur radio operators ("hams") are usually referred to in meters instead of the actual frequencies. The number of meters approximates the wavelength at the band of frequencies being designated. The meter

TABLE 2-14
"Ham" Bands

Frequency band limits	Types of emission	Band (meters)
1800–2000 kHz	A1, A3	160
3500–4000 kHz	A1	80
3500–3750 kHz	F1	80
3750–4000 kHz	A3, A4, A5, F3, F4, F5	80
5167.5 kHz	A3A, A3J	80
7000–7300 kHz	A1	40
7000–7150 kHz	F1	40
7075–7100 kHz	A3, F3	40
7150–7300 kHz	A3, A4, A5, F3, F4, F5	40
10,100–10,109 kHz	A1, F1	30
10,115–10,150 kHz	A1, F1	30
14,000–14,350 kHz	A1	20
14,000–14,150 kHz	F1	20
14,150–14,350 kHz	A3, A4, A5, F3, F4, F5	20
21.000–21.450 MHz	A1	15
21.000–21.200 MHz	F1	15
21.200–21.450 MHz	A3, A4, A5, F3, F4, F5	15
28.000–29.700 MHz	A1	
28.000–28.500 MHz	F1	
28.500–29.700 MHz	A3, A4, A5, F3, F4, F5	
50.000–54.000 MHz	A2, A3, A4, A5, F1, F2, F3, F4, F5	
51.000–54.000 MHz	A0, F0	
144–140 MHz	A1	2
144.100–148.000 MHz	A0, A2, A3, A4, A5, F0, F1, F2, F3, F4, F5	2
220–225 MHz	A0, A1, A2, A3, A4, A5, F0, F1, F2, F3, F4, F5	1¼
420–450 MHz	A0, A1, A2, A3, A4, A5, F0, F1, F2, F3, F4, F5	¾
1215–1300 MHz	A0, A1, A2, A3, A4, A5, F0, F1, F2, F3, F4, F5	
2300–2450 MHz	A0, A1, A2, A3, A4, A5, F0, F1, F2, F3, F4, F5	
3300–3500 MHz	A0, A1, A2, A3, A4, A5, F0, F1, F2, F3, F4, F5, P	
5650–5925 MHz	A0, A1, A2, A3, A4, A5, F0, F1, F2, F3, F4, F5, P	
10.0–10.5 GHz	A0, A1, A2, A3, A4, A5, F0, F1, F2, F3, F4, F5	
24.0–24.25 GHz	A0, A1, A2, A3, A4, A5, F1, F2, F3, F4, F5, P	
48–50, 71–76 GHz	A0, A1, A2, A3, A4, A5, F0, F1, F2, F3, F4, F5, P	
Above 300 GHz	A0, A1, A2, A3, A4, A5, F0, F1, F2, F3, F4, F5, P	

bands and their frequency limits are given in Table 2-14.

Note. Frequencies between 220 and 225 MHz are sometimes referred to as 1¼ m and between 420 and 450 MHz as ¾ m.

The maximum DC plate input power in watts for the 160-m band (1.8–2.0 MHz) is shown in Table 2-15 for all states and U.S. possessions.

TYPES OF EMISSIONS

Emissions are classified according to their modulation, type of transmission, and supplementary characteristics. These classifications are given in Table 2-16. When a full designation of the emissions—including bandwidth—is necessary, the symbols in Table 2-16 are prefixed by a number indicating the bandwidth in kilohertz. Below 10 kHz, this number is given to two significant figures.

TABLE 2-15
Maximum Power for the 160-m Band

	Maximum DC plate input power in watts							
	1800–1825 kHz	*1825–1850 kHz*	*1850–1875 kHz*	*1875–1900 kHz*	*1900–1925 kHz*	*1925–1950 kHz*	*1950–1975 kHz*	*1975–2000 kHz*
Area	*Day/Night*	*Day/Night*	*Day/Night*	*Day/Night*	*Day/Night*	*Day/Night*	*Day/Night*	*Day/Night*
Alabama	500/100	100/25	0	0	0	0	100/25	500/100
Alaska	1000/200	500/100	500/100	100/25	0	0	0	0
Arizona	1000/200	500/100	500/100	0	0	0	0	0
Arkansas	1000/200	500/100	100/25	0	0	100/25	100/25	500/100
California	1000/200	500/100	500/100	100/25	0	0	0	0
Colorado	1000/200	500/100	200/50	0	0	0	0	200/50
Connecticut	500/100	100/25	0	0	0	0	0	0
Delaware	500/100	100/25	0	0	0	0	0	100/25
District of Columbia	500/100	100/25	0	0	0	0	0	100/25
Florida	500/100	100/25	0	0	0	0	100/25	500/100
Georgia	500/100	100/25	0	0	0	0	0	200/50
Hawaii	0	0	0	0	200/50	100/25	100/25	500/100
Idaho	1000/200	500/100	500/100	100/25	100/25	100/25	100/25	500/100
Illinois	1000/200	500/100	100/25	0	0	0	0	200/50
Indiana	1000/200	500/100	100/25	0	0	0	0	200/50
Iowa	1000/200	500/100	200/50	0	0	100/25	100/25	500/100
Kansas	1000/200	500/100	100/25	0	0	100/25	100/25	500/100
Kentucky	1000/200	500/100	100/25	0	0	0	0	200/50
Louisiana	500/100	100/25	0	0	0	0	100/25	500/100
Maine	500/100	100/25	0	0	0	0	0	0
Maryland	500/100	100/25	0	0	0	0	0	100/25
Massachusetts	500/100	100/25	0	0	0	0	0	0
Michigan	1000/200	500/100	100/25	0	0	0	0	100/25
Minnesota	1000/200	500/100	500/100	100/25	100/25	100/25	100/25	500/100
Mississippi	500/100	100/25	0	0	0	0	100/25	500/100
Missouri	1000/200	500/100	100/25	0	0	100/25	100/25	500/100
Montana	1000/200	500/100	500/100	100/25	100/25	100/25	100/25	500/100
Nebraska	1000/200	500/100	200/50	0	0	100/25	100/25	500/100
Nevada	1000/200	500/100	500/100	100/25	0	0	0	0
New Hampshire	500/100	100/25	0	0	0	0	0	0
New Jersey	500/100	100/25	0	0	0	0	0	0
New Mexico	1000/200	500/100	100/25	0	0	100/25	500/100	1000/200
New York	500/100	100/25	0	0	0	0	0	0
North Carolina	500/100	100/25	0	0	0	0	0	100/25
North Dakota	1000/200	500/100	500/100	100/25	100/25	100/25	100/25	500/100
Ohio	1000/200	500/100	100/25	0	0	0	0	100/25
Oklahoma	1000/200	500/100	100/25	0	0	100/25	100/25	500/100
Oregon	1000/200	500/100	500/100	100/25	0	0	0	0
Pennsylvania	500/100	100/25	0	0	0	0	0	0
Rhode Island	500/100	100/25	0	0	0	0	0	0
South Carolina	500/100	100/25	0	0	0	0	0	200/50
South Dakota	1000/200	500/100	500/100	100/25	100/25	100/25	100/25	500/100
Tennessee	1000/200	500/100	100/25	0	0	0	0	200/50
Texas	500/100	100/25	0	0	0	0	0	200/50
Utah	1000/200	500/100	500/100	100/25	100/25	0	0	100/25
Vermont	500/100	100/25	0	0	0	0	0	0
Virginia	500/100	100/25	0	0	0	0	0	100/25

TABLE 2-15 Cont.
Maximum Power for the 160-m Band

	Maximum DC plate input power in watts							
	1800–1825 kHz	*1825–1850 kHz*	*1850–1875 kHz*	*1875–1900 kHz*	*1900–1925 kHz*	*1925–1950 kHz*	*1950–1975 kHz*	*1975–2000 kHz*
Area	***Day/Night***	***Day/Night***	***Day/Night***	***Day/Night***	***Day/Night***	***Day/Night***	***Day/Night***	***Day/Night***
Washington	1000/200	500/100	500/100	100/25	0	0	0	0
West Virginia	1000/200	500/100	100/25	0	0	0	0	100/25
Wisconsin	1000/200	500/100	200/50	0	0	0	0	200/50
Wyoming	1000/200	500/100	500/100	100/25	100/25	0	0	200/50
Puerto Rico	500/100	100/25	0	0	0	0	0	200/50
Virgin Islands	500/100	100/25	0	0	0	0	0	200/50
Swan Island	500/100	100/25	0	0	0	0	100/25	500/100
Serrana Bank	500/100	100/25	0	0	0	0	100/25	500/100
Roncador Key	500/100	100/25	0	0	0	0	100/25	500/100
Navassa Island	500/100	100/25	0	0	0	0	0	200/50
Baker, Canton, Enderbury, Howland	100/25	0	0	100/25	100/25	0	0	100/25
Guam, Johnston, Midway	0	0	0	0	100/25	0	0	100/25
American Samoa	200/50	0	0	200/50	200/50	0	0	200/50
Wake	100/25	0	0	100/25	0	0	0	0
Palmyra, Jarvis	0	0	0	0	200/50	0	0	200/50

TABLE 2-16
Types of Emission

Type of modulation	Type of transmission	Supplementary characteristics	Symbol
1. amplitude	absence of any modulation	—	A0
	telegraphy without the use of modulating audiofrequency (on–off keying)	—	A1
	telegraphy by the keying of a modulating audiofrequency or audiofrequencies or by the keying of the modulated emission (special case: an unkeyed modulated emission)	—	A2
	telephony	double sideband, full carrier	A3
		single sideband, reduced carrier	A3a
		two independent sidebands, reduced carrier	A3b
	facsimile	—	A4
	television	—	A5

TABLE 2-16 Cont.
Types of Emission

Type of modulation	Type of transmission	Supplementary characteristics	Symbol
1. amplitude	composite transmissions, and cases not covered by the above	—	A9
	composite transmissions	reduced carrier	A9c
2. frequency (or phase) modulated	absence of any modulation	—	F0
	telegraphy without the use of modulating audiofrequency (frequency shift keying)	—	F1
	telegraphy by the keying of a modulating audiofrequency or audiofrequencies or by the keying of the modulated emission (special case: an unkeyed emission modulated by audiofrequency)	—	F2
	telephony	—	F3
	facsimile	—	F4
	television	—	F5
	composite transmissions and cases not covered by the above	—	F9
3. pulsed emissions	absence of any modulation-carrying information	—	P0
	telegraphy without the use of modulating audiofrequency	—	P1
	telegraphy by the keying of a modulating audiofrequency or of the modulated pulse (special case: an unkeyed modulated pulse)	audiofrequency or audiofrequencies modulating the pulse in amplitude	P2d
		audiofrequency or audiofrequencies modulating the width of the pulse	P2c
		audiofrequency or audiofrequencies modulating the phase (or position) of the pulse	P2f
	telephony	amplitude-modulated pulse	P3d
		width-modulated pulse	P3e
		phase-(or position-)modulated pulse	P3f
	composite transmissions and cases not covered by the above	—	P9

TELEVISION SIGNAL STANDARDS

The signal standards for television broadcasting are given in Fig. 2-8.

Note. The standards given here are for color transmission. For monochrome transmission, the standards are the same except the color burst signal is omitted.

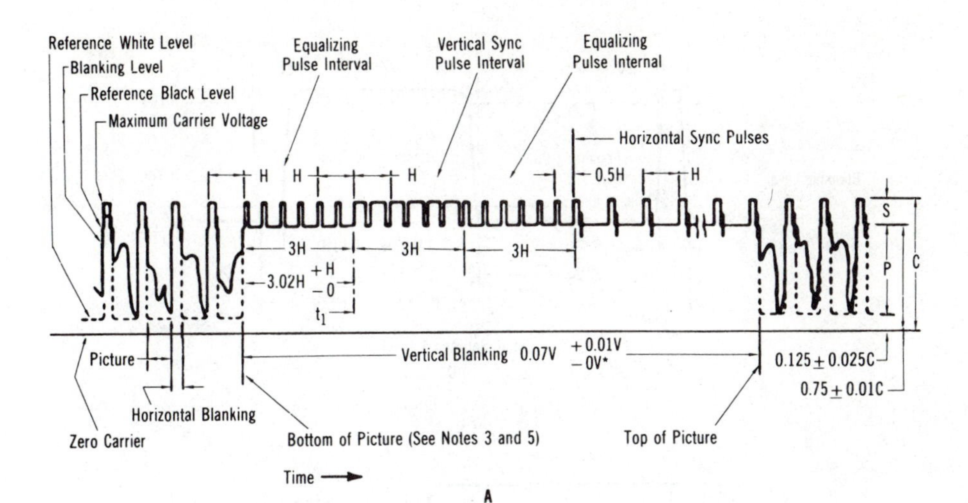

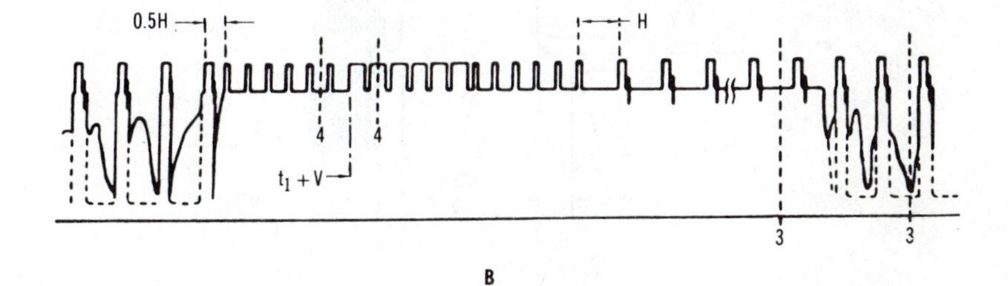

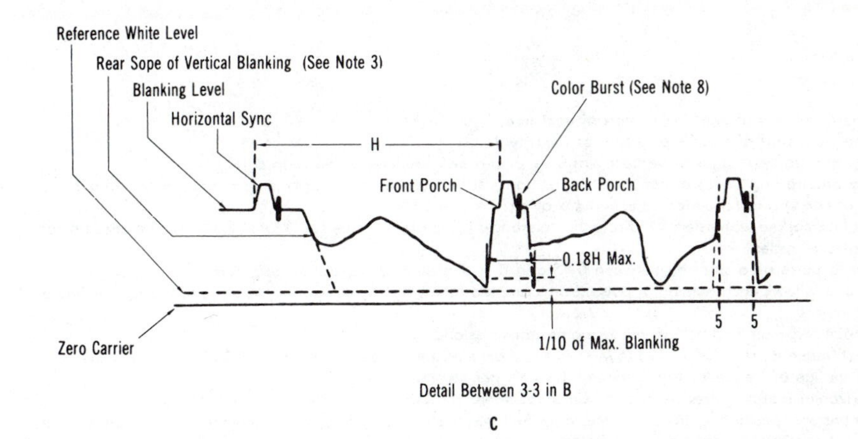

Horizontal Dimensions Not to Scale in A, B, and C

Fig. 2-8. Television signal standards.

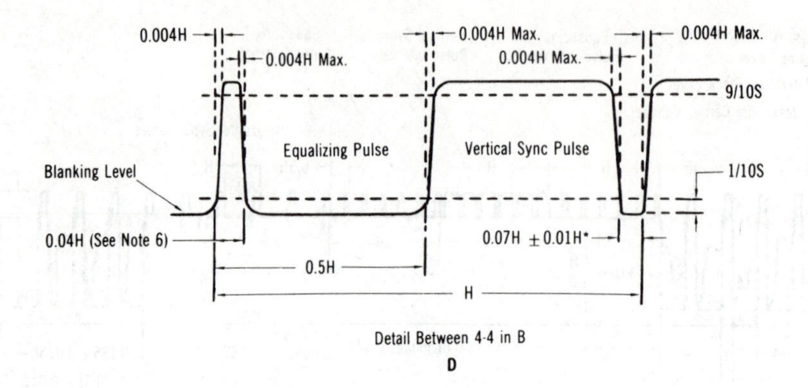

Detail Between 4-4 in B

D

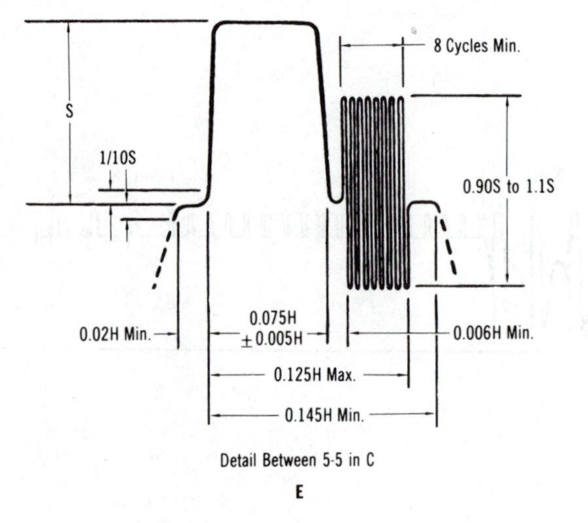

Detail Between 5-5 in C

E

NOTES

1. H = Time from start of one line to start of next line.
2. V = Time from start of one field to start of next field.
3. Leading and trailing edges of vertical blanking should be complete in less than 0.1H.
4. Leading and trailing slopes of horizontal blanking must be steep enough to preserve minimum and maximum values of (x + y) and (z) under all conditions of picture content.
5. Dimensions marked with asterisk indicate that tolerances given are permitted only for long time variations and not for successive cycles.
6. Equalizing pulse area shall be between 0.45 and 0.5 of area of a horizontal sync pulse.
7. Color burst follows each horizontal pulse, but is omitted following the equalizing pulses and during the broad vertical pulses.
8. Color burst to be omitted during monochrome transmissions.
9. The burst frequency shall be 3.579545 MHz. The tolerance on the frequency shall be ±0.0003% with a maximum rate of change of frequency not to exceed 1/10 Hz per second.
10. The horizontal scanning frequency shall be 2/455 times the burst frequency.
11. The dimensions specified for the burst determine the times of starting and stopping the burst but not its phase. The color burst consists of amplitude modulation of a continuous sine wave.
12. Dimension "P" represents the peak excursion of the luminance signal at blanking level but does not include the chrominance signal. Dimension "S" is the sync amplitude above blanking level. Dimension "C" is the peak carrier amplitude.

Fig. 2-8. Television signal standards. Cont.

TELEVISION CHANNEL FREQUENCIES

Table 2-17 lists the broadcast frequency limits of all television channels and the frequency of the video, color, and sound carriers of each channel. The frequencies of the signals are altered on most cable systems. Table 2-18 lists the cable channel frequency assignments generally used.

TABLE 2-17
Television Channel Frequencies*

Channel	Freq	Carriers			Channel	Freq	Carriers		
no.	range	*Video*	*Color*	*Sound*	no.	range	*Video*	*Color*	*Sound*
2	54-60	55.25	58.83	59.75	43	644-650	645.25	648.83	649.75
3	60-66	61.25	64.83	65.75	44	650-656	651.25	654.83	655.75
4	66-72	67.25	70.83	71.75	45	656-662	657.25	660.83	661.75
5	76-82	77.25	80.83	81.75	46	662-668	663.25	666.83	667.75
6	82-88	83.25	86.83	87.75	47	668-674	669-25	672.83	673.75
7	174-180	175.25	178.83	179.75	48	674-680	675.25	678.83	679.75
8	180-186	181.25	184.83	185.75	49	680-686	681.25	684.83	685.75
9	186-192	187.25	190.83	191.75	50	686-692	687.25	690.83	691.75
10	192-198	193.25	196.83	197.75	51	692-698	693.25	696.83	697.75
11	198-204	199.25	202.83	203.75	52	698-704	699.25	702.83	703.75
12	204-210	205.25	208.83	209.75	53	704-710	705.25	708.83	709.75
13	210-216	211.25	214.83	215.75	54	710-716	711.25	714.83	715.75
14	470-476	471.25	474.83	475.75	55	716-722	717.25	720.83	721.75
15	476-482	477.25	480.83	481.75	56	722-728	723.25	726.83	727.75
16	482-488	483.25	486.83	487.75	57	728-734	729.25	732.83	733.75
17	488-494	489.25	492.83	493.75	58	734-740	735.25	738.83	739.75
18	494-500	495.25	498.83	499.75	59	740-746	741.25	744.83	745.75
19	500-506	501.25	504.83	505.75	60	746-752	747.25	750.83	751.75
20	506-512	507.25	510.83	511.75	61	752-758	753.25	756.83	757.75
21	512-518	513.25	516.83	517.75	62	758-764	759.25	762.83	763.75
22	518-524	519.25	522.83	523.75	63	764-770	765.25	768.83	769.75
23	524-530	525.25	528.83	529.75	64	770-776	771.25	774.83	775.75
24	530-536	531.25	534.83	535.75	65	776-782	777.25	780.83	781.75
25	536-542	537.25	540.83	541.75	66	782-788	783.25	786.83	787.75
26	542-548	543.25	546.83	547.75	67	788-794	789.25	792.83	793.75
27	548-554	549.25	552.83	553.75	68	794-800	795.25	798.83	799.75
28	554-560	555.25	558.83	559.75	69	800-806	801.25	804.83	805.75
29	560-566	561.25	564.83	565.75	70†	806-812	807.25	810.83	811.75
30	566-572	567.25	570.83	571.75	71†	812-818	813.25	816.83	817.75
31	572-578	573.25	576.83	577.75	72†	818-824	819.25	822.83	823.75
32	578-584	579.25	582.83	583.75	73†	824-830	825.25	828.83	829.75
33	584-590	585.25	588.83	589.75	74†	830-836	831.25	834.83	835.75
34	590-596	591.25	594.83	595.75	75†	836-842	837.25	840.83	841.75
35	596-602	597.25	600.83	601.75	76†	842-848	843.25	846.83	847.75
36	602-608	603.25	606.83	607.75	77†	848-854	849.25	852.83	853.75
37	608-614	609.25	612.83	613.75	78†	854-860	855.25	858.83	859.75
38	614-620	615.25	618.83	619.75	79†	860-866	861.25	864.83	865.75
39	620-626	621.25	624.83	625.75	80†	866-872	867.25	870.83	871.75
40	626-632	627.25	630.83	631.75	81†	872-878	873.25	876.83	877.75
41	632-638	633.25	636.83	637.75	82†	878-884	879.25	882.83	883.75
42	638-644	639.25	642.83	643.75	83†	884-890	885.25	888.83	889.75

* All frequencies in megahertz.

† Channels 70-83 (806-890 MHz), formerly allocated to television broadcasting, are now allocated to the land mobile services. Operation, on a secondary basis, of some television translators may continue on these frequencies.

FREQUENCY SPECTRUM—SOUND AND ELECTROMAGNETIC RADIATION

The spectrum of electromagnetic waves, shown in Fig. 2-9, covers a range of 10^8 m to about 10^{-5} nm. The sound or audiofrequencies start about 8 Hz and the top of the range is around 20 kHz. The FCC allocation chart starts just below 10 kHz and ends at about 100 GHz. All of the different classes of radiowaves are in this region. Following the allocation chart frequencies are infrared

TABLE 2-18
Cable TV Channel Frequencies*

Channel no.	Freq range	Carriers *Video*	*Color*	*Sound*	Channel no.	Freq range	Carriers *Video*	*Color*	*Sound*
T-7	5.75-11.75	7	10.58	11.5	28	246-252	247.25	250.83	251.75
T-8	11.75-17.75	13	16.58	17.5	29	252-258	253.25	256.83	257.75
T-9	17.75-23.75	19	22.58	23.5	30	258-264	259.25	262.83	263.75
T-10	23.75-29.75	25	34.58	35.5	31	264-270	265.25	268.83	269.75
T-11	29.75-35.75	31	40.58	41.5	32	270-276	271.25	274.83	275.75
T-12	35.75-41.75	37	40.58	41.5	33	276-282	277.25	280.83	281.75
T-13	41.75-47.55	43	46.58	47.5	34	282-288	283.25	286.83	287.75
2	54-60	55.25	58.83	59.75	35	288-294	289.25	292.83	293.75
3	60-66	61.25	64.83	65.75	36	294-300	295.25	298.83	299.75
4	66-72	67.25	70.83	71.75	37	300-306	301.25	304.83	305.75
5	76-82	77.25	80.83	81.75	38	306-312	307.25	310.83	311.75
6	82-88	83.25	86.83	87.75	39	312-318	313.25	316.83	317.75
7	174-180	175.25	178.83	179.75	40	318-324	319.25	322.83	323.75
8	180-186	181.25	184.83	185.75	41	324-330	325.25	328.83	329.75
9	186-192	187.25	190.83	191.75	42	330-336	331.25	334.83	335.75
10	192-198	193.25	196.83	197.75	43	336-342	337.25	340.83	341.75
11	198-204	199.25	202.83	203.75	44	342-348	343.25	346.83	347.75
12	204-210	205.25	208.83	209.75	45	348-354	349.25	352.83	353.75
13	210-216	211.25	214.83	215.75	46	354-360	355.25	358.83	359.75
FM	88-108	—	—	—	47	360-366	361.25	364.83	365.75
14	120-126	121.25	124.83	125.75	48	366-372	367.25	370.83	371.75
15	126-132	127.25	130.83	131.75	49	372-378	373.25	376.83	377.75
16	132-138	143.25	136.83	137.75	50	378-384	379.25	382.83	383.75
17	138-144	139.25	142.83	143.75	51	384-390	385.25	388.83	389.75
18	144-150	145.25	148.83	149.75	52	390-396	391.25	394.83	395.75
19	150-156	151.25	154.83	155.75	53	396-402	397.25	400.83	401.75
20	156-162	157.25	160.83	161.75	54	72-78	73.25	76.83	77.75
21	162-168	163.25	166.83	167.75	55	78-84	79.25	82.83	83.75
22	168-174	169.25	172.83	173.75	56	84-90	85.25	88.83	89.75
23	216-222	217.25	220.83	221.75	57	90-96	91.25	94.83	95.75
24	222-228	223.25	226.83	227.75	58	96-102	97.25	100.83	101.75
25	228-234	229.25	232.83	233.75	59	102-108	103.25	106.83	107.75
26	234-240	235.25	238.83	239.75	60	108-114	109.25	112.83	113.75
27	240-246	241.25	244.83	245.75	61	114-120	115.25	118.83	119.75

* All frequencies in megahertz.

frequencies, visible light frequencies, X-rays, and gamma rays. Little is known beyond the gamma-ray frequencies. These are known as cosmic rays. The visible light spectrum covers a very small area, but thousands of color frequencies are present in this region.

AUDIOFREQUENCY SPECTRUM

The audiofrequency spectrum is generally accepted as extending from 15 to 20,000 Hz. Figure 2-10 presents the frequencies for each tone of the standard organ keyboard, based on the current musical pitch of A = 440 Hz. Figure 2-11 shows the frequency range of various musical instruments and of other sounds. The frequency range shown for each sound is the range needed for faithful reproduction and includes the fundamental frequency and the necessary harmonic frequencies. The frequency range of the human ear and the various broadcasting and recording media are also included in Fig. 2-11.

Unisons have a 1:1 frequency ratio, and octaves have a 2:1 frequency ratio. The perfect fifth has a 3:2 frequency ratio, and its complement (the perfect fourth) has a 4:3 frequency ratio. The additive numerical measure for intervals is a logarithmic function wherein the octave is divided into 1200 cents:

$$\text{Cents} = \frac{1200}{\log 2} \times \text{log of frequency ratio}$$

Example. The ratio and cents for the perfect fifth are 3:2 and 701.955. For the major third, they are 5:4 and 386.314.

All musical intervals are based on ratios of products of the prime numbers 2, 3, and 5. The prevailing musical temperament is the result of a long history of experimentation with various temperaments (an infinite spectrum is possible), and further evolution can be anticipated.

RADIOFREQUENCY SPECTRUM

The radiofrequency spectrum of 3 kHz to 3,000,000 MHz is divided into the various bands shown in Table 2-19 for easier identification.

TABLE 2-19
Frequency Classification

Frequency	Band no.	Classification	Abbreviation
30–300 Hz	2	extremely low frequencies	ELF
300–3000 Hz	3	voice frequencies	VF
3–30 kHz	4	very low frequencies	VLF
30–300 kHz	5	low frequencies	LF
300–3000 kHz	6	medium frequencies	MF
3–30 MHz	7	high frequencies	HF
30–300 MHz	8	very high frequencies	VHF
300–3000 MHz	9	ultrahigh frequencies	UHF
3–30 GHz	10	super-high frequencies	SHF
30–300 GHz	11	extremely high frequencies	EHF
300 GHz–3 THz	12	—	—

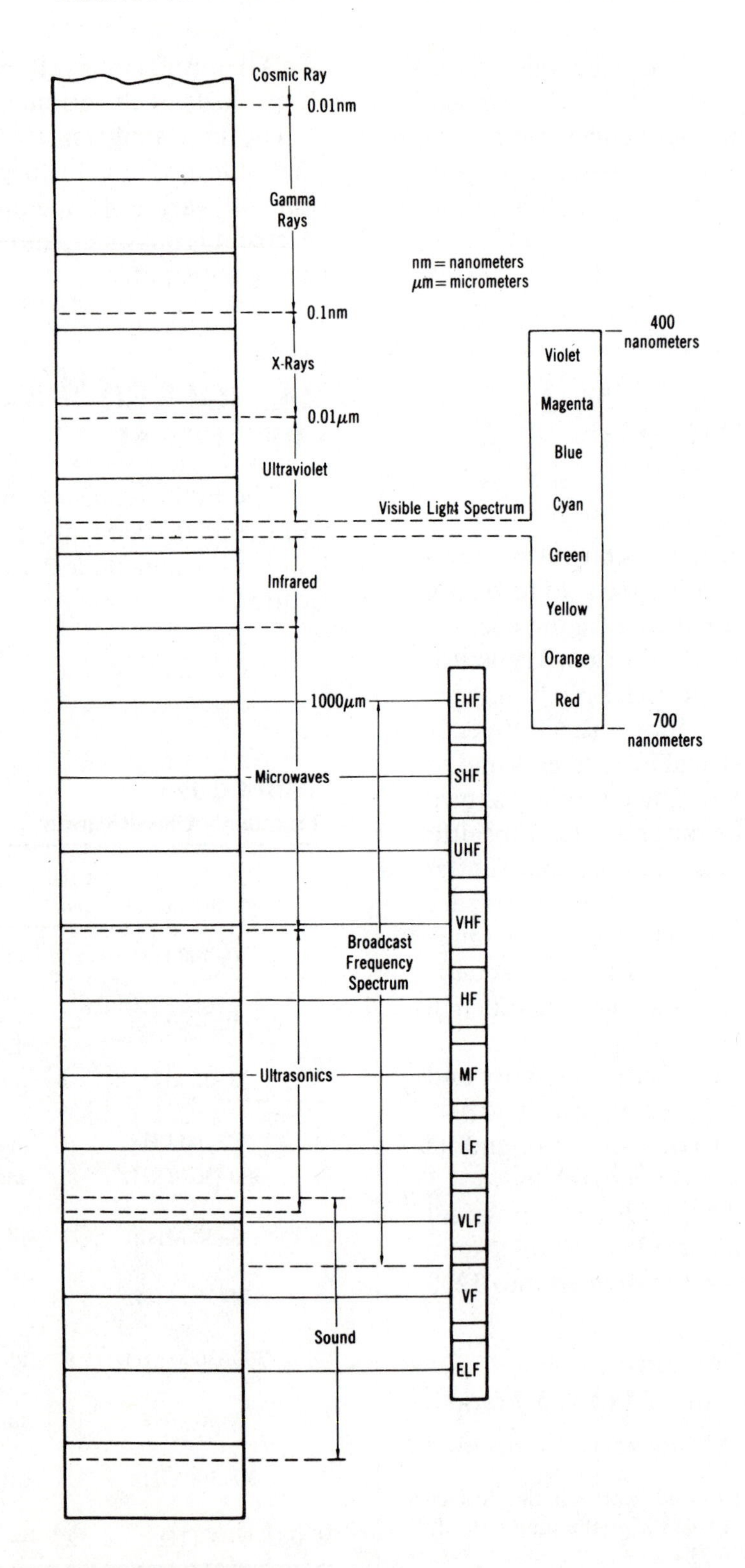
Cosmic Ray
0.01nm
Gamma Rays
nm = nanometers
μm = micrometers
0.1nm
400 nanometers
X-Rays
Violet
Magenta
0.01μm
Blue
Ultraviolet
Visible Light Spectrum
Cyan
Green
Infrared
Yellow
Orange
1000μm
EHF
Red
700 nanometers
Microwaves
SHF
UHF
VHF
Broadcast Frequency Spectrum
HF
Ultrasonics
MF
LF
VLF
VF
Sound
ELF

Fig. 2-9

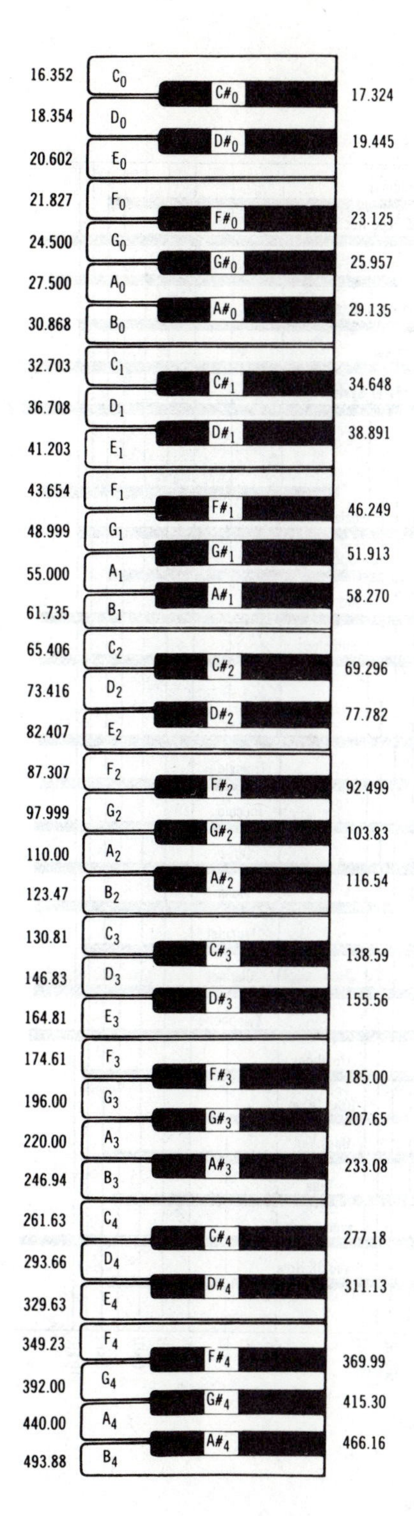
16.352 C0
C#0 17.324
18.354 D0
D#0 19.445
20.602 E0
21.827 F0
F#0 23.125
24.500 G0
G#0 25.957
27.500 A0
A#0 29.135
30.868 B0
32.703 C1
C#1 34.648
36.708 D1
D#1 38.891
41.203 E1
43.654 F1
F#1 46.249
48.999 G1
G#1 51.913
55.000 A1
A#1 58.270
61.735 B1
65.406 C2
C#2 69.296
73.416 D2
D#2 77.782
82.407 E2
87.307 F2
F#2 92.499
97.999 G2
G#2 103.83
110.00 A2
A#2 116.54
123.47 B2
130.81 C3
C#3 138.59
146.83 D3
D#3 155.56
164.81 E3
174.61 F3
F#3 185.00
196.00 G3
G#3 207.65
220.00 A3
A#3 233.08
246.94 B3
261.63 C4
C#4 277.18
293.66 D4
D#4 311.13
329.63 E4
349.23 F4
F#4 369.99
392.00 G4
G#4 415.30
440.00 A4
A#4 466.16
493.88 B4
523.25 C5
C#5 554.37
587.33 D5
D#5 622.25
659.26 E5
698.46 F5
F#5 739.99
783.99 G5
G#5 830.61
880.00 A5
A#5 932.33
987.77 B5
1,046.5 C6
C#6 1.108.7
1,174.7 D6
D#6 1.244.5
1,318.5 E6
1,396.9 F6
F#6 1.480.0
1,568.0 G6
G#6 1.611.2
1,760.0 A6
A#6 1.864.7
1,975.0 B6
2,093.0 C7
C#7 2.217.5
2,349.3 D7
D#7 2.489.0
2,637.0 E7
2,793.8 F7
F#7 2.960.0
3,136.0 G7
G#7 3.322.4
3,520.0 A7
A#7 3.729.3
3,591.0 B7
4,186.0 C8
C#8 4.434.9
4,698.6 D8
D#8 4.978.0
5,274.0 E8
5,587.7 F8
F#8 5.919.9
6,271.9 G8
G#8 6.644.9
7,040.0 A8
A#8 7.458.6
7,902.1 B8

Fig. 2-10

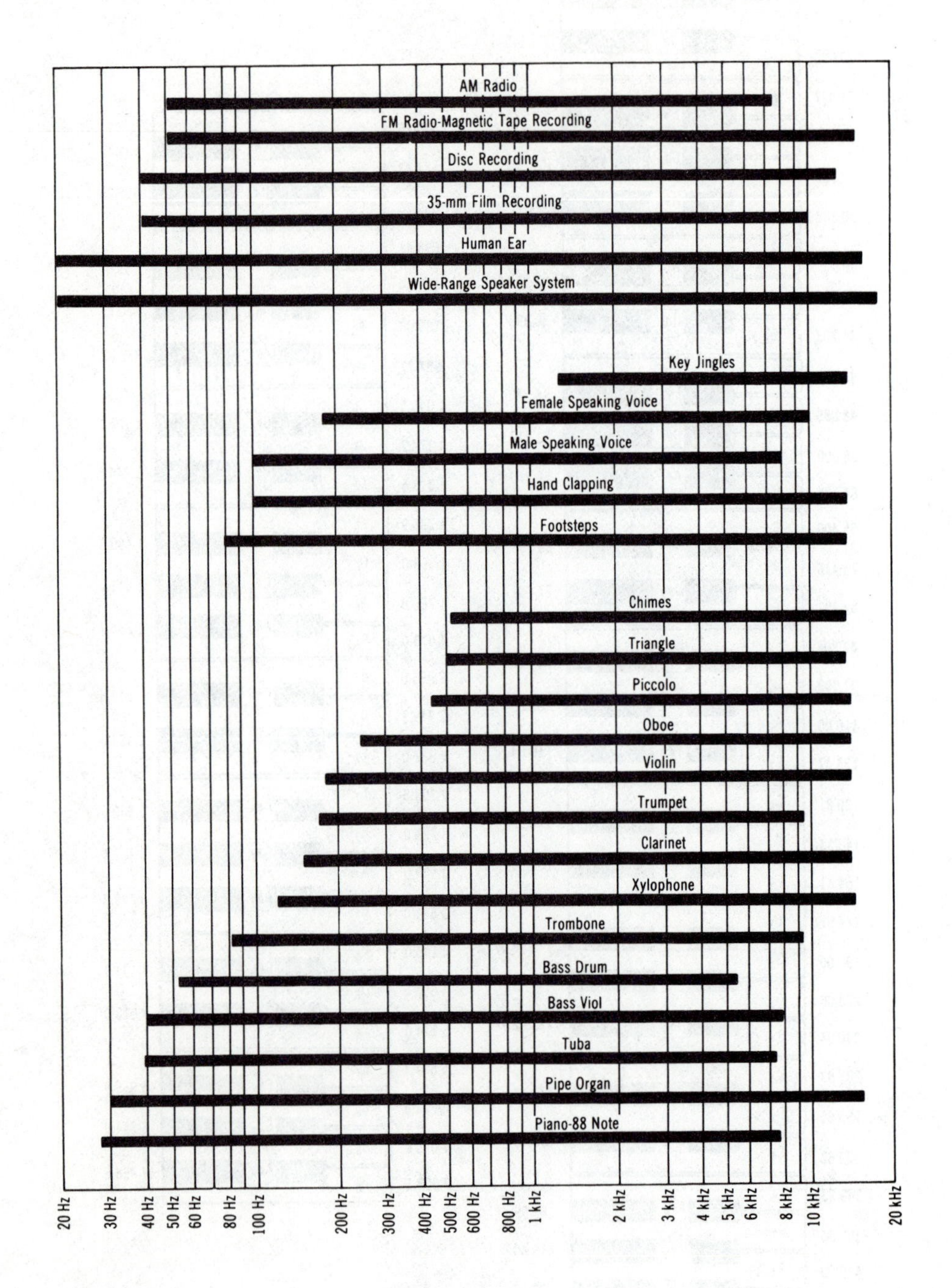

AM Radio
FM Radio-Magnetic Tape Recording
Disc Recording
35-mm Film Recording
Human Ear
Wide-Range Speaker System
Key Jingles
Female Speaking Voice
Male Speaking Voice
Hand Clapping
Footsteps
Chimes
Triangle
Piccolo
Oboe
Violin
Trumpet
Clarinet
Xylophone
Trombone
Bass Drum
Bass Viol
Tuba
Pipe Organ
Piano-88 Note
20 Hz
30 Hz
40 Hz
50 Hz
60 Hz
80 Hz
100 Hz
200 Hz
300 Hz
400 Hz
500 Hz
600 Hz
800 Hz
1 kHz
2 kHz
3 kHz
4 kHz
5 kHz
6 kHz
8 kHz
10 kHz
20 kHz

Fig. 2-11

NOAA WEATHER FREQUENCIES

The FCC has allocated three frequencies to the U.S. Department of Commerce, National Oceanic and Atmospheric Administration (NOAA), National Weather Service for the dissemination of weather information to the public. The frequencies assigned are:

162.40 MHz
162.475 MHz
162.55 MHz

Chapter 3

Symbols and Codes

INTERNATIONAL Q SIGNALS

The international Q signals were first adopted to enable ships at sea to communicate with each other or to contact foreign shores without experiencing language difficulties. The signals consist of a series of three-letter groups starting with Q and having the same meaning in all languages. Today, Q signals serve as a convenient means of abbreviation in communications between amateurs. Each Q signal has both an affirmative and an interrogative meaning. The question is designated by the addition of the question mark after the Q signal. The most common Q signals are listed in Table 3-1.

TABLE 3-1
Q Signals

Signals	Question	Answer or advice
QRA	What station are you?	My station name is _____.
QRB	How far are you from me?	I am _____ from your station.
QRD	Where are you headed and from where?	I am headed for _____ from _____.
QRE	What is your estimated time of arrival?	My ETA is _____ hours.
QRF	Are you returning to _____?	I am returning to _____.
QRG	What is my exact frequency?	Your exact frequency is _____ kHz.
QRH	Does my frequency vary?	Your frequency varies.
QRI	How is the tone of my transmission?	Your tone is _____.
QRJ	Do you receive me badly?	I cannot receive you. Your signals are too weak.
QRK	How do your read my signals?	The legibility of your signal is _____.
QRL	Are you busy?	I am busy (or I am busy with _____). Do not interfere.
QRM	Are you being interfered with?	I am being interfered with.
QRN	Are you troubled by static?	I am troubled by static.
QRO	Shall I increase power?	Increase power.
QRP	Shall I decrease power?	Decrease power.
QRQ	Shall I speak faster?	Speak faster.
QRS	Shall I speak slower?	Speak slower.
QRT	Shall I stop transmitting?	Stop sending.

TABLE 3-1 Cont.
Q Signals

Signals	Question	Answer or advice
QRU	Have you anything for me?	I have nothing for you.
QRV	Are you ready?	I am ready.
QRW	Shall I inform _____ that you are calling him on _____ kHz?	Please tell _____ that I am calling him on _____ kHz.
QRX	When will you call me again?	Wait _____ I will call you at _____ hours.
QRY	What is my turn?	Your turn is _____.
QRZ	Who is calling me?	You are being called by _____.
QSA	What is the strength of my signals?	The strength of your signals is _____.
QSB	Are my signals fading?	Your signals are fading.
QSD	Are my signals mutilated?	Your keying is incorrect; your signals are bad.
QSG	Shall I send _____ messages at a time?	Send _____ messages.
QSK	Can you hear me between your signals?	I can hear you.
QSL	Will you send me a confirmation of our communication?	I give you acknowledgment of receipt.
QSM	Shall I repeat the last message?	Repeat the last telegram you have sent me.
QSN	Did you hear me on _____ kHz?	I heard you on _____ kHz.
QSO	Can you communicate with _____ direct or by relay?	I can communicate with _____ direct (or through the medium of _____).
QSP	Will you relay to _____ free of charge?	I will relay.
QSQ	Have you a doctor aboard?	Yes. Or no, we have no doctor.
QRR	Have the distress calls from _____ been cleared?	Distress calls from _____ have cleared.
QSU	Shall I send reply on this frequency or on _____ kHz?	Reply on this frequency, or reply on _____ kHz.
QSV	Shall I send a series of Vs on this frequency?	Send a series of Vs.
QSW	Will you send on this frequency?	Yes, I will send on this frequency.
QSX	Will you listen to _____ on _____ kHz?	I will listen to _____ on _____ kHz.
QSY	Shall I change to transmission on another frequency?	Change to transmission on _____ kHz without changing the type of wave.
QSZ	Shall I send each word or group more than once?	Say each word or group of words twice or _____ times.
QTA	Shall I cancel message number?	Cancel message number _____ as if not sent.
QTB	Do you agree with my counting words?	I agree, or word count is _____.
QTC	How many messages do you have for me?	I have _____ messages for you.
QTE	What is my true bearing from you?	True bearing from me is _____ degrees.
QTF	Will you give me the position of my station according to the bearings of your direction finding station?	Your bearing is _____.
QTG	Will you send two dashes of ten seconds each followed by our call sign—repeated _____ times on kHz?	I am sending two dashes of ten seconds each with my call sign _____ times on _____ kHz at _____ hours.
QTH	What is your location in latitude and longitude?	My location is _____.
QTI	What is your true track—in degrees?	My true track is _____ degrees.
QTJ	What is your true speed?	My true speed is _____.
QTL	What is your true heading—in degrees?	My true heading is _____ degrees.
QTM	Send signals to enable me to fix my bearing and distance.	Fix your bearing and distance on my radio signal.
QTN	At what time did you depart from _____?	I departed from _____ (place) at _____ hours.
QTO	Have you left port/dock?	I left port at _____ hours.
QTO	Are you going to enter port/dock?	I am entering port.

TABLE 3-1 Cont.
Q Signals

Signals	Question	Answer or advice
QTQ	Can you communicate with my station by means of the International Code of Signals?	I will communicate with you by International Code of Signals.
QTR	What is the correct time?	The correct time is _____ hours.
QTS	Will you send your call sign for _____ minutes now, or at _____ hours on _____ kHz so that your frequency may be measured?	I will send your call sign for _____ now, or at _____ hours on _____ kHz so you can measure my frequency.
QTU	During what hours is your station open?	My station is open from _____ to _____ hours.
QTV	Shall I stand guard for you on _____ kHz?	Listen for me on channel _____ (from _____ to _____ hours).
QTX	Will you keep your station open for further communication with me for _____ hours?	I will keep my station open for further communication with you until further notice (or until _____ hours).
QTY	Are you proceeding to the position of incident and if so when do you expect to arrive?	I am proceeding to the position of incident and expect to arrive at _____ hours.
QTZ	Are you continuing the search?	I am continuing the search for _____.
QUA	Do you have news of _____?	I have news of _____.
QUB	Can you give me, in the following order, information concerning visibility, height of clouds, direction and velocity of ground wind at _____ (place of observation)?	Information desired follows: visibility is _____ clouds are _____ wind is _____ knots from _____ at _____ latitude _____ longitude.
QUC	What is the number (or other) of the last message you received from me?	The number of my last message to you is _____.
QUD	Have you received the urgency signal sent by _____?	I have urgency signal sent by _____.
QUF	Have you received the distress signal sent by _____?	I have received the distress signal sent by _____.
QUG	Will you be forced to alight (or land)?	I must land now.
QUH	Will you give me the present barometric pressure at sea level?	Barometric pressure at sea level is now _____.
QUJ	Will you indicate the true course for me to follow?	Follow course _____ degrees true.
QUM	Is the distress traffic ended?	Distress traffic is ended.
QRRR	(Official ARRL land distress call)	This is a special signal, and if you hear it keep off the frequency except to listen, unless you are in a position to help. It is the official ARRL land distress call, for emergency use only. It is the equivalent of SOS as used by ships at sea and must receive the same attention and priority.

Z-SIGNALS

The Z-code signals shown in Table 3-2 are used to communicate at sea. The U.S. military also uses these codes.

TABLE 3-2
Z Code for Point-to-Point Service*

Signal	Message
	A
*ZAA	YOU ARE NOT OBSERVING CIRCUIT DISCIPLINE.
*ZAB	YOUR SPEED KEY IMPROPERLY ADJUSTED.
ZAC	Advise (Call sign of) frequency you are reading.
*ZAC	Cease using speed key.
*ZAD	Signal (1) Not understood, (2) Not held.
*ZAE	Unable to receive you, try, via _____.
*ZAF	Reroute the circuit by patching.
*ZAH	Unable to relay. We file.
*ZAI	Run (foxes, RYs, mk, etc.).
*ZAJ	Have been unable to break you.
ZAL	Alter your wavelength.
*ZAL	Closing down, due to _____.
ZAN	WE CAN RECEIVE ABSOLUTELY NOTHING.
*ZAN	Transmit only messages of above precedence _____.
*ZAO	CAN'T UNDERSTAND VOICE. USE TELEGRAPH.
ZAP	ACKNOWLEDGE, PLEASE.
*ZAP	Work (simplex, duplex, mux, sb).
*ZAQ	Last word received (sent) was _____.
ZAR	Revert to Automatic Relay.
*ZAS	Rerun tapes run on _____ since _____.
*ZAT	Punching tape for transmission.
*ZAV	Send blind until advised.
*ZAX	You are causing interference.
*ZAY	Send on _____ (kHz). Will confirm later.
	B
*ZBA	Cause of delay is _____.
*ZBD	Following was sent _____ (time).
*ZBE	Retransmit message _____ to _____.
*ZBG	You are sending uppercase.
*ZBH	Make call before transmitting traffic.
*ZBI	Listen for telephony.
*ZBL	Do not use break-in.
*ZBM	Put _____ (speed opr.) this frequency.
ZBN	Break and go ahead with New slip.
*ZBN	Your tape reversed.
*ZBO	I HAVE TRAFFIC.
*ZBP	(1) Characters indistinct, (2) Spacing bad.
*ZBQ	When and on what frequency was message received.
ZBR	Break circuit. Retuning.
*ZBR	Shall I send by _____ (method).
ZBS	Blurring Signals. (1) Dots heavy, (2) Dots light.
*ZBT	Count _____ as _____ groups.
*ZBU	Report when in communication with _____.
*ZBV	Answer on _____ MHz.
*ZBW	I shift (or ask _____) to _____ (kHz).
ZBY	Pull Back your tape one Yard.
	C
ZCA	Circuit affected. Make readable signals.
ZCB	Circuit broken. Signal unheard.
ZCC	Collate code.
ZCD	Your Collation is Different.
ZCF	CHECK YOUR CENTER FREQUENCY, PLEASE.
ZCI	Circuit Interrupted. Running and available.
ZCK	Check Keying.
ZCL	TRANSMIT CALL LETTERS INTELLIGIBLY.
ZCO	Your Collation Omitted.
ZCP	Conditions poor, increase to maximum.
ZCR	Using concentrator, make warning signals.
ZCS	Cease Sending.
ZCT	Send Code Twice.
ZCW	Are you in direct Communication With _____?
	D
*ZDA	Hr. formal message, priority _____.
*ZDB	Expedite reply to my _____.
ZDC	Diagnosing Circuit faults, will advise.
ZDE	Message undelivered. (1) Will keep trying, (2) Advise disposal, (3) Canceling, (4) Btr. ads.
*ZDE	Message _____ undelivered.
ZDF	Frequency is Drifting to degree indicated, 1–5.
*ZDF	Message _____ received by addressee _____ (time).
*ZDG	Accuracy of following doubtful.
ZDH	Your Dots are too Heavy (long).

TABLE 3-2 Cont.
Z Code for Point-to-Point Service*

Signal	Message
ZDL	Your Dots are too Light (short).
ZDM	Your Dots are Missing.
*ZDM	This is a multiple-address message.
*ZDN	Report disposal of message _____.
*ZDQ	Message _____ relayed to _____ at _____ by _____.
*ZDS	Message just transmitted erroneous. Correct version is _____.
ZDT	Following transmitters running dual.
*ZDT	Don't transmit exercise messages until advised.
ZDV	Your Dots Varying length, please remedy.
*ZDV	Private message received for _____. Advise.
*ZDY	No private messages until ordered.
	E
*ZEC	Have you received message _____?
ZED	We are Experiencing Drop-outs, 1-5.
ZEF	We are Experiencing Fill-ins, 1-5.
ZEG	We are Experiencing Garbles, 1-5.
*ZEI	Accuracy of heading doubtful.
*ZEK	No answer required.
*ZEL	This message is correction to _____.
*ZEN	This message is classified.
*ZEP	Parts marked ZEP coming later.
*ZEU	Exercise _____ (drill message).
	F
ZFA	Failing Auto.
*ZFA	Message intercepted or copied blind.
ZFB	SIGNALS ARE FADING BADLY.
*ZFB	Pass this message to _____.
ZFC	Check your FSK shift, please.
ZFD	Depth of Fading is as indicated, 1-5.
*ZFD	This message is a suspected duplicate.
*ZFF	Advise when message received by _____.
*ZFH	Message for (1) Action, (2) Info, (3) Comment.
*ZFI	Reply message? There is no reply.
ZFK	Revert to FSK.
ZFO	SIGNALS FADED OUT.
ZFQ/x	Frequency shift your signal is _____ Hz.
ZFR	Rapidity of Fading is as indicated, 1-5.
*ZFR	Cancel transmission _____.
ZFS	Signals are Fading Slightly.
	G
*ZGB	Send _____ (answer).
ZGF	_____ signals Good For _____ wpm.
*ZGF	Make call signs more distinctly.
ZGP	Please Give Priority.
ZGS	YOUR SIGNALS GETTING STRONGER.
ZGW	YOUR SIGNALS GETTING WEAKER.
	H
ZHA	How are conditions for Auto reception?
ZHC	HOW ARE YOUR RECEIVING CONDITIONS?
ZHM/x	Harmonic radiation from transmitter.
ZHS	Send High Speed auto _____ wpm.
ZHY	We are Holding Your _____.
	I
ZIM	Industrial or Medical interference, 1-5.
ZIP	Increase Power.
ZIR	You have strong Idle Radiation.
ZIS	Atmospheric Interference, 1-5.
	J
ZJF	Frequency Jumping to degree indicated, 1-5.
	K
*ZKA	Who is controlling station? or I am _____.
*ZKB	Permission necessary before transmitting messages.
*ZKD	Take control of net _____ or shall I _____.
*ZKE	I (or _____) reports into circuit (net).
*ZKF	Station leaves net temporarily.
*ZKJ	Closing down (until _____).
ZKO	REVERT TO ON-OFF KEYING.
ZKQ	Say when ready to resume.
*ZKS	What Stations Keeping watch on _____?
ZKW	Keying weight is _____ (percent).
	L
ZLB	Give Long Breaks, please.
ZLD	We are getting Long Dash from you.
ZLL	Distorted Land Line control signals.
ZLP	Low (minimum) Power.
ZLS	WE ARE SUFFERING FROM LIGHTNING STORM.
	M
ZMG	Magnetic activity.
ZMO	STAND BY MOMENT.
ZMP	MisPunch or Perforator failures.
ZMQ	Stand by for _____.
ZMU/x	MUltipath making _____ mark bias.
	N
ZNB	No Breaks, we send twice.
*ZNB	Authentication is _____.
ZNC	NO COMMUNICATIONS WITH _____.
*ZNC	All stations authenticate.
*ZND	You are misusing authenticator.
ZNG	Receiving conditions No Good for code.
ZNI	NO CALL LETTERS (IDENTIFICATION) HEARD.
ZNN	ALL CLEAR OF TRAFFIC.
ZNO	Not On the air.
ZNR	Not Received.

TABLE 3-2 Cont.
Z Code for Point-to-Point Service*

Signal	Message
ZNS	Here New Slip.
	O
ZOA	Have checked (call letters) _____ OK.
*ZOC	Relay to your substations.
ZOD	Observing _____ will transfer when better.
*ZOD	Act as radio link between me and _____.
*ZOE	Can you accept message? (or) Give me message.
*ZOF	Relay this message.
*ZOG	Transmit your message (give info.).
ZOH	What traffic have you On Hand?
ZOK	WE ARE RECEIVING OK.
*ZOK	Relay this message via _____.
ZOL	OK. On Line.
*ZOM	Mail delivery permissible.
ZOR	Transmit Only Reversals.
*ZOU	Give instructions for routing traffic.
*ZOZ	Obtain retransmission of message _____.
	P
ZPA	Printer line Advance not received.
*ZPA	Your speech distorted.
ZPC	Printer Carriage-return not received.
*ZPC	Signals fading, 1–5.
ZPE	Punch Everything.
ZPF	Printer motor Fast.
ZPO	Send Plain Once.
ZPP	Punch Plain only.
ZPR	Reruns slip at Present Running.
ZPS	Printer motor Slow.
ZPT	Send Plain Twice.
	R
ZRA	Reverse Auto tape.
*ZRA	My frequency OK? Your frequency is _____.
ZRB	Relayed signal Bad, adjust receiver.
*ZRB	Check your frequency on this circuit.
ZRC	Can you Receive Code?
*ZRC	Shall I, or tune your transmitter to _____.
*ZRE	Hear you best on _____ (kHz).
*ZRF	Send tuning signals on present frequency.
ZRK	Reversed Keying.
ZRL	Rerun slip before one now running.
ZRM	Please Remove Modulation from _____.
*ZRM	I receive _____ (usb, lsb, isb).
ZRN	ROUGH NOTE.
ZRO	ARE YOU RECEIVING OK?
ZRR	Run Reversals.
ZRS	Rerun message No. _____.
ZRT	Revert to Traffic.
ZRY	Run test slip, please.
	S
ZSF	SEND FASTER.
ZSH	STATIC HEAVY HERE.
ZSI/x	Please furnish Signal Intensity.
ZSM/x	Microvolt input to receiver is _____.
ZSN	Give SINPO report on _____.
	S Signal strength.
	I Interference.
	N Noise.
	P Propagation.
	O Overall readability.
ZSO	Transmit Slips Once.
ZSR	YOUR SIGNALS STRONG AND READABLE.
ZSS	Send Slower.
ZST	Transmit Slips Twice.
ZSU	YOUR SIGNALS ARE UNREADABLE.
ZSV	Your Speed Varying.
	T
ZTA	Transmit by Auto.
ZTH	Transmit by Hand.
ZTI	Transmission temporarily Interrupted.
	U
ZUA	Conditions Unsuitable for Automatic recording.
*ZUA	Timing signal will be transmitted.
ZUB	WE HAVE BEEN UNABLE TO BREAK YOU.
ZUC	UNABLE TO COMPLY. WILL DO SO AT _____.
*ZUE	Affirmative (Yes).
*ZUG	Negative (No).
*ZUH	Unable to comply.
*ZUJ	Wait. Stand by.
	V
ZVB	Varying Bias.
ZVF	Signals Varying in Frequency.
ZVP	Send Vs Please.
*ZVR	Pass at once to substations.
ZVS	Signals Varying in intensity.
	W
*ZWB	Name of operator on watch.
ZWC	Wipers or Clicks here.
ZWO	SEND WORDS ONCE.
ZWR	YOUR SIGNALS WEAK BUT READABLE.
ZWS	Wavelength (frequency) is Swinging, 1-5.
ZWT	SEND WORDS TWICE.
	Y
ZYS	WHAT IS YOUR SPEED OF TRANSMISSION?
	Z
*ZZF	Incorrect.
*ZZG	You are correct.

TABLE 3-2 Cont.
Z Code for Point-to-Point Service*

Signal	Message	Signal	Message
*ZZH	TRY AGAIN.		**Radiophoto and Facsimile**
	Multiplex	ZXA	Adjust to receive speeds _____.
ZYA	Cease traffic; send As on A channel.	ZXC	PIX _____ Conditionally accepted.
ZYC	Cycling on ARQ, errors stored your end.	ZXD	Send Dashes, please.
ZYK	Check Your Keying on channel _____.	ZXF	You are Floating Fast.
ZYM	Change from single printer to Multiplex.	ZXH	Limits High, reduce _____ Hz.
ZYN	Make bias Neutral.	ZXJ	You are Jumping out of phase.
ZYP	Change from multiplex to single Printer.	ZXK	Is your synchronizing correct?
ZYR	Please put _____ on MUX revolutions.	ZXL	Limits are Low, increase _____ Hz.
ZYT	Check Your Thyratrons.	ZXO	Last run defaced due to _____.
ZYX/x	Revert to MUX frames _____ channels.	ZXP	Go ahead with Pix.
		ZXS	You are floating Slow.
		ZXV	Your modulation is Varying.

*Asterisk indicates U.S. military usage. Numbers 1–5 following the "Z" signal mean: (1) very slight, (2) slight, (3) moderate, (4) severe, (5) extreme.

Sources: Cable and Wireless Ltd., U.S. Army Communications Manual SIG 439-2, ACP-131 (A), Allied Communication Procedures, W3AFM.

10-Signals

Numerous versions of 10-signals are in use. The one in Table 3-3, adopted by the Associated Public Safety Communications Officers, Inc. (APCO), is the result of an in-depth study to develop a uniform code that could be used by all radio services. Containing only 34 signals, it is easier to memorize than the others, yet most of the needed signals are included. Two other versions, one used primarily by CBers and the other by police agencies, are given in Tables 3-4 and 3-5.

11-CODE SIGNALS

Table 3-6 is the 11-code, also sometimes used by law enforcement agencies.

TABLE 3-3
APCO 10-Signals

Number	Meaning	Number	Meaning	Number	Meaning
10-1	signal weak	10-13	existing conditions	10-25	report to (meet)
10-2	signal good	10-14	message information	10-26	estimated arrival time
10-3	stop transmitting	10-15	message delivered	10-27	license/permit information
10-4	affirmative (OK)	10-16	reply to message		
10-5	relay (to)	10-17	enroute	10-28	ownership information
10-6	busy	10-18	urgent	10-29	records check
10-7	out of service	10-19	(in) contact	10-30	danger/caution
10-8	in service	10-20	location	10-31	pick-up
10-9	say again	10-21	call (_____) by phone	10-32	_____ units needed specify/number/type
10-10	negative	10-22	disregard		
10-11	_____ on duty	10-23	arrived at scene	10-33	help me quick
10-12	stand by (stop)	10-24	assignment completed	10-34	time

TABLE 3-4
CBers 10-Code

Number	Meaning	Number	Meaning
10-1	Receiving poorly; signal weak.	10-39	Your message delivered.
10-2	Receiving well; signal good.	10-41	Please tune to channel _____.
10-3	Stop transmitting.	10-42	Traffic accident at _____.
10-4	OK, message received; acknowledgment.	10-43	Traffic tie-up at _____.
10-5	Relay message; or relay to _____.	10-44	I have a message for you (or _____).
10-6	Busy, stand by.	10-45	All units within range, please report.
10-7	Out of service (leaving air).	10-46	Assist motorist.
10-8	In service (subject to call).	10-50	Break channel.
10-9	Repeat, or repeat message.	10-51	Wrecker needed.
10-10	Transmission completed; standing by.	10-52	Ambulance needed.
10-11	Transmitting (talking) too rapidly.	10-60	What is next message number?
10-12	Visitors (or officials) present.	10-62	Unable to copy, use phone.
10-13	Weather and/or road conditions.	10-63	Net directed to _____.
10-16	Make pickup at _____.	10-64	Net clear.
10-17	Urgent business.	10-65	Awaiting your next message/assignment.
10-18	Anything (message) for us?	10-67	All units comply.
10-19	Nothing for you, return to base.	10-70	Fire at _____.
10-20	Location.	10-71	Proceed with transmission in sequence.
10-21	Call _____ by phone; or number _____.	10-73	Speed trap at _____.
10-22	Report in person to _____.	10-75	You are causing interference.
10-23	Stand by.	10-77	Negative contact.
10-24	Completed last assignment.	10-81	Reserve hotel room for _____.
10-25	Can you contact _____?	10-82	Reserve lodging (room).
10-26	Disregard last information.	10-84	Telephone number.
10-27	I am moving to channel _____.	10-85	My address is _____.
10-28	Identify your station.	10-89	Radio repairmen needed at _____.
10-29	Time is up for contact.	10-90	I have TVI.
10-30	Does not conform to FCC Rules.	10-91	Talk closer to mike.
10-32	I will give you a radio check.	10-92	Your transmitter is out of adjustment.
10-33	Emergency traffic at this station.	10-93	Frequency check.
10-34	Trouble at this station, need help.	10-94	Please give me a long count.
10-35	Confidential information.	10-95	Transmit dead carrier for 5 seconds.
10-36	Correct time.	10-99	Mission completed, all units secure.
10-37	Wrecker needed at _____.	10-100	Rest room pause.
10-38	Ambulance needed at _____.	10-200	Police needed at _____.

TABLE 3-5
Police 10-Code

Number	Meaning	Number	Meaning
10-0	Caution.	10-49	Traffic light out.
10-1	Unable to copy, change location.	10-50	Accident _____ (F, PI, PD).
10-2	Signal good.	10-51	Wrecker needed.
10-3	Stop transmitting.	10-52	Ambulance needed.
10-4	Acknowledgment.	10-53	Road blocked.
10-5	Relay (to).	10-54	Livestock on highway.
10-6	Busy, stand by unless urgent.	10-55	Intoxicated driver.
10-7	Out of service (give location).	10-56	Intoxicated pedestrian.
10-8	In service.	10-57	Hit and run.
10-9	Repeat.	10-58	Direct traffic.
10-10	Fight in progress.	10-59	Convoy or escort.
10-11	Dog case.	10-60	Squad in vicinity.
10-12	Stand by.	10-61	Personnel in area.
10-13	Weather and road report.	10-62	Reply in message.
10-14	Report of prowler.	10-63	Prepare to make written copy.
10-15	Civil disturbance.	10-64	Message for local delivery.
10-16	Domestic trouble.	10-65	Net message assignment.
10-17	Meet complainant.	10-66	Message cancellation.
10-18	Complete assignment quickly.	10-67	Clear to read next message.
10-19	Nothing for you, return to _____.	10-68	Dispatch information.
10-20	Location.	10-70	Fire alarm.
10-21	Call _____ by phone.	10-71	Advise nature of fire.
10-22	Disregard.	10-72	Report progress on fire.
10-23	Arrived at scene.	10-73	Smoke report.
10-24	Assignment completed.	10-74	Negative.
10-25	Report in person to _____.	10-75	In contact with.
10-26	Detaining subject, expedite.	10-76	Enroute.
10-27	Drivers license information.	10-77	ETA (Estimated Time of Arrival).
10-28	Vehicle registration information.	10-78	Need assistance.
10-29	Check records for wanted.	10-79	Notify coroner.
10-30	Illegal use of radio.	10-80	Chase in progress.
10-31	Crime in progress.	10-81	Breathalizer report.
10-32	Man with gun.	10-82	Reserve lodging.
10-33	Emergency.	10-83	Work school crossing at _____.
10-34	Riot.	10-84	If meeting, advise ETA.
10-35	Major crime alert.	10-85	Delayed, due to _____.
10-36	Correct time.	10-86	Officer/operator on duty.
10-37	Investigate suspicious vehicle.	10-87	Pick up/distribute checks.
10-38	Stopping suspicious vehicle.	10-88	Advise present telephone number.
10-39	Urgent—use light and siren.	10-89	Bomb threat.
10-40	Silent run, no light and siren.	10-90	Bank alarm at _____.
10-41	Beginning tour of duty.	10-91	Pick up prisoner/subject.
10-42	Ending tour of duty.	10-92	Improperly parked vehicle.
10-43	Information.	10-93	Blockade.
10-44	Request permission to leave patrol for _____.	10-94	Drag racing.
		10-95	Prisoners/subject in custody.
10-45	Animal carcass in _____ lane at _____.	10-96	Mental subject.
10-46	Assist motorist.	10-97	Check (test) signal.
10-47	Emergency road repairs needed, _____.	10-98	Prison/jail break.
10-48	Traffic standard needs repair.	10-99	Records indicate wanted or stolen.

TABLE 3-6
Law Enforcement 11-Code

Signal	Meaning	Signal	Meaning
11-6	Illegal discharge of firearms.	11-43	Doctor required.
11-7	Prowler.	11-44	Coroner required.
11-8	Person down.	11-45	Attempted suicide.
11-10	Take a report.	11-46	Death report.
11-12	Dead animal.	11-47	Injured person.
11-13	Injured animal.	11-48	Provide transportation.
11-14	Animal bite.	11-65	Traffic signal light out.
11-15	Ball game in street.	11-66	Traffic signal out of order.
11-17	Wires down.	11-70	Fire alarm.
11-24	Abandoned vehicle.	11-71	Fire report.
11-25	Vehicle—traffic hazard.	11-79	Traffic accident—ambulance sent.
11-25X	Female motorist needs assistance.	11-80	Traffic accident—serious injury.
11-27	Subject has felony record but is not wanted.	11-81	Traffic accident—minor injury.
11-28	Rush vehicle registration information—driver is being detained.	11-82	Traffic accident—no injury.
11-29	Subject has no record and is not wanted.	11-83	Traffic accident—no details.
11-30	Incomplete phone call.	11-84	Direct traffic.
11-31	Person calling for help.	11-85	Dispatch tow truck.
11-40	Advise if ambulance is needed.	11-86	Special detail.
11-41	Request ambulance.	11-87	Assist other unit.
11-42	Ambulance not required.	11-98	Meet officer.
		11-99	Officer needs help.

THE INTERNATIONAL CODE

A	*di dah*	J	*di dah dah dah*	S	*di di dit*
B	*dah di di dit*	K	*dah di dah*	T	*dah*
C	*dah di dah dit*	L	*di dah di dit*	U	*di di dah*
D	*dah di dit*	M	*dah dah*	V	*di di di dah*
E	*dit*	N	*dah dit*	W	*di dah dah*
F	*di di dah dit*	O	*dah dah dah*	X	*dah di di dah*
G	*dah dah dit*	P	*di dah dah dit*	Y	*dah di dah dah*
H	*di di di dit*	Q	*dah dah di dah*	Z	*dah dah di dit*
I	*di dit*	R	*di dah dit*		

1	*di dah dah dah dah*	6	*dah di di di dit*
2	*di di dah dah dah*	7	*dah dah di di dit*
3	*di di di dah dah*	8	*dah dah dah di dit*
4	*di di di di dah*	9	*dah dah dah dah dit*
5	*di di di di dit*	0	*dah dah dah dah dah*

question mark	*di di dah dah di dit*	period	*di dah di dah di dah*
error	*di di di di di di di dit*	comma	*dah dah di di dah dah*
wait	*di dah di di dit*	end of message	*di dah di dah dit*

SINPO RADIO-SIGNAL REPORTING CODE

SINPO is an acronym for **S**ignal Strength, **I**nterference, **N**oise, **P**ropagation, and **O**verall merit. The code has a five number rating scale, as shown in Table 3-7, and provides a rapid and fairly accurate means for evaluating and reporting the quality of a received radio signal.

GREEK ALPHABET

The Greek alphabet is given in Table 3-8. The items for which each letter is a symbol are listed in Table 3-9.

TABLE 3-8
Greek Alphabet

Letter			Letter		
Small	*Capital*	**Name**	*Small*	*Capital*	**Name**
α	A	alpha	ν	N	nu
β	B	beta	ξ	Ξ	xl
γ	Γ	gamma	ο	O	omicron
δ	Δ	delta	π	Π	pi
ϵ	E	epsilon	ϱ	P	rho
ζ	Z	zeta	σ	Σ	sigma
η	H	eta	τ	T	tau
θ	Θ	theta	υ	Υ	upsilon
ι	I	iota	ϕ	Φ	phi
ϰ	K	kappa	χ	X	chi
λ	Λ	lambda	ψ	Ψ	psi
μ	M	mu	ω	Ω	omega

TABLE 3-7
SINPO Signal-Reporting Code

	S	I	N	P	O
		Degrading effect of			
Rating scale	*Signal strength*	*Interference (QRM)*	*Noise (QRN)*	*Propagation disturbance*	*Overall readability (QRK)*
5	excellent	nil	nil	nil	excellent
4	good	slight	slight	slight	good
3	fair	moderate	moderate	moderate	fair
2	poor	severe	severe	severe	poor
1	barely audible	extreme	extreme	extreme	unusable

TABLE 3-9
Greek Symbol Designations

Symbol	Designates	Symbol	Designates
α	angles, angular acceleration, coefficients, absorptance, linear current density	λ_e	critical wavelengths
β	angles, phase coefficient, transfer ratio	λ_g	wavelength in a guide
B	magnetic induction	λ_f	resonance wavelength
γ	electrical conductivity, propagation coefficient, specific quantity	Λ	logarithmic decrement, magnetic flux linkage
Γ	reciprocal inductance, propagation constant	μ	amplification factor, Poisson's ratio, permeability
Γ_e	electric constant	μ_i	magnetic susceptibility
Γ_m	magnetic constant	μ_o	initial (relative) permeability
δ	density, damping coefficient, angles, sign of variation	μ_r	relative (magnetic) permeability
Δ	permittivity, determinant	μ_v	permeability of vacuum
ϵ	linear strain, capacitivity, permittivity, base of natural logarithms	ν	frequency, Poisson's ratio, reluctivity
ϵ_b	complex dielectric constant	N_A	Avogadro's constant
ϵ_r	relative capacitivity, relative permittivity, dielectric constant	ξ	coordinates
ϵ_t	total emissivity	π	3.1416 . . . (circumference divided by diameter)
ϵ_l	electric susceptibility	ϱ	volume density of charge, resistivity, cylindrical coordinates
ϵ_λ	emissivity (a function of wavelength)	ϱ_0	thermal resistivity
E	energy, electric field strength	σ	wavenumber, normal stress, surface density of charge
ζ	coordinate, coefficients	Σ	sign of summation
Z_0	characteristic impedance, surge impedance	τ	time constant, shear stress, transmittance
η	permeability, efficiency, modulation index (FM)	$1/\tau$	signaling speed
H	magnetic field strength	ϕ	angle (plane), electrostatic potential, phase angle
θ	temperature, volume strain, angular phase displacement, angle	Φ	heat flow, radiant power, luminous flux, magnetic flux
ι	length	Φ_r	radiant power
$\varkappa$	thermal conductivity, coupling coefficient, Boltzmann's constant, circular and angular wave number	Φ_v	luminous flux
$\varkappa_e$	eddy-current coefficient	χ	cartesian coordinate
$\varkappa_f$	form factor	χ_e	electric susceptibility
$\varkappa_h$	hysteresis coefficient	χ_m	magnetic susceptibility
K	electric field strength, luminous efficiency, susceptibility	ψ	angle (plane)
λ	conductivity, wavelength, linear density charge	Ψ	electric flux
		ω	angle, angular frequency, angular velocity
		ω_e	critical angular frequency
		ω_l	synchronous angular frequency
		ω_r	resonance angular frequency
		Ω	resistance in ohms, angles (solid)

LETTER SYMBOLS AND ABBREVIATIONS

Although letter symbols are often regarded as being abbreviations, each has a separate and distinct use. A symbol represents a unit or quantity, and is the same in all languages. An abbreviation is a letter or a combination of letters (with or without punctuation marks) that represent a word or name in a particular language. An abbreviation, therefore, may be different for different languages. Letter symbols, with few exceptions, are restricted to the Greek and English alphabets. Unit symbols are most commonly written in lowercase letters except when the unit name is derived from a proper name. The distinction between capital and lowercase letters is part of the symbol and should be followed. The terms marked with an asterisk (*) in the following list are not the preferred unit in SI. They have been included because they are used frequently. The preferred SI unit may be included in parentheses ( ) following the previously used unit.

Letter Symbols

A—ampere; ampere turn
Ah—ampere-hour
A/m—ampere per meter
Å—angstrom* (micrometer)
a—atto (10^{-18})
B—bel
b—bit; barn
Bd—baud
Bq—becquerel
Btu—British thermal unit
C—coulomb
°C—degree Celsius
c—centi (10^{-2}); cycle
cd—candela
cd/m²—candela per square meter
Ci—curie* (becquerel)
cm—centimeter
cm³—cubic centimeter
cmil—circular mil
c/s—cycle per second* (hertz)
d—deci (10^{-1}); day
da—deka (10)
dB—decibel
dyn—dyne* (newton)
E—exa (10^{18})
erg—erg* (joule)
eV—electronvolt
F—farad
°F—degree Fahrenheit
f—femto (10^{-15})
fc—footcandle* (lux)
fL—footlambert* (candela per square meter)
ft—foot
ft²—square foot
ft³—cubic foot
ft³/min—cubic foot per minute
ft³/s—cubic foot per second
ft/min—foot per minute
ft/s—foot per second
ft/s²—foot per second squared
ft·lbf—foot-pound (force)
G—giga (10^{9}); gauss* (tesla)
g—gram
g/cm³—gram per cubic centimeter
gal—gallon
Gb—gilbert* (ampere turn)
GeV—gigaelectronvolt
GHz—gigahertz
gr—gram
Gy—gray
H—henry
h—hecto (10^{2}); hour
hp—horsepower* (watt)
Hz—hertz
in—inch
in²—square inch
in³—cubic inch
in/s—inch per second
J—joule

J/K—joule per kelvin
K—kelvin
k—kilo (10^3)
kg—kilogram
kHz—kilohertz
kΩ—kilohm
km—kilometer
km/h—kilometer per hour
kV—kilovolt
kVA—kilovoltampere
kW—kilowatt
kWh—kilowatthour
L—liter; lambert* (candela per square meter)
lb—pound
lm—lumen
lm/m^2—lumen per square meter
lm/w—lumen per watt
lm·s—lumen second
lx—lux (lm/m^2)
M—mega (10^6)
MeV—megaelectronvolt
MHz—megahertz
MΩ—megohm
Mx—maxwell* (Weber)
My—myria* (10^4)
m—meter; milli (10^{-3})
m^2—square meter
m^3—cubic meter
m^3/s—cubic meter per second
mA—milliampere
mH—millihenry
mho—mho* (siemens)
mL—milliliter
mm—millimeter
ms—millisecond
mV—millivolt
mW—milliwatt
mi/h—mile per hour
mil—mil (0.001 in)
min—minute (time)
mol—mole
N—newton
n—nano (10^{-9})
N·m—newton-meter
N/m^2—newton per square meter* (pascal)
N·s/m^2—newton-second per square meter
nA—nanoampere
nF—nanofarad
nm—nanometer
ns—nanosecond
Oe—oersted* (ampere per meter)
P—peta (10^{15}); poise* (pascal second)
Pa—pascal
p—pico (10^{-12})
pF—picofarad
pt—pint
pW—picowatt
qt—quart
R—roentgen
°R—degree Rankine
rad—radian
rd—rad* (gray)
r/min—revolutions per minute
r/s—revolutions per second
S—siemens (Ω^{-1})
s—second time
sr—steradian
T—tera (10^{12}); tesla
u—(unified) atmoic mass unit
V—volt
V/A—voltampere
V/m—volt per meter
var—var
W—watt
W/(m·K)—watt per meter kelvin
W/sr—watt per steradian
W/(sr·m^2)—watt per steradian-square meter
Wb—weber (V·s)
Wh—watt-hour
yd—yard
yd^2—square yard
yd^3—cubic yard
μ—micro (10^{-6})
μA—microampere
μF—microfarad
μH—microhenry
μm—micrometer
μs—microsecond

°—degree (plane angle)
′—minute (plane angle)
″—second (plane angle)

Abbreviations

AC—alternating current
AF—audiofrequency
AFC—automatic frequency control
AGC—automatic gain control
AM—amplitude modulation
amm—ammeter
amp—ampere
amp hr—ampere hour
ampl—amplifier
amptd—amplitude
ANL—automatic noise limiter
ant—antenna
APC—automatic phase control
ASCII—American Standard Code for Information Interchange
assy—assembly
atten—attenuation; attenuator
aud—audible; audio
auto—automatic; automobile
aux—auxiliary
AVC—automatic volume control
avg—average
AWG—American Wire Gage
BA—Buffer amplifier
bal—balance
BC—broadcast
BFO—beat frequency oscillator
bnd—band
BO—blocking oscillator
bp—bandpass
buz—buzzer
bw—bandwidth
byp—bypass
B&S—Brown & Sharpe Wire Gage
Btu—British thermal unit
cal—calibrate
cap—capacitor
carr—carrier
cath—cathode
CB—common base
CC—color code
CCW—counterclockwise
CE—common emitter
cermet—ceramic metal element
CF—cathode follower
chan—channel
ckt—circuit
CRT—cathode-ray tube
c/s—cycle per second
CT—center tap
C to C—center to center
CW—continuous wave; clockwise
cy—cycle
C—capacitance; capacitor; collector; coulomb
°C—Celsius temperature scale
DB—double break
dblr—doubler
dB—decibel
DC—direct current; double contact
DCC—double cotton-covered
deg—degree
degusg—degaussing
demod—demodulator
det—detail; detach
DF—direction finder
disc—disconnect
disch—discharge
DL—delay line
dly—delay
dmgz—demagnetize
dmr—dimmer
DP—double-pole
DPBC—double-pole, back-connected
DPDT—double-pole, double-throw
DPFC—double-pole, front-connected
dplxr—duplexer
DPST—double-pole, single-throw
DPSW—double-pole switch
DSB—double-sideband
DSC—double silk-covered
DSSB—double single-sideband
DT—double throw

DTVM—differential thermocouple voltmeter
dty cy—duty cycle
dyn—dynamo
dynm—dynamotor
dynmt—dyamometer
D—drain; duty factor; electric flux density
EBCDIC—extended binary-coded decimal interchange code
EC—enamel covered
ECO—electronic checkout
EDT—electronic discharge tube
EF—emitter follower
elctd—electrode
elec—electric
elek—electronic
elex—electronics
EM—electromagnetic-epitaxial mesa
EMF—electromotive force
emsn—emission
EMT—electrical metallic tubing
emtr—emitter
engy—energy
env—envelope
EP—epitaxial planar
ER—electrical resistance
ERP—effective radiated power
es—electrostatic
EVM—electronic voltmeter
EVOM—electronic voltohmmeter
exctr—exciter
E—east; emitter; voltage
f—frequency; force
FATR—fixed autotransformer
FB—fuse block
fdbk—feedback
FF—flip-flop
fil—filament
FM—frequency modulation
foc—focus
freq chg—frequency changer
freq con—frequency converter
freqm—frequency meter
FSC—full scale
fu—fuse
fubx—fuse box
fuhlr—fuse holder
FV—full voltage
FW—full wave
°F—Fahrenheit temperature scale
FET—field-effect transistor
g—grounded
ga—gage; graphic ammeter
galvnm—galvanometer
gdlk—grid leak
glpg—glowplug
gnd—ground
G—gain; gate
GMT—Greenwich mean time
hdst—headset
HF—high frequency
HFO—high-frequency oscillator
hifi—high fidelity
hkp—hook-up
hndst—handset
hp—high pass; high pressure; horsepower
HSD—hot side
HT—high tension
HV—high voltage
HVR—high-voltage regulator
hyb—hybrid
hyp—hypotenuse
ID—inside diameter
IF—intermediate frequency
illum—illuminate
impd—impedance
imprg—impregnate
incand—incandescent
incr—increase, increment
ind—indicate
ind lp—indicating lamp
inf—infinite; infinity
inp—input
insp—inspect
inst—install; installation
instm—instrumentation
instr—instrument
insul—insulate; insulation

intercom—intercommunicating; intercommunication
inv—inverter
I/O—input-output
IR—insulation resistance
IT—insulating transformer
I—current
IC—integrated circuit
IGFET—insulated-gate field-effect transistor
JB—junction box
jct—junction
jk—jack
JAN—Joint Army-Navy
JANAF—Joint Army-Navy-Air Force
JFET—junction field-effect transistor
K—dielectric constant
kn sw—knife switch
KO—knock out
kWhm—kilowatt-hour meter
l—length; inductance; inductor; luminance
lam—laminate
lc—line-carrying
LF—low frequency
LFO—low-frequency oscillator
LH—left hand
lim—limit
lim sw—limit switch
LIRLY—load-indicating relay
lkg—leakage
lkrot—locked rotor
llres—load-limiting resistor
LO—local oscillator
loff—leakoff
LP—low pass
LPO—low power output
LR—load resistor (relay)
LSB—lower sideband
LSR—load-switching resistor
lt sw—light switch
lyr—layer
LSHI—large-scale hybrid integration
LSI—large-scale integration
m—magnaflux; mode
MA—mecury arc
mag—magnet; magnetic
mag amp—magnetic amplifier
mag mod—magnetic modulator
MC—momentary contact; multichip
mdl—module
melec—microelectronics
MF—microfilm
mg—magnetic armature
mgn—magneto; magnetron
mic—microphone
mom—momentary
mt—mount
MOS—metal-oxide semiconductor
MOSFET—metal-oxide semiconductor field-effect transistor
NC—No coil; no connection; normally closed
nelec—nonelectric
neut—neutral
NF—noise figure; noise frequency
nfsd—nonfused
nmag—nonmagnetic
NO—normally open
NOL—normal overload
nom—nominal
norm—normal
ntn—neutron
nyl—nylon
N—north
OC—over current
OCO—open-close-open
OCR—over-current relay
ohm—ohmmeter
opr—operate
ORLY—overload relay
osc—oscillate; oscillator
OSMV—one-shot multivibrator
out—output
ovld—overload
ovrd—override
p—pole; probe
PA—pulse amplifier
PAM—pulse-amplitude modulation
PB SW—pull-button switch
PC—printed circuit

PCM—pulse-code modulation; pulse-count modulation
pct—percent
PDM—pulse-duration modulation
PEC—photoelectric cell
pelec—photoelectric
pent—pentode
permb—permeability
PF—power factor; pulse frequency
PFM—pulse-frequency modulation
pF—picofarad
ph—phase
phen—phenolic
phm—phase meter
PIM—pulse-interval modulation
pk—peak
PLB—pull button
plnr—planar
pls—pulse
plyph—polyphase
plz—polarize
plzn—polarization
PNP—positive-negative-positive
pos—positive
pot—potentiometer
pr—pair
preamp—preamplifier
pri—primary
PRV—peak reverse voltage
psiv—passive
PU—pickup
PVC—polyvinyl chloride
pwr—power
pwr sply—power supply
qtz—quartz
Q—merit of a capacitor or coil; quantity of electricity
r—radius
rad—radio
rcdr—recorder
rcv—receive
rcvr—receiver
rechrg—recharge
rect—rectifier
ref—reference
reg—regenerate
res—resistor
resn—resonant
rev cur—reverse current
RF—radiofrequency
RFC—radiofrequency choke
RFI—radiofrequency interference
rgltr—regulator
RH—right hand
RIFI—radio interference field intensity
rinsul—rubber insulation
RLY—relay
rms—root mean square
rmt—remote
rot—rotate
rpm—revolutions per minute
rps—revolutions per second
rpt—repeat
rtr—rotor
RTTY—radio teletypewriter
R—resistance; resistor
RC—resistance-capacitance
RC cpld—resistance-capacitance coupled
RL—resistance-inductance
RLC—resistance-inductance-capacitance
RTL—resistor-transistor logic
SB—sideband
SC—single contact
SCC—single-conductor cable; single cotton-covered
SCE—single cotton enamel
schem—schematic
SCR—short-circuit ratio
scrterm—screw terminal
sec—second; secondary
sel—selector
semicond—semiconductor
sens—sensitive; sensitivity
seq—sequence
servo—servomechanism
sft—shaft
sh—shunt
SHF—superhigh frequency

shld—shield; shielding
short—short circuit
SHTC—short time constant
sig—signal
sig gen—signal generator
slp—slope
slv—sleeve
SLWL—straight-line wavelength
SNR—signal-to-noise ratio
snsr—sensor
sol—solenoid
SP—single pole
spdr—spider
SPDT—single-pole, double-throw
SPDT SW—single-pole, double-throw switch
spk—spike
spkr—speaker
SPST—single-pole, single-throw
SPST SW—single-pole, single-throw switch
SP SW—single-pole switch
sq—square
sqcq—squirrel cage
sqw—square wave
SR—slip ring; split ring
SRLY—series relay
SS—subsystem
SSB—single-sideband
SSBO—single swing blocking oscillator
SSC—single silk-covered
SSW—synchro switch
ST—sawtooth; Schmitt trigger; single throw
STALO—stabilized local oscillator
STAMO—stabilized master oscillator
st & sp—start and stop
stbscp—stroboscope
stby—standby
stdf—standoff
subassy—subassembly
submin—subminiature
substr—substrate
sup cur—superimposed current
suppr—suppressor
svmtr—servomotor
svo—servo
sw—shortwave; switch
swbd—switchboard
swgr—switchgear
swp—sweep
swp exp—sweep expand
swp gen—sweep generator
swp integ—sweep integrator
SWR—standing-wave ratio (voltage)
sym—symbol
syn—synchronous
sync—synchronize
sys—system
syncap—synchroscope
S—signal power; voltage standing-wave ratio
SCR—semiconductor-controlled rectifier
SH—shield (electronic device)
SWG—Stubs Wire Gage
t—temperature; time
tach—tachometer
TB—terminal board
TC—thermocouple; time constant
TCU—tape control unit
tel—telephone
telecom—telecommunications
temp—temperature
templ—template
term—terminal
tet—tetrode
TF—thin film
TFT—thin-film transistor
thermo—thermostat
thms—thermistor
thrm—thermal
thymo—thyratron motor
thyr—thyristor
tlg—telegraph
tlm—telemeter
tlmy—telemetry
TM—temperature meter
TMX—telemeter transmitter
tpho—telephotograph
tpr—teleprinter
TRF—tuned radiofrequency
tsteq—test equipment

T SW—temperature switch; test switch
TT—teletype
TTY—teletypewriter
TV—television
TVM—tachometer voltmeter
TS—telegraph system
UF—ultrasonic frequency
UHF—ultrahigh frequency
undc—undercurrent
undf—underfrequency
unf—unfused
unrgltd—unregulated
USB—upper sideband
util—utility
UJT—unijunction transistor
USG—United States Gage
v—vertical, voltage
vac—vacuum
vam—voltammeter
var—variable; varistor
varhm—var-hour meter
varistor—variable resistor
VC—voice coil
VCO—voltage-controlled oscillator
VD—voltage drop
vdet—voltage detector
vern—vernier
VF—variable frequency; voice frequency
VFO—variable-frequency oscillator
vfreq clk—variable-frequency clock
VHF—very high frequency
vib—vibrate; vibration
vid—video; visual
vidamp—video amplifier
VF—video frequency
VLF—very low frequency
vm—voltmeter
vo—voice
vol—volume
VOM—volt-ohm-milliammeter
VR—voltage regulator
VRLY—voltage relay
VSM—vestigial-sideband modulation
VSWR—voltage standing-wave ratio
VT—vacuum tube
VTVM—vacuum-tube voltmeter
VOX—voice-operated transmitter keyer
VU—volume unit
w—wide
wb—wide band
wd—watt demand meter
wdg—winding
wfr—wafer
WG—waveguide; wire gage
WHDM—watt-hour demand meter
WHM—watt-hour meter
WL—wavelength
WM—wattmeter
wnd—wound
wpg—wiping
WR—wall receptacle
wrg—wiring
wtrprf—waterproof
WV—working voltage
ww—wire-wound
X—reactance
X_c—capacitive reactance
X_L—inductive reactance
y—admittance
yr—year
ZA—zero adjusted
Z—impedance; zone
1/c—single conductor
1 PH—single-phase
3/C—three-conductor
3P—three-pole
3PDT—three-pole, double-throw
3PDT SW—three-pole, double-pole switch
3PH—three-phase
3PST—three-pole, single-throw
3PST SW—three-pole, single-throw switch
3W—three-wire
3way—three-way
4/C—four conductor
4P—four-pole
4PDT—four-pole, double-throw
4PDT SW—four-pole, double-throw switch
4PST—four-pole, single-throw

4PST SW—four-pole, single-throw switch
4W—four-wire
4way—four-way

SEMICONDUCTOR ABBREVIATIONS

The following abbreviations have been adopted for use with semiconductor devices.

α—alpha, common-base short-circuit current gain
B, b—base electrode for units employing a single base
b_1, b_2, **etc.**—base electrodes for more than one base
B—beta, common-emitter short-circuit current gain
BV_R—breakdown voltage, reverse
C, c—collector electrode
C_{cb}—interterminal capacitance, collector-to-base
C_{ce}—interterminal capacitance, collector-to-emitter
C_{ds}—drain-source capacitance, with gate connected to the guard terminal of a three-terminal bridge
C_{dso}—open-circuit drain-source capacitance
C_{du}—drain-substrate capacitance, with gate and source connected to the guard terminal of a three-terminal bridge
C_{eb}—interterminal capacitance, emitter-to-base
C_{gdo}—open-circuit gate-drain capacitance
C_{gso}—open-circuit gate-source capacitance
C_{ibo}—open-circuit input capacitance (common base)
C_{ibs}—short-circuit input capacitance (common base)
C_{ieo}—open-circuit input capacitance (common emitter)
C_{ies}—short-circuit input capacitance (common emitter)
C_{iss}—gate-source capacitance, with drain short-circuited to source
C_{obo}—open-circuit output capacitance (common base)
C_{obs}—short-circuit output capacitance (common base)
C_{ods}—short-circuit output capacitance (gate-drain short-circuited to AC)
C_{oeo}—open-circuit output capacitance (common emitter)
C_{oes}—short-circuit output capacitance (common emitter)
C_{oss}—drain-source capacitance, with gate short-circuited to source
C_{rbs}—short-circuit reverse transfer capacitance (common base)
C_{res}—short-circuit reverse transfer capacitance (common collector)
C_{res}—short-circuit reverse transfer capacitance (common emitter)
C_{rss}—drain-gate capacitance, with the source connected to the guard terminal of a three-terminal bridge
D—duty cycle
d—damping coefficient
E, e—emitter electrode
f_{hfb}—small-signal, short-circuit, forward-current, transfer-ratio cutoff frequency (common base)
f_{hfc}—small-signal, short-circuit, forward-current, transfer-ratio cutoff frequency (common collector)
f_{hfe}—small-signal, short-circuit, forward-current, transfer-ratio cutoff frequency (common emitter)
f_{max}—maximum frequency of oscillation
f_T—transition frequency
g_{MB}—static transconductance (common base)
g_{mb}—small-signal transconductance (common base)
g_{MC}—static transconductance (common collector)
g_{mc}—small-signal transconductance (common collector)

g_{ME}—static transconductance (common emitter)

g_{me}—small-signal transconductance (common emitter)

Ge—germanium

G_{PB}—large-signal average power gain (common base)

G_{pb}—small-signal average power gain (common base)

G_{PC}—large-signal average power gain (common collector)

G_{pc}—small-signal average power gain (common collector)

G_{PE}—large-signal average power gain (common emitter)

G_{pe}—small-signal average power gain (common emitter)

G_{pg}—small-signal insertion power gain, common gate

G_{ps}—small-signal insertion power gain, common source

G_{TB}—large-signal transducer power gain (common base)

G_{tb}—small-signal transducer power gain (common base)

G_{TC}—large-signal transducer power gain (common collector)

G_{tc}—small-signal transducer power gain (common collector)

G_{TE}—large-signal transducer power gain (common emitter)

G_{te}—small-signal transducer power gain (common emitter)

G_{tg}—small-signal transducer power gain, common gate

G_{ts}—small-signal transducer power gain, common source

h_{FB}—static value of the forward-current transfer ratio (common base)

h_{fb}—small-signal, short-circuit, forward-current transfer ratio (common base)

H_{FC}—static value of the forward-current transfer ratio (common collector)

h_{fc}—small-signal, short-circuit, forward-current transfer ratio (common collector)

h_{FE}—static value of the forward-current transfer ratio (common emitter)

h_{fe}—small-signal, short-circuit, forward-current transfer ratio (common emitter)

h_{FEL}—inherent large-signal, forward-current, transfer ratio

h_{IB}—static value of the input resistance (common base)

h_{ib}—small-signal value of short-circuit input impedance (common base)

h_{IC}—static value of the input resistance (common collector)

h_{ic}—small-signal value of short-circuit input impedance (common collector)

h_{IE}—static value of the input resistance (common emitter)

h_{ie}—small-signal value of short-circuit input impedance (common emitter)

h_{ie} **(real)**—real part of small-signal value of short-circuit input impedance (common emitter)

h_{OB}—static value of open-circuit output conductance (common base)

h_{ob}—small-signal value of open-circuit output admittance (common base)

h_{OC}—static value of open-circuit output conductance (common collector)

h_{oc}—small-signal value of open-circuit output admittance (common collector)

h_{OE}—static value of open-circuit output conductance (common emitter)

h_{oe}—small-signal value of open-circuit output admittance (common emitter)

h_{rb}—small-signal value of open-circuit, reverse-voltage transfer ratio (common base)

h_{rc}—small-signal value of open-circuit, reverse-voltage transfer ratio (common collector)

h_{re}—small-signal value of open-circuit, reverse-voltage transfer ratio (common emitter)

I, i—intrinsic region of a device (where neither holes nor electrons predominate)

I_B—base current (DC)

I_b—base current (rms)
i_B—base current (instantaneous)
I_{BO}—breakover current, direct
I_C—collector current (DC)
I_c—collector current (rms)
i_C—collector current (instantaneous)
I_{CBO}—current cutoff current (DC), emitter open
I_{CEO}—collector cutoff current (DC), base open
I_{CER}—collector cutoff current (DC), with specified resistance between base and emitter
I_{CES}—collector cutoff current (DC), with base short-circuited to emitter
I_{CEV}—collector cutoff current with specified voltage between base and emitter
I_{CEX}—collector current (DC), with specified circuit between base and emitter
I_{CO}—collector leakage current (cutoff current)
I_D—drain current (DC)
$I_{D(off)}$—drain cutoff current
I_{DSR}—drain current, (external) gate-source resistance specified
I_{DSS}—drain current, zero gate voltage
I_{DSX}—drain current, gate-source condition specified
I_E—emitter current (DC)
I_e—emitter current (rms)
i_E—emitter current (instantaneous)
I_{EBO}—emitter cutoff current (DC), collector open
$I_{EC(ofs)}$—emitter-collector offset current
I_{ECS}—emitter cutoff current (DC), base short-circuited to collector
I_{E1E2}—emitter cutoff current (double-emitter transistors)
I_F—forward current (DC)
I_f—forward current, alternating component
i_F—forward current (instantaneous)
$I_{F(av)}$—forward current, DC value with alternating component
I_{FG}—forward gate current (DC)
I_{FGM}—peak forward gate current
I_{FM}—forward current, peak total value
$I_{F(OV)}$—forward current, overload
I_{FRM}—forward current, peak repetitive
I_{FSM}—forward current, peak surge
I_G—gate current (DC)
I_{GF}—forward gate current
I_{GR}—reverse gate current
I_H—holding current (DC)
I_i—infection-point current
I_O—average output rectified current
I_{OV}—overload on-state current
I_P—peak-point current (double-base transistor)
I_R—reverse current (DC)
I_r—alternating component of reverse current (rms value)
i_R—reverse current (instantaneous)
$i_{R(REC)}$—reverse recovery current
I_{RRM}—peak reverse current, repetitive
$I_{R(rms)}$—reverse current, total rms value
I_S—source current
I_{SDS}—source current, zero gate voltage
I_{SDX}—source current, gate-drain condition specified
I_{TRM}—peak on-state current, repetitive
I_{TSM}—on-state current surge (nonrepetitive)
I_U—substate current
I_V—valley-point current (double-base transistor)
I_Z—regulator current, reference current (DC)
I_{ZK}—regulator current, reference current (DC nar breakdown knee)
I_{ZM}—regulator current, reference current (DC maximum rated current)
K_O—thermal derating factor
L_c—conversion loss
M—figure of merit
N, n—region of a device where electrons are the majority carriers
NF—noise figure
NF_o—overall noise figure
NR_o—output noise ratio
P, p—region of a device where holes are the majority carriers
P_{BE}—total power input (DC or average) to the base electrode with respect to the emitter electrode

P_{BE}—total power input (instantaneous) to the base electrode with respect to the emitter electrode
P_{CB}—total power input (DC or average) to the collector electrode with respect to the base electrode
P_{CB}—total power input (instantaneous) to the collector electrode with respect to the base electrode
P_{CE}—total power input (DC or average) to the collector electrode with respect to the emitter electrode
P_{CE}—total power input (instantaneous) to the collector electrode with respect to the emitter electrode
P_{DS}—drain-source power dissipation
P_{EB}—total power input (DC or average) to the emitter electrode with respect to the base electrode
P_{EB}—total power input (instantaneous) to the emitter electrode with respect to the base electrode
P_F—forward power loss (DC)
p_F—forward power loss (instantaneous)
P_{FM}—forward power loss, total peak value
P_{IB}—large-signal input power (common base)
P_{ib}—small-signal input power (common base)
P_{IC}—large-signal input power (common collector)
P_{ic}—small-signal input power (common collector)
P_{IE}—large-signal input power (common emitter)
P_{ie}—small-signal input power (common emitter)
P_{OB}—large-signal output power (common base)
P_{ob}—small-signal output power (common base)
P_{OC}—large-signal output power (common collector)
P_{oc}—small-signal output power (common collector)
P_{OE}—large-signal output power (common emitter)
P_{oe}—small-signal output power (common emitter)
P_R—reverse power loss
p_R—reverse power loss (instantaneous)
P_{SM}—surge nonrepetitive power
P_T—total power input (DC or average) to all electrodes
p_T—total power input (instantaneous) to all electrodes
Q_s—recovered charge (stored charge)
R_B—external base resistance
r_{BB}—resistance between two bases, emitter zero (double-base transistor)
$r_b{}'C_c$—collector-base time constant
R_C—external collector resistance
$r_{CE(sat)}$—collector-to-emitter saturation resistance
r_d—damping resistance
R_E—external emitter resistance
R_{EB}—emitter-base junction resistance (assume 4 Ω average)
r_{e1e2}—small-signal emitter-emitter on-state resistance (double emitter transistors)
r_i—dynamic resistance at inflection point
R_L—load resistance
R_θ—thermal resistance
$R_{\theta CA}$—thermal resistance, case-to-ambient
$R_{\theta JA}$—thermal resistance, junction-to-ambient
$R_{\theta Jc}$—thermal resistance, junction-to-case
r_T—slope resistance
Si—silicon
T—temperature
T_A—ambient temperature
T_C—case temperature
t_d—delay time
t_f—fall time
t_{fr}—forward recovery time
T_j—junction temperature
t_{off}—turn-off time
t_{on}—turn-on time
T_{opr}—operating temperature
t_p—pulse time
t_r—rise time
t_{rr}—reverse recovery time

t_s—storage time
TSS—tangential signal sensitivity
T_{stg}—storage temperature
t_w—pulse average time
V_B—base voltage (DC)
V_{BB}—base supply voltage (DC)
V_{BC}—base-to-collector voltage (DC)
V_{bc}—base-to-collector voltage (rms)
v_{bc}—base-to-collector voltage (instantaneous)
V_{BE}—base-to-emitter voltage (DC)
$V_{BE(sat)}$—saturation voltage, base to emitter
V_{be}—base-to-emitter voltage (rms)
v_{be}—base-to-emitter voltage (instantaneous)
V_{BO}—breakover voltage (instantaneous)
$V_{(BR)CBO}$—breakdown voltage, collector-to-base, emitter open
$V_{(BR)CEO}$—breakdown voltage, collector-to-emitter, base open
$V_{(BR)CER}$—breakdown voltage, collector-to-emitter, with specified resistance between base and emitter
$V_{(BR)CES}$—breakdown voltage, collector-to-emitter, with base short-circuited to emitter
$V_{(BR)CEV}$—breakdown voltage, collector-to-emitter, voltage between base and emitter
$V_{(BR)CEX}$—breakdown voltage, collector-to-emitter, circuit between base and emitter
$V_{(BR)EBO}$—breakdown voltage, emitter-to-base, collector open
$V_{(BR)ECO}$—breakdown voltage, emitter-to-collector, base open (formerly BV_{ECO})
$V_{(BR)E1E2}$—breakdown voltage, emitter-to-emitter (double-emitter transistor)
$V_{(BR)GSS}$—breakdown voltage, gate-to-source, drain short-circuited to source
$V_{(BR)GSSF}$—breakdown voltage, forward voltage applied to gate-source, drain short-circuited to source
$V_{(BR)GSSR}$—breakdown voltage, reverse voltage to gate-source, drain short-circuited to source
$V_{(BR)R}$—reverse breakdown voltage
V_{B2B1}—bias DC voltage between base 2 and base 1 (double-base transistor)
V_C—collector voltage (DC)
V_{CB}—collector-to-base voltage (DC)
$V_{CB(fl)}$—DC open-circuit voltage, floating potential, collector-to-base
V_{cb}—collector-to-base voltage (rms)
v_{cb}—collector-to-base voltage (instantaneous)
V_{CBO}—collector-to-base voltage (DC), with emitter open
V_{CC}—collector supply voltage (DC)
V_{CE}—collector-to-emitter voltage (DC)
V_{ce}—collector-to-emitter voltage (rms)
$V_{CE(fl)}$—DC open-circuit voltage, floating potential, collector-to-emitter
V_{CEO}—collector-to-emitter voltage (DC), with base open
V_{CER}—collector-to-emitter voltage (DC), with specified resistance between base and emitter
V_{CES}—collector-to-emitter voltage (DC), with base short-circuited to emitter
$V_{CE(sat)}$—saturation voltage, collector to emitter
V_{CEV}—collector-to-emitter voltage (DC), with voltage between base and emitter
V_{CEX}—collector-to-emitter voltage (DC), with circuit between base and emitter
V_D—off-state voltage (direct)
V_{DD}—drain supply voltage (DC)
V_{DG}—drain-to-gate voltage (DC)
V_{DM}—peak off-state voltage
V_{DRM}—peak off-state voltage repetitive
V_{DS}—drain-to-source voltage (DC)
V_{DSM}—peak off-state voltage, nonrepetitive
V_{DU}—drain-to-substrate voltage (DC)
V_{DWM}—peak off-state voltage, working
V_E—emitter voltage (DC)
V_{EB}—emitter-to-base voltage (DC)
$V_{EB(fl)}$—DC open-circuit voltage, floating potential, emitter-to-base
V_{EBO}—emitter-to-base voltage (DC), with collector open
V_{EC}—emitter-to-collector voltage (DC)
$V_{EC(fl)}$—DC open-circuit voltage, floating potential, emitter-to-collector
V_{EE}—emitter supply voltage (DC)
V_F—forward voltage (DC)

V_f—alternating component of forward voltage (rms value)
v_F—forward voltage (instantaneous)
V_{FG}—forward gate voltage (direct)
V_{FGM}—peak forward gate voltage
V_{FM}—forward voltage, peak total value
$V_{F(rms)}$—forward voltage, total rms value
V_{GD}—gate nontrigger (direct) voltage
V_{GG}—gate supply voltage (DC)
V_{GO}—gate turn-off voltage (direct)
V_{GS}—gate-to-source voltage (DC)
$V_{GS(off)}$—gate-to-source cutoff voltage
$V_{GS(th)}$—gate-to-source theshold voltage
V_{GSF}—forward gate-to-source voltage (DC), of such polarity that an increase in its magnitude causes the channel resistance to decrease
V_{GSR}—reverse gate-to-source voltage (DC), of such polarity that an increase in its magnitude causes the channel resistance to increase
V_{GT}—gate trigger voltage (direct)
$V_{GT(min)}$—minimum gate trigger voltage
V_{GU}—gate-to-substrate voltage (DC)
V_I—inflection-point voltage
V_P—peak-point voltage (double-base transistor)
V_{PP}—projected peak-point voltage
V_{PT}—punch-through voltage
V_R—reverse voltage (DC)
V_r—alternating component of reverse voltage (rms value)
v_R—reverse voltage (instantaneous)
$V_{R(rms)}$—reverse voltage, total rms value
V_{RRM}—reverse voltage, maximum recurrent
V_{RSM}—reverse voltage, peak transient
V_{RT}—reverse collector-to-base voltage, reach-through voltage
V_{RWM}—reverse voltage, (peak) working
V_{SB}—source-substrate voltage
V_{SS}—source supply voltage (DC)
V_{SU}—source-to-substrate voltage (DC)
V_T—on-state voltage, direct
$V_{T(min)}$—minimum on-state voltage
V_{TO}—threshold voltage
V_V—valley-point voltage (double-base transistor)
V_Z—regulator voltage, reference voltage (DC working voltage)
V_{ZM}—regulator voltage, reference voltage (DC at maximum rated current)
Y_{fb}—small-signal, short-circuit forward transfer admittance, common base
Y_{fe}—small-signal, short-circuit forward transfer admittance, common emitter
Y_{fs}—small-signal, short-circuit forward transfer admittance, common source
Y_{ib}—small-signal, short-circuit input admittance, common base
Y_{ic}—small-signal, short-circuit input admittance, common collector
Y_{ie}—small-signal, short-circuit input admittance, common emitter
Y_{ob}—small-signal, short-circuit output admittance, common base
Y_{oc}—small-signal, short-circuit output admittance, common collector
Y_{oe}—small-signal, short-circuit output admittance, common emitter
Y_{rb}—small-signal, short-circuit reverse transfer admittance, common base
Y_{rc}—small-signal, short-circuit reverse transfer admittance, common collector
Y_{re}—small-signal, short-circuit reverse transfer admittance, common emitter
Z_m—impedance, modulator frequency load
z_{RF}—impedance, radio frequency
$Z_{\theta(t)}$—transient thermal impedance
$Z_{\theta JA(t)}$—transient thermal impedance, junction-to-ambient
$Z_{\theta JC(t)}$—transient thermal impedance, junction-to-case
z_v—video impedance
z_z—regulator impedance, reference impedance (small-signal at I_Z)
z_{ZK}—regulator impedance, reference impedance (small-signal at I_{zk})

RESISTOR COLOR CODES

Both composition resistors and the smaller types of wirewound resistors are color-coded for values. The various methods of marking the resistors are shown in Fig. 3-1. Table 3-10 gives the significance of each color.

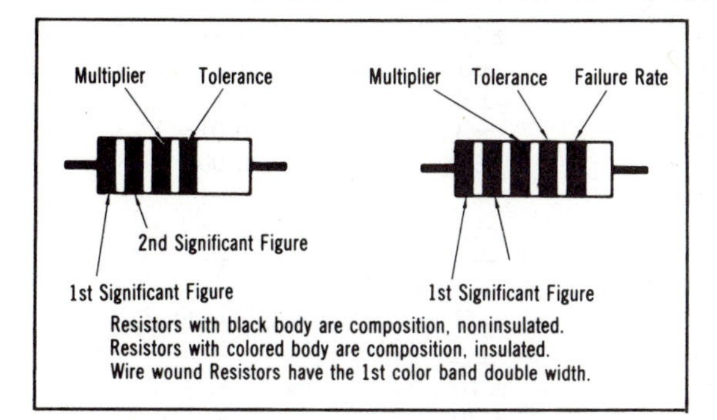

Color-band system (two significant figures)

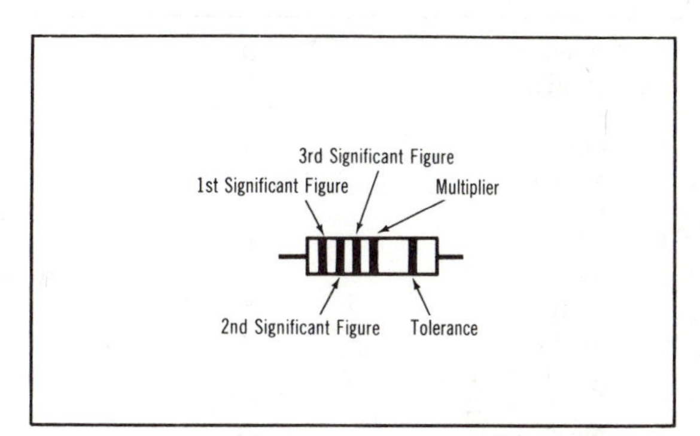

Color-band system (three significant figures)

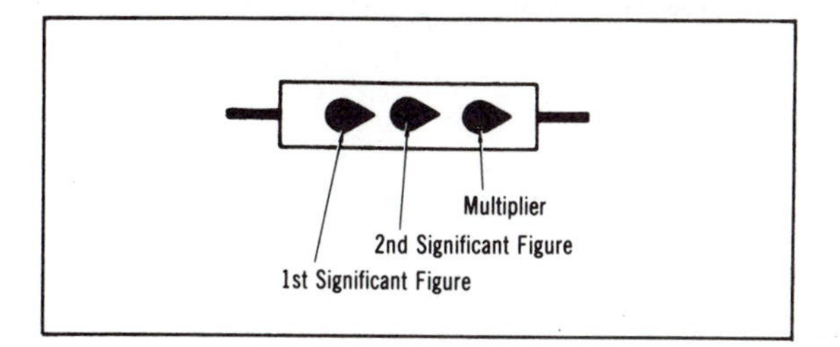

Body-dot system

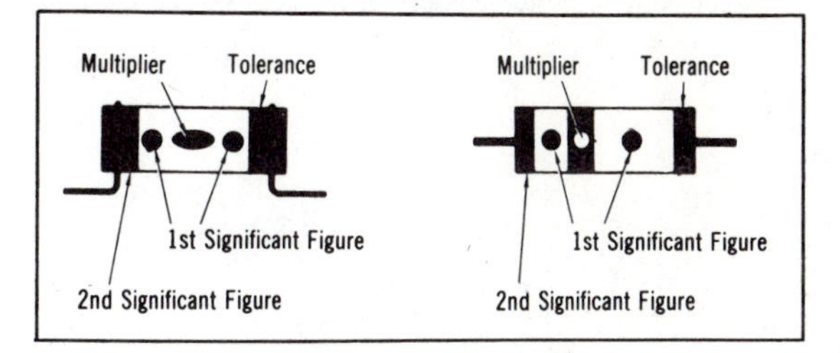

Dot-band system

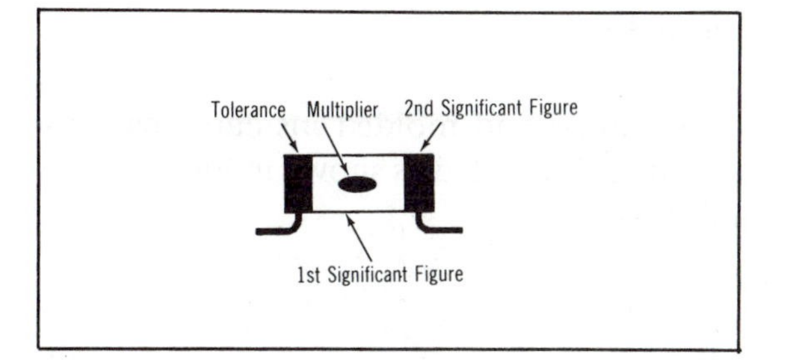

Body-end-dot system

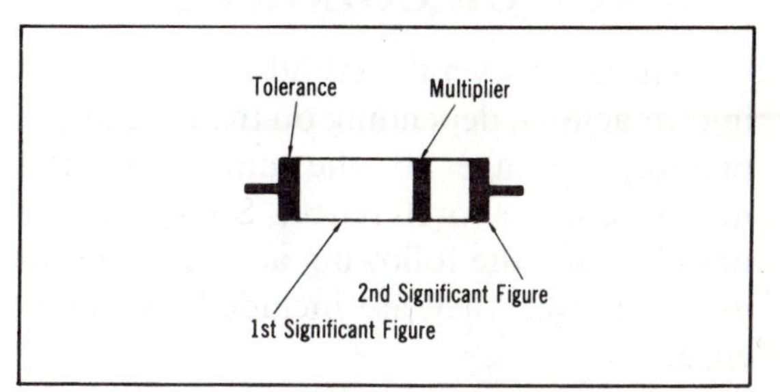

Body-end band system

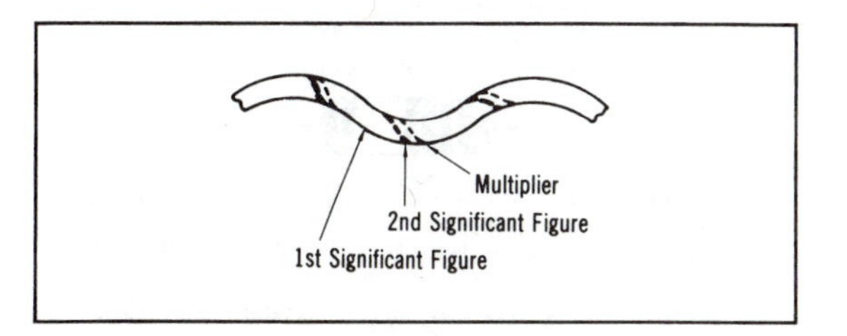

Dash-band system

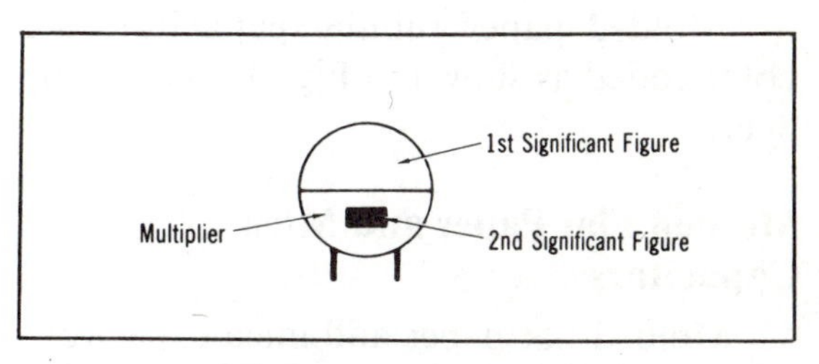

Miniature resistor code

Fig. 3-1

TABLE 3-10
Resistor Color Code

Color	Significant figures	Multiplier	Tolerance (%)	Failure rate*
Black	0	1	± 20	—
Brown	1	10	± 1	1.0
Red	2	100	± 2	0.1
Orange	3	1,000	± 3	0.01
Yellow	4	10,000	± 4	0.001
Green	5	100,000	—	—
Blue	6	1,000,000	—	—
Violet	7	10,000,000	—	—
Gray	8	100,000,000	—	—
White	9		—	Solderable*
Gold	—	0.1	± 5	—
Silver	—	0.01	± 10	—
No Color	—		± 20	—

*On composition resistor indicates failures per 1000 hours. On film resistor indicates solderable terminal.

CAPACITOR COLOR CODES

There are several methods of color coding capacitors, depending on the type of capacitor, the age of the unit, and the manufacturer's preference. Some of the ones listed in the following are no longer in use. However, they are included for reference.

Molded Paper Tubular Capacitors

Molded paper tubular capacitors are color coded as shown in Fig. 3-2 and Table 3-11.

Molded Flat Paper and Mica Capacitors

Molded flat paper and mica capacitors are color coded as shown in Fig. 3-3 and Table 3-12.

Ceramic and Molded Insulated Capacitors

Ceramic and molded insulated capacitors are color coded as shown in Fig. 3-4 and Table 3-13.

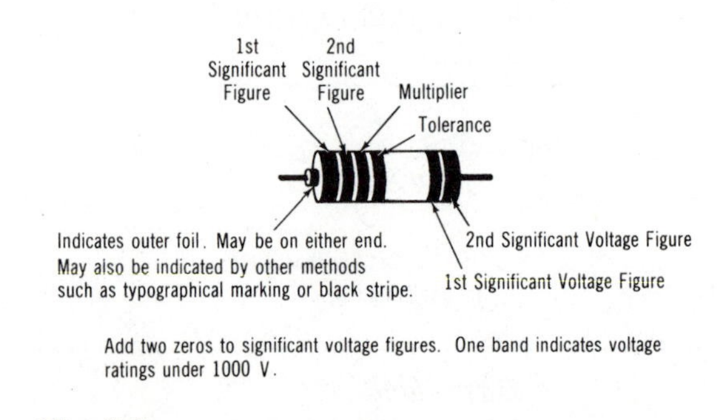

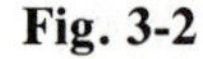
Fig. 3-2

TABLE 3-11
Molded Paper Tubular Capacitor Color Code*

Color	First & second significant figures	Multiplier	Tolerance (%)
Black	0	1	±20
Brown	1	10	—
Red	2	100	—
Orange	3	1,000	—
Yellow	4	10,000	—
Green	5	100,000	±5
Blue	6	1,000,000	—
Violet	7	—	—
Gray	8	—	—
White	9	—	±10
Gold	—	—	±5
Silver	—	—	±10
No Color	—	—	±20

*All values in picofarads.

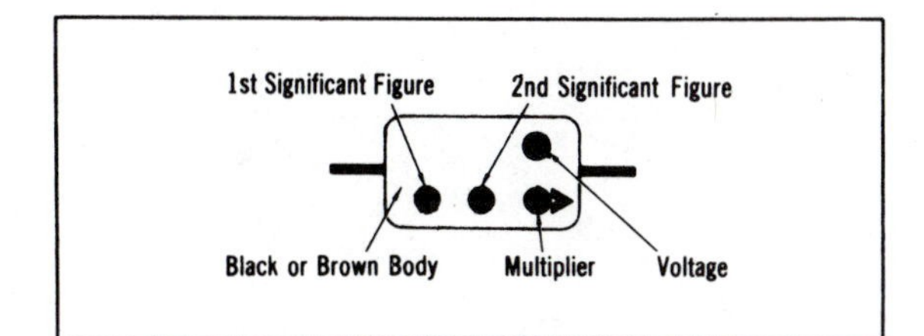

Molded flat paper (commercial grade)

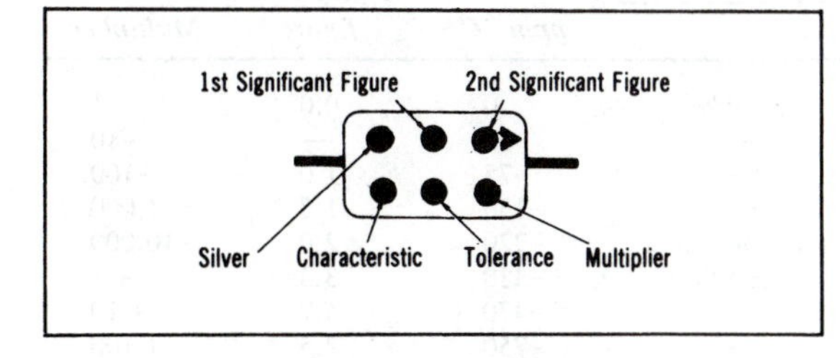

Molded flat paper (military grade)

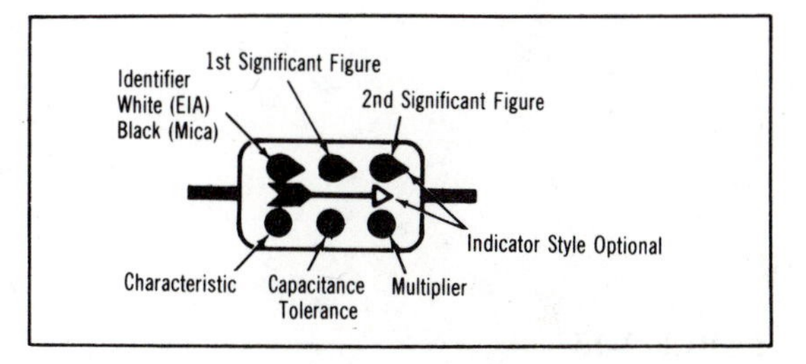

Molded mica (6-dot)

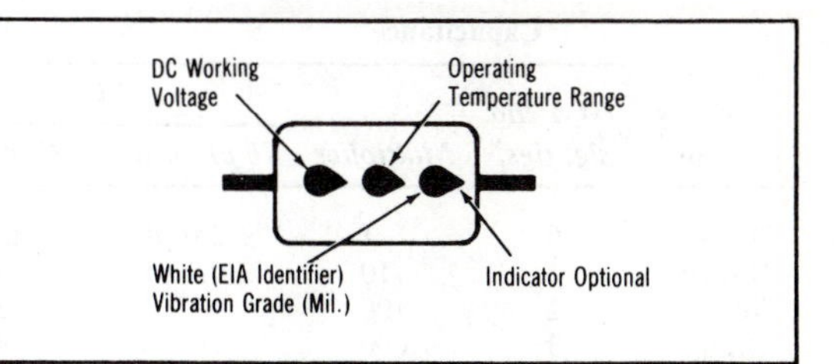

Molded mica (9 dot rear—front is same as for 6-dot code)

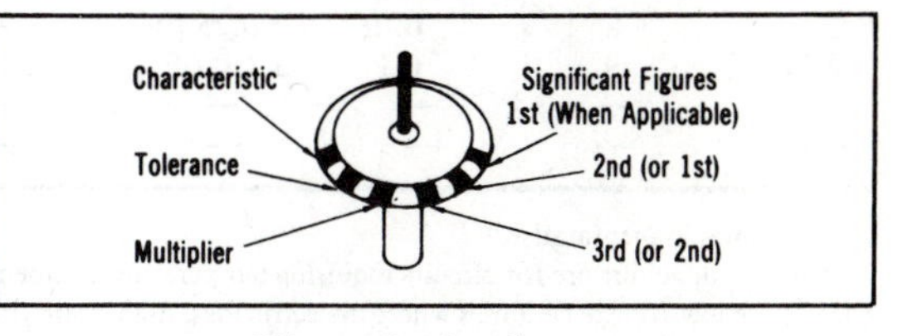

Silvered mica button

Fig. 3-3

TABLE 3-12
Molded Flat Paper and Mica Capacitor Color Code*

Color	Charac-teristic†	Capacitance: *1st & 2nd sig. figs.*	Capacitance: *Multiplier*	Tolerance (%)	DC working voltage	Operating temperature range	Vibration grade (MIL)
Black	A (EIA)	0	1	± 20	—	-55 to + 70 °C (MIL)	10–55 Hz
Brown	B	1	10	± 1	100 (EIA)	—	—
Red	C	2	100	± 2	—	-55 to + 85 °C	—
Orange	D	3	1,000	—	300	—	—
Yellow	E	4	10,000 (EIA)	—	—	-55 to + 125 °C	10–2000 Hz
Green	F	5	—	± 5	500	—	—
Blue	—	6	—	—	—	-55 to + 150 °C (MIL)	—
Violet	—	7	—	—	—	—	—
Gray	—	8	—	—	—	—	—
White	—	9	—	—	—	—	—
Gold	—	—	0.1	± 0.5 (EIA)‡	1000 (EIA)	—	—
Silver	—	—	0.01 (EIA)	± 10	—	—	—

*All values in picofarads.
†Denotes specifications of design involving Q factors, temperature coefficients, and production test requirements.
‡Or + 5.0 pF, whichever is greater. All others are specified tolerance or + 1.0 pF, whichever is greater.

TABLE 3-13
Ceramic Capacitor Color Codes*

Color	Capacitance: *1st & 2nd sig. figs.*	Capacitance: *Multiplier*	Tolerance† Class 1: *10 pF or less*	Tolerance† Class 1: *Over 10 pF*	Tolerance† *Class 2*	Temperature coefficient: *ppm °C*	Temperature coefficient: *Significant figure*	Temperature coefficient: *Multiplier*
Black	0	1	± 2.0 pF	± 20%	± 20%	0	0.0	-1
Brown	1	10	± 0.1 pF	± 1%	—	-33	—	-10
Red	2	100	—	± 2%	—	-75	1.0	-100
Orange	3	1,000	—	± 3%	—	-150	1.5	-1,000
Yellow	4	10,000	—	—	+ 100%, -0%	-220	2.0	-10,000
Green	5	—	± 0.5 pF	± 5%	± 5%	-330	3.3	+ 1
Blue	6	—	—	—	—	-470	4.7	+ 10
Violet	7	—	—	—	—	-750	7.5	+ 100
Gray	8	0.01	± 0.25 pF	—	+ 80%, -20%	+ 150 to -1500	—	+ 1,000
White	9	0.1	± 1.0 pF	± 10%	± 10%	+ 100 to -750	—	+ 10,000
Silver	—	—	—	—	—	—	—	—
Gold	—	—	—	—	—	—	—	—

*All values in picofarads.
†Class 1 capacitors are for circuits requiring temperature compensation and high Q. Class 2 are for circuits where Q and stability are not required. Class 3 are low voltage ceramics where dielectric loss, high insulation resistance and stability are not of major importance. Tolerance of Class 3 ceramics is typographically marked with code M for ± 20% or code Z for + 80%, -20% where space permits.

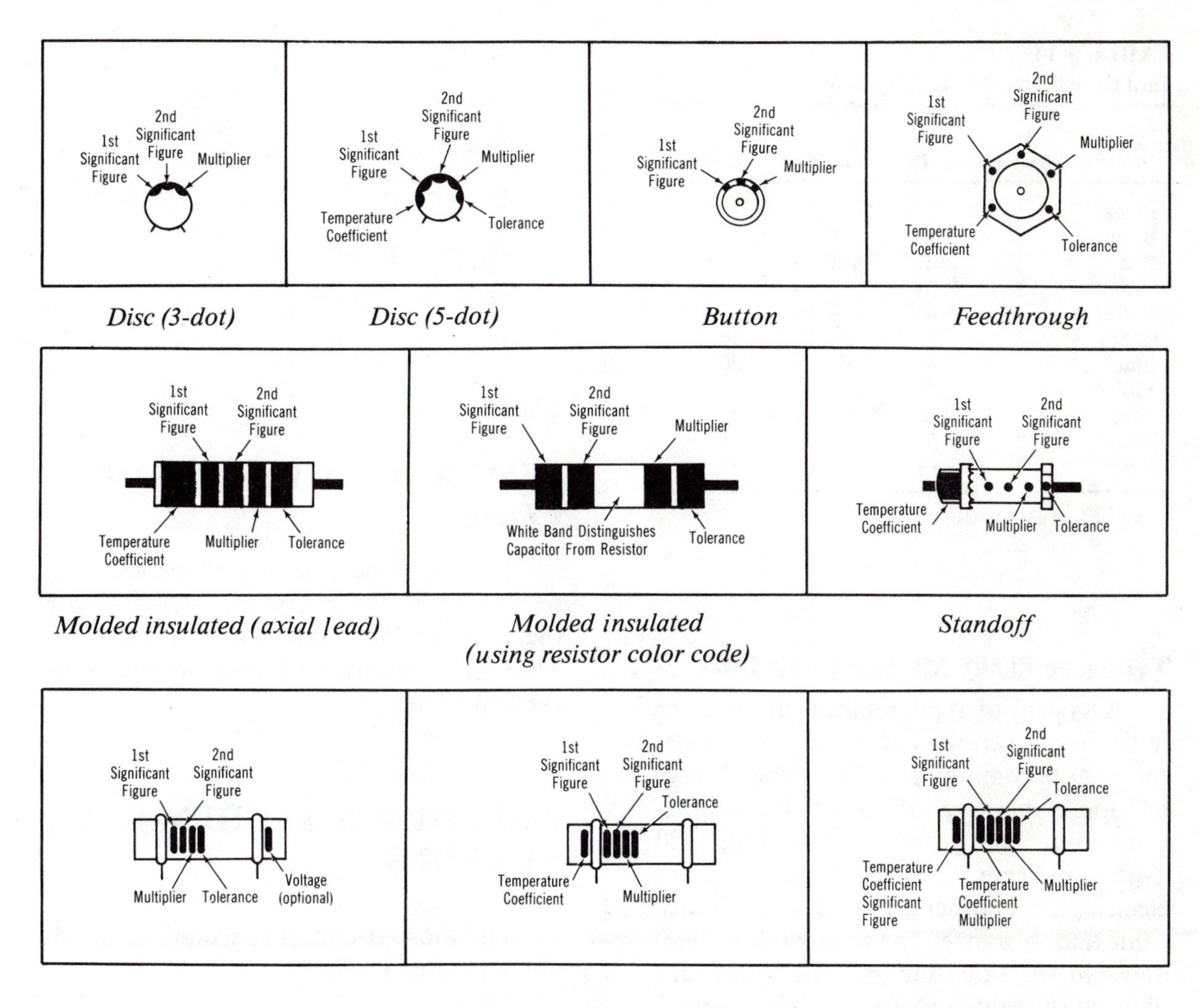

Fig. 3-4

Tantalum Capacitors

Tantalum capacitors are color coded as shown in Fig. 3-5 and Table 3-14.

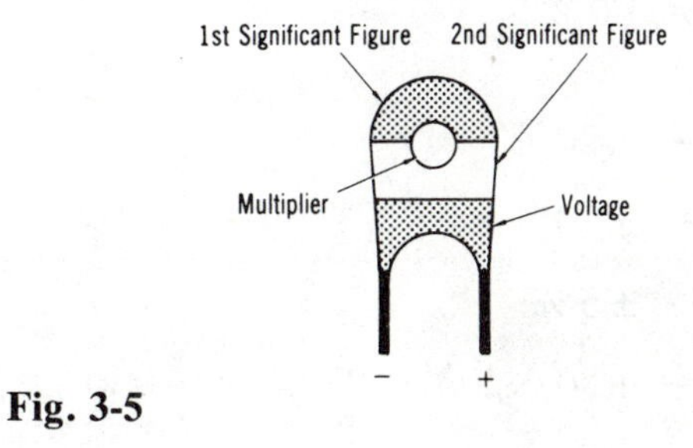

Fig. 3-5

TABLE 3-14
Tantalum Capacitor Color Codes

Color	1st sig. fig.	2nd sig. fig.	Multiplier	Rated DC voltage
Black	—	0	1	10
Brown	1	1	10	—
Red	2	2	100	—
Orange	3	3	—	—
Yellow	4	4	—	6.3
Green	5	5	—	16
Blue	6	6	—	20
Violet	7	7	—	—
Gray	8	8	0.01	25
White	9	9	0.1	3
Pink	—	—	—	35

*All values in microfarads.

Typographically Marked Capacitors

A system of typographical marking to indicate the various parameters of capacitors is becoming popular. The actual method of marking will vary with manufacturers but one group of markings will usually indicate the type, voltage, and dielectric, and another group the capacitance value and tolerance. The first two (or three) digits in the value indicate the significant digits of the value and the last digit, the multiplier or number of zeros to add to obtain the value in picofarads. An R included in the digits indicates a decimal point. A letter following the value indicates the tolerance. The significance of these letters is as follows:

M — ±20%

K — ±10%

J — ±5%

Z — +80, -20%

P — GMV

X — Special

H — ±3%

G — ±2%

F — ±1%

B — ±0.1 pF

C — ±0.25 pF

D — ±0.5 pF

SEMICONDUCTOR COLOR CODE

The sequence numbers of semiconductor type numbers and suffix letters may use the color-coding indicated in Table 3-15. The colors conform to EIA standard for numerical values.

ELECTRONICS SCHEMATIC SYMBOLS

The most common schematic symbols are illustrated in Fig 3-6.

TABLE 3-15
Semiconductor Color Code

Number	Color	Suffix letter
0	black	not applicable
1	brown	A
2	red	B
3	orange	C
4	yellow	D
5	green	E
6	blue	F
7	violet	G
8	gray	H
9	white	J

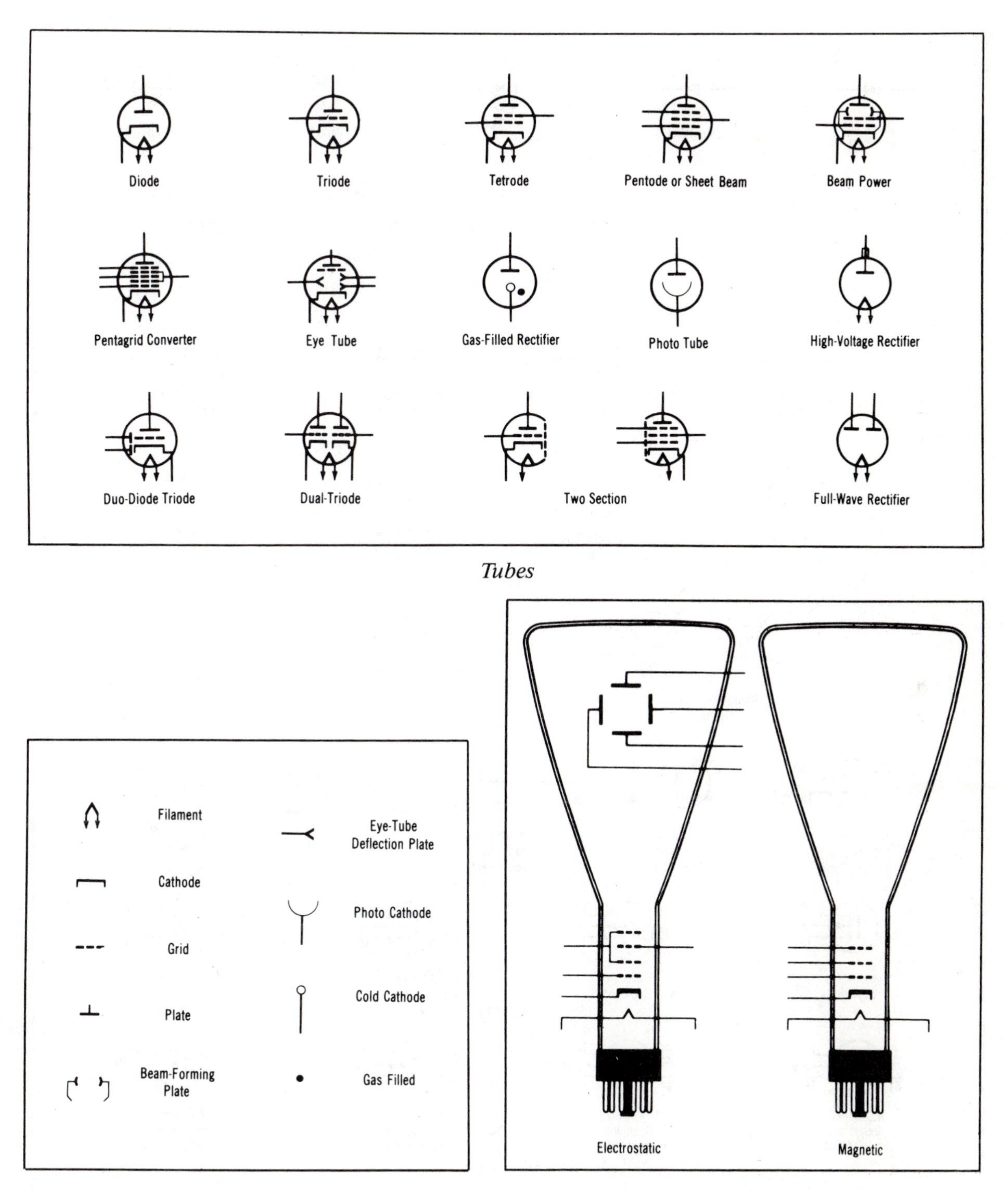

Tubes

Tube elements

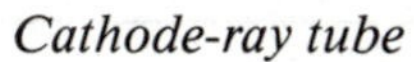

Cathode-ray tube

Fig. 3-6A. Electronics schematic symbols.

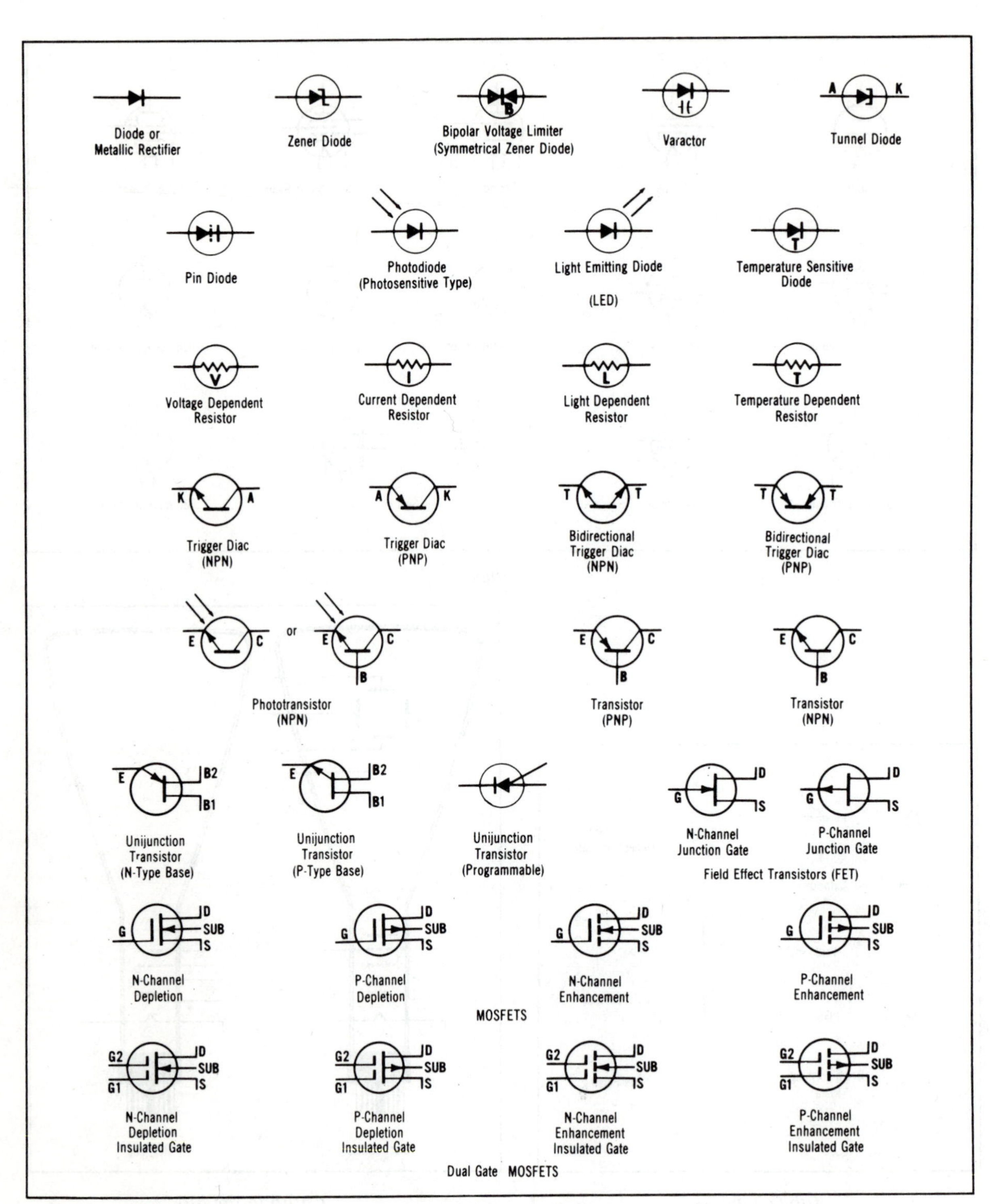

Semiconductor devices

Fig. 3-6B. Electronics schematic symbols.

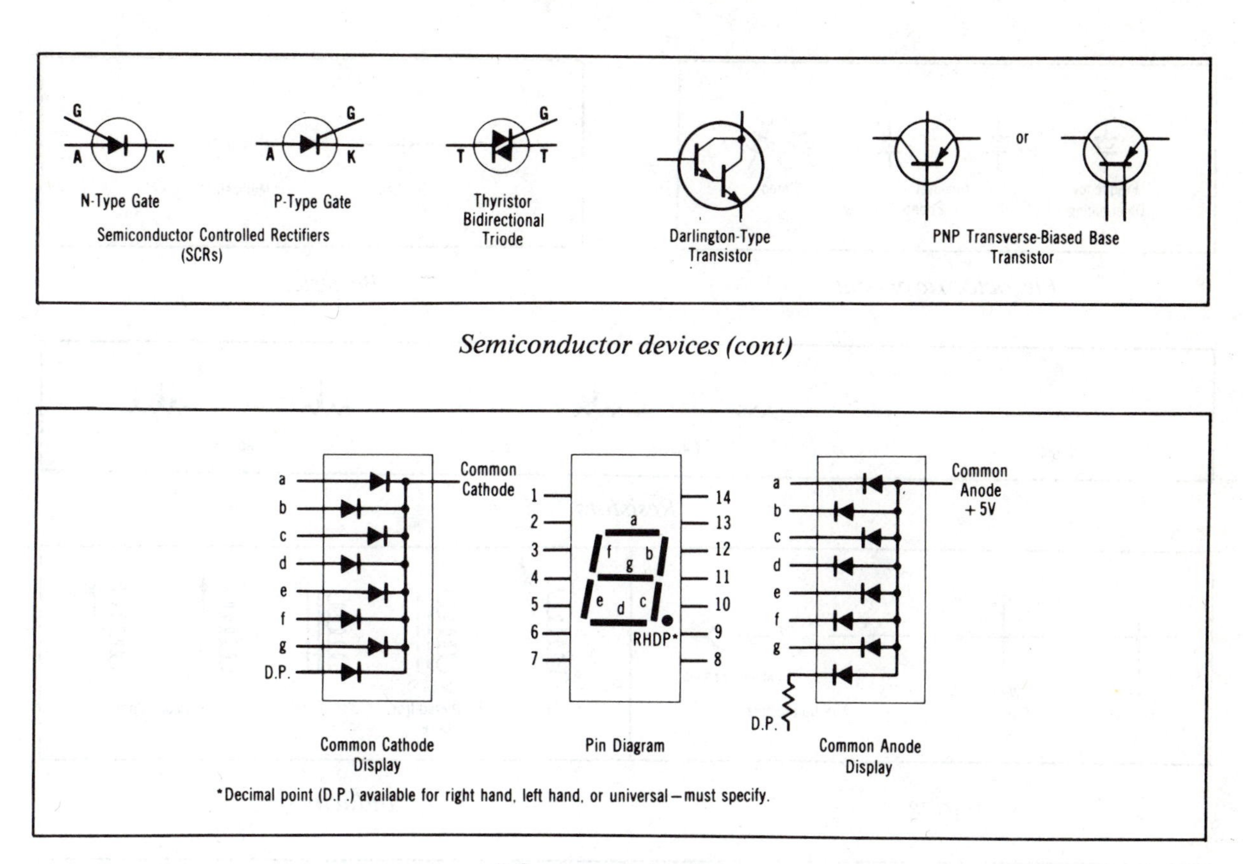

Semiconductor devices (cont)

7-Segment led indicator

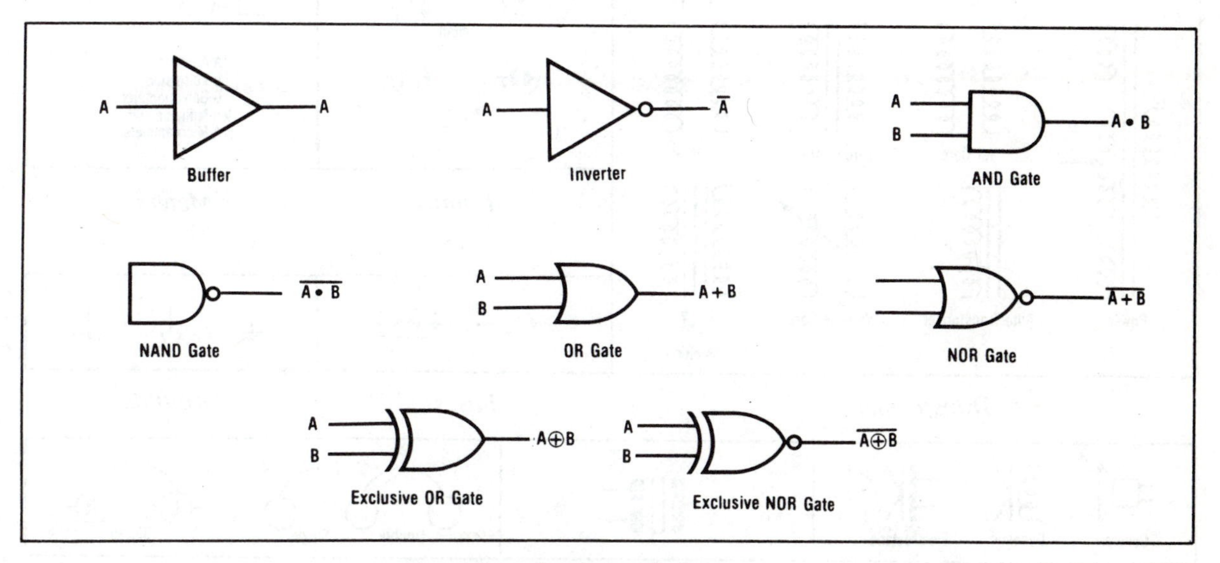

Logic symbols

Fig. 3-6C. Electronics schematic symbols.

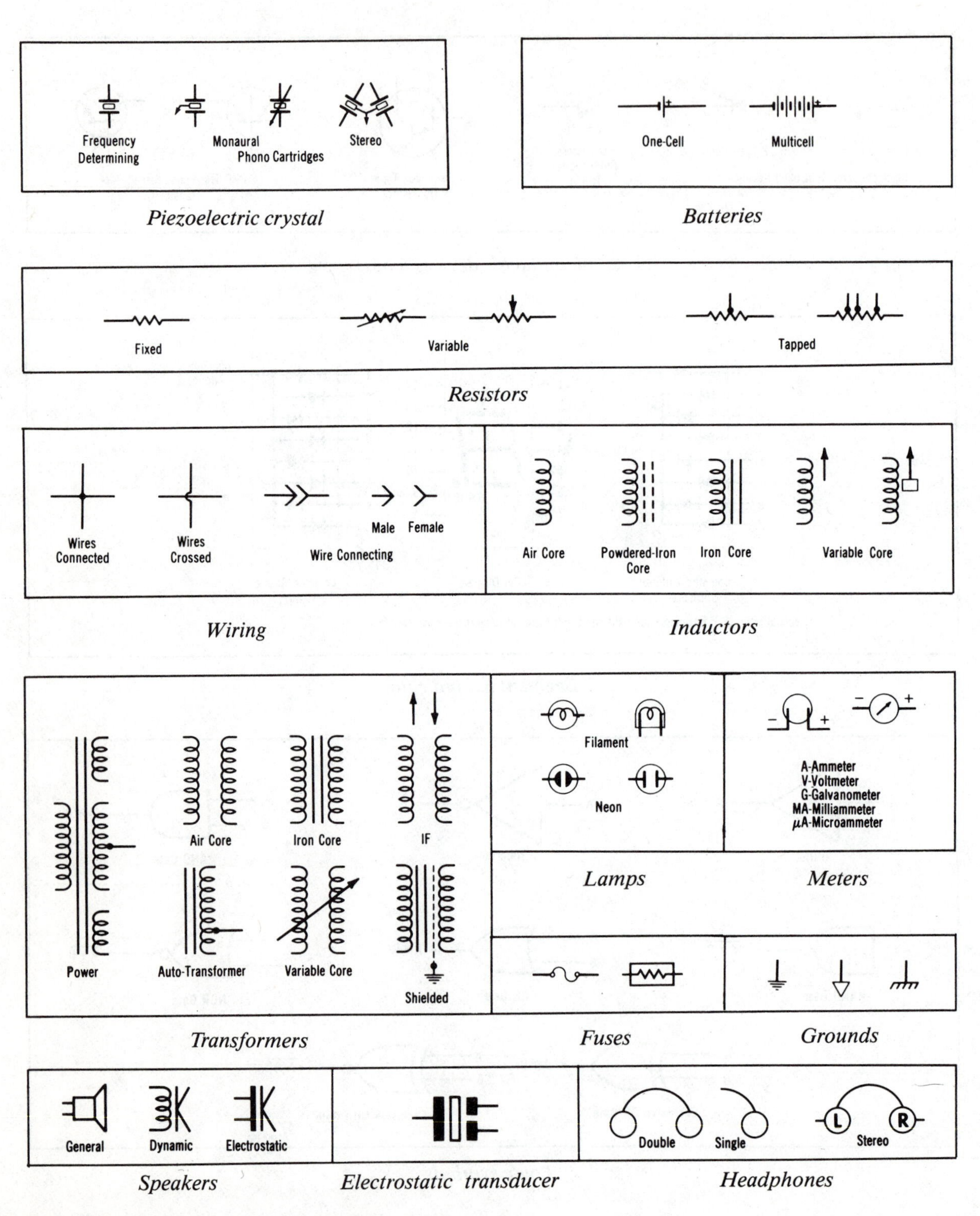

Fig. 3-6D. Electronics schematic symbols.

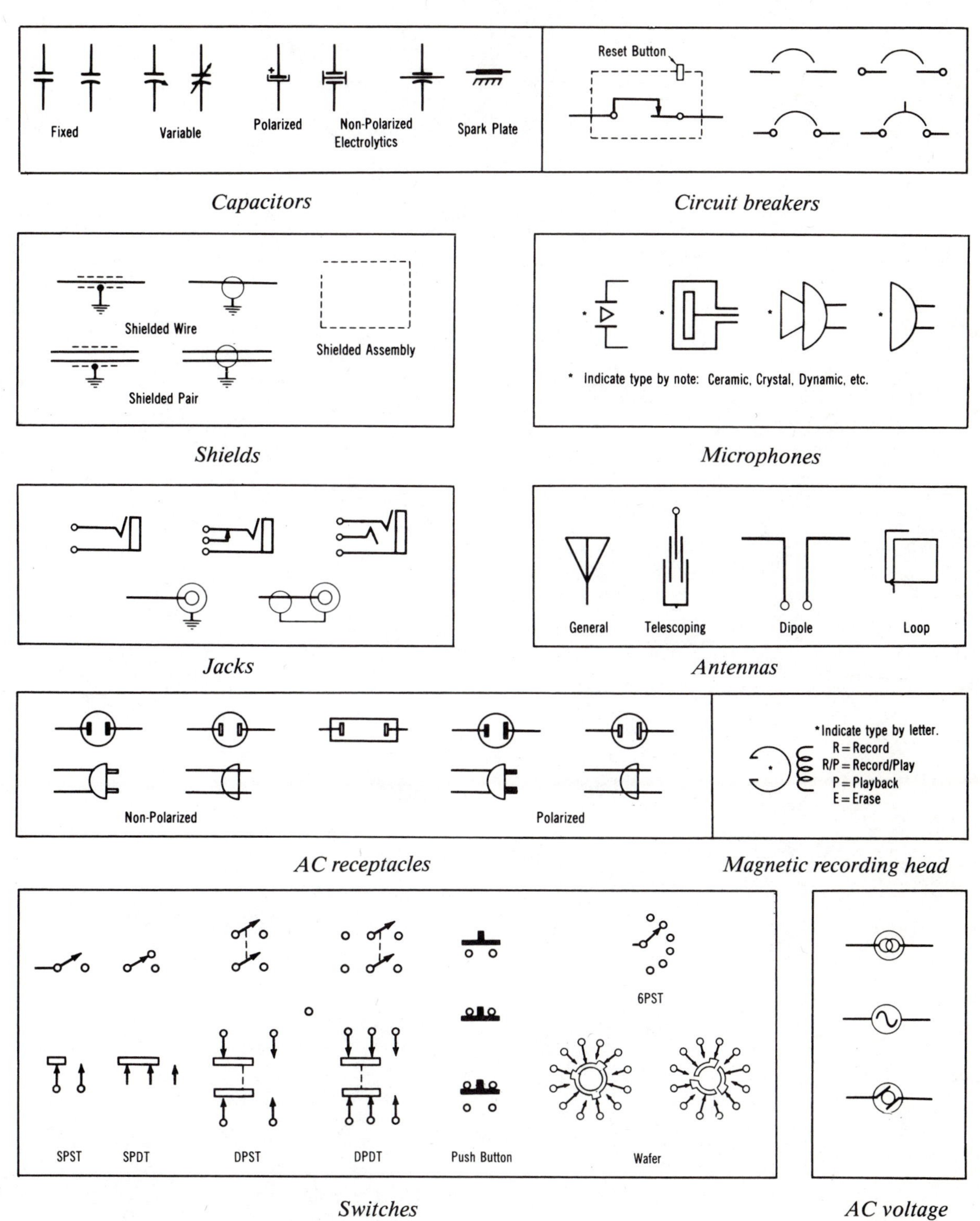

Fig. 3-6E. Electronics schematic symbols.

Chapter 4

Service and Installation Data

COAXIAL CABLE CHARACTERISTICS

Table 4-1 lists the most frequently used coaxial cables. The electrical specifications include the impedance in ohms, capacitance in picofarads per foot, attenuation in decibels per 100 ft and 100 m, and the outside diameter in inches or millimeters. (See page 30 for formulas.)

TEST-PATTERN INTERPRETATION

Many television stations transmit a test pattern, a color bar pattern, or a combination of color bars and test pattern. Generally, the test pattern is transmitted before the station starts its broadcasting day. The test pattern is broadcast as a "station check of performance," indicating proper operation of the transmitter equipment. It is also a check of performance for the receiver. A person trained in electronics can see at a glance if a receiver is operating properly, and appropriate adjustments can be made on the receiver.

In the following explanation, the significance of various test patterns is given. The test pattern broadcast from the television station follows the characteristics of the Indian Head test pattern (Fig. 4-1), which has been in use since the start of television broadcasting.

The roundness of the circles (A and G) in the test pattern provide a quick check on the width, height, and linearity. Horizontal and vertical lines (B) may be used to check linearity, and diagonal lines (C) can be used to check interlace. The vertical wedges (E) or any other pattern details in the vertical plane are used to determine horizontal resolution. Hence, they serve to check the overall video-amplifying circuits and receiver alignment. There should not be any black or white trailing edges from the vertical wedge or circle. That would indicate a problem associated with the receiver. Also, if the test pattern has a vertical wedge, the wedge has separate lines that seem to come together at a certain point and become one wide vertical line. The point where the vertical lines are no longer clear indicates the extent of horizontal resolution.

TABLE 4-1
Coaxial Cable Characteristics

Type RG . . .	Nominal impedance	Nominal capacitance		Nominal OD		Nominal attenuation								
						100 MHz		200 MHz		400 MHz		900 MHz		
/U	Ω	pF/ft	pF/m	in	mm	dB/100 ft	dB/100 m	dB/100 ft	dB/100 m	dB/100 ft	dB/100 m	dB/100 ft	dB/100 m	Remarks
														CATV—MATV
6	75	17.3	56.8	0.290	7.37	2.1	6.9	3.1	10.2	—	—	6.9	22.6	IF & video
6A	75	20.5	67.3	0.336	7.53	2.9	9.5	4.3	14.1	6.5	21.3	10.1	33.1	small, IF/video
8	52	29.5	95.7	0.405	10.29	2.0	6.6	3.0	9.8	4.7	15.4	7.8	25.6	gen. purpose
8A	52	29.5	95.7	0.405	10.29	2.0	6.6	3.0	9.8	4.7	15.4	7.8	25.6	gen. purpose, mil.
9	51	30.0	97.3	0.420	10.67	1.9	6.2	2.8	9.2	4.1	13.5	6.5	21.3	gen. purpose
9B	50	30.0	97.3	0.430	10.92	1.9	6.2	2.8	9.2	4.1	13.5	6.5	21.3	gen. purpose, mil
11	75	20.5	66.5	0.405	10.29	2.0	6.6	2.9	9.5	4.2	13.8	6.5	21.3	commun. tv
11A	75	20.5	66.5	0.405	10.29	2.0	6.6	2.9	9.5	4.2	13.8	6.5	21.3	mil. spec.
12A	75	20.5	66.5	0.475	12.07	2.1	6.9	3.2	10.5	4.7	15.4	7.8	25.6	with armor
14A	52	29.5	95.7	0.558	14.17	1.5	4.8	2.3	7.8	3.5	11.6	—	—	RF power
17A	52	29.5	95.7	0.885	22.48	0.95	3.1	1.5	4.8	2.4	7.9	—	—	gen. purpose
19A	52	29.5	95.7	1.135	28.83	0.69	2.3	1.1	3.2	1.8	5.8	—	—	gen. purpose
22B	95	16.0	52.5	0.420	10.67	3.0	9.8	4.5	14.8	6.8	22.3	11.0	36.1	double shield
55	53.5	28.5	92.5	0.206	5.23	4.8	15.8	7.0	23.1	10.5	34.4	16.0	52.9	flexible, small
55B	53.5	28.5	92.5	0.206	5.23	4.8	15.8	7.0	23.1	10.5	34.4	16.0	52.9	double braid
58	53.5	28.5	92.5	0.195	4.95	4.1	13.5	6.2	20.3	9.5	31.2	14.5	47.6	U/L listed
58	53.5	28.5	92.5	0.206	5.23	4.9	16.1	6.6	21.7	9.2	30.2	13.4	44.2	double shield
58A	50	30.8	101.1	0.195	4.95	5.3	17.4	8.2	26.9	12.6	41.3	20.0	65.6	test leads
58A	50	30.8	101.1	0.195	4.95	4.8	15.8	6.9	22.6	10.1	33.1	15.5	50.9	double shield
58C	50	30.8	101.1	0.195	4.95	5.3	17.4	8.2	26.9	12.6	41.3	20.0	65.6	mil. spec.
59	73	21.0	68.9	0.242	6.15	3.4	11.2	4.9	16.1	7.1	23.3	—	—	gen. purpose, TV
59B	75	20.5	67.3	0.242	6.15	3.4	11.2	4.9	16.1	7.1	23.3	11.1	36.4	mil. spec.
62	93	13.5	44.3	0.242	6.15	3.1	10.2	4.4	14.4	6.3	20.7	11.0	36.1	low capacity, small
62A	93	13.5	44.3	0.242	6.15	3.1	10.2	4.4	14.4	6.3	20.7	11.0	36.1	mil. spec.
62A	93	14.5	47.6	0.260	6.60	3.1	10.2	4.4	14.4	6.3	20.7	11.0	36.1	U/L listed
62B	93	13.5	44.3	0.242	6.15	3.1	10.2	4.4	14.4	6.3	20.7	11.0	36.1	transmission
63B	125	10.0	33.1	0.415	10.54	2.0	6.6	2.9	9.5	4.1	13.5	—	—	low capacitance
71B	93	13.5	44.3	0.250	6.28	2.7	8.9	3.9	12.7	5.8	18.9	—	—	transmission
122	50	30.8	101.1	0.160	4.06	7.0	23.0	11.0	36.1	16.5 max	54.1	28.0	91.9	—
141A	50	29.0	95.1	0.190	4.83	—	—	—	—	9.0 max	29.5	—	—	Teflon, Fiberglas
142B	50	29.0	95.1	0.195	4.95	—	—	—	—	9.0	29.5	—	—	Teflon, 2 shield
174	50	30.8	101.1	0.100	2.54	8.8	28.9	13.0	42.7	20.0 max	65.6	—	—	miniature
178B	50	29.0	95.1	0.070	1.78	—	—	—	—	29.0 max	95.1	—	—	Teflon, trans.
179B	75	19.5	64.0	0.100	2.54	—	—	—	—	21.0 max	68.9	—	—	Teflon, trans.
180B	95	15.0	49.21	0.140	3.56	—	—	—	—	17.0	55.8	—	—	Teflon, trans.
187A	75	19.5	64.0	0.110	2.81	—	—	—	—	21.0 max	68.9 max	—	—	Teflon, trans., miniature
188A	50	27.5	88.5	0.110	2.81	—	—	—	—	20.0 max	65.6 max	—	—	Teflon, trans., miniature
195A	95	14.5	57.6	0.155	3.96	—	—	—	—	17.0 max	55.8 max	—	—	pulse, low cap.
196A	50	28.5	92.4	0.080	2.03	—	—	—	—	29.0	95.1	—	—	Teflon, miniature
212	50	29.5	95.7	0.336	7.53	2.4	7.9	3.6	11.9	5.2	17.0	—	—	double braid
213	50	30.8	101.1	0.405	10.29	2.0	6.6	3.0	9.8	4.7	15.4	7.8	25.6	gen. purpose
214	50	30.8	101.1	0.425	10.80	2.0	6.6	3.0	9.8	4.7	15.4	7.8	25.6	gen. purpose
215	50	30.5	99.4	0.412	10.46	2.1	6.9	3.1	10.2	5.0	16.5	—	—	gen. purpose
217	50	30.0	97.3	0.555	14.14	1.5	4.8	2.3	7.8	3.5	11.5	—	—	double braid
218	50	30.0	97.3	0.880	22.45	0.95	3.1	1.5	4.8	2.4	7.9	—	—	low attenuation
219	50	30.0	97.3	0.880	22.45	0.95	3.1	1.5	4.8	2.4	7.9	—	—	low attenuation/ armor
223	50	30.0	97.3	0.216	5.30	4.8	15.8	7.0	23.1	10.5 max	34.4 max	—	—	double braid, miniature
316	50	29.0	95.2	0.098	2.49	—	—	—	—	20.0 max	65.6 max	—	—	Teflon, mil. spec.

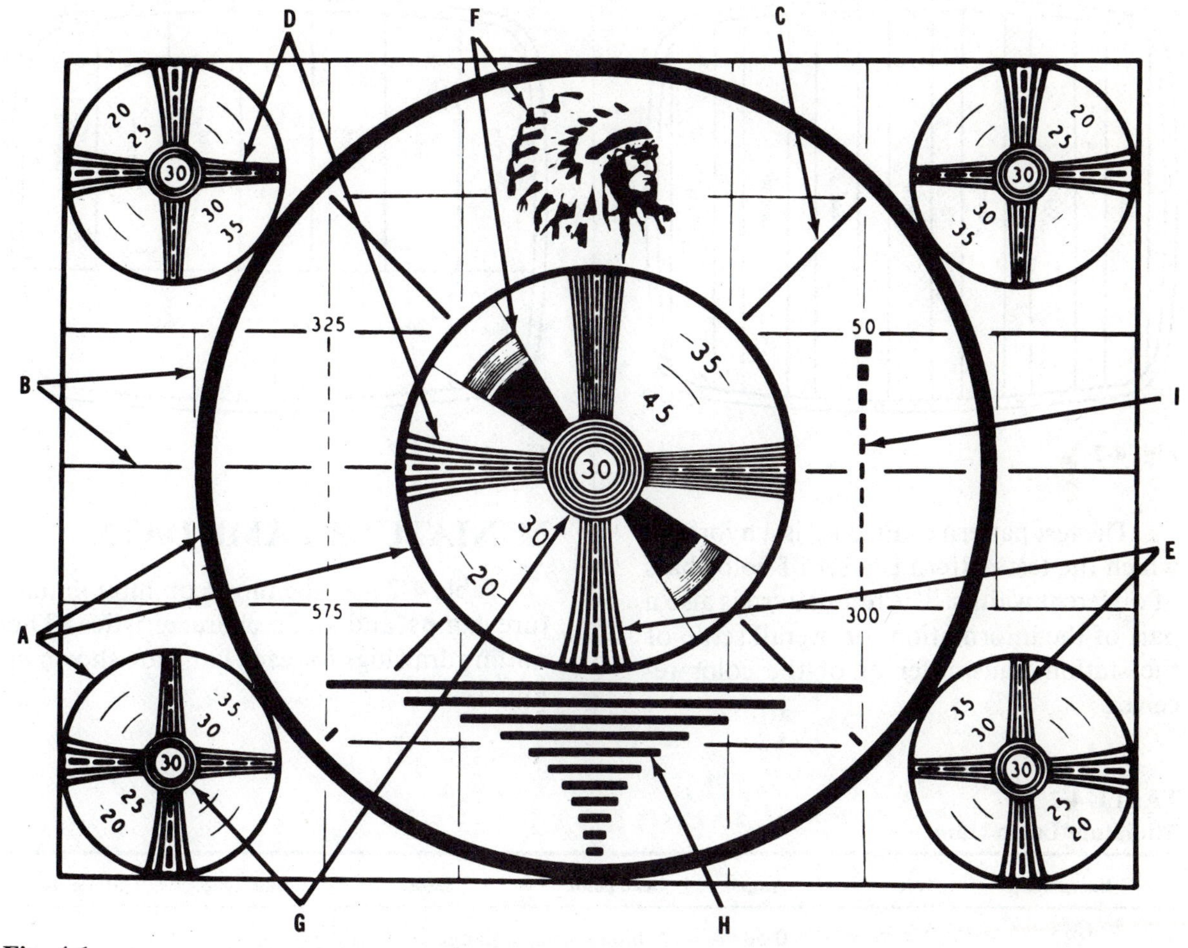

Fig. 4-1

A horizontal wedge (D) in the test pattern is used to indicate the vertical resolution and interlace of the receiver. Generally these wedges have numbers. Various breaks in the lines indicate the number of lines the receiver is capable of producing.

There are one or two diagonal wedges (F) that indicate the contrast ratio. Therefore, they can be used to check the adjustments of the contrast, brightness, and automatic gain controls, as well as the video-amplifying and picture-tube circuits. When video-amplifying and picture-tube circuits are operating properly and the controls are properly adjusted, four degrees of shading should be observed, ranging from black at the center to light gray at the outermost point on the wedge.

The horizontal bars (H) are used to check for low-frequency phase shift. High-frequency ringing can be checked using the single resolution lines (I).

Another test pattern is the color bar pattern shown in Fig. 4-2. The color pattern consists of red to yellow to green, then to blue, and is used for a station check of the color transmitter. It is also used for "receiver color setup."

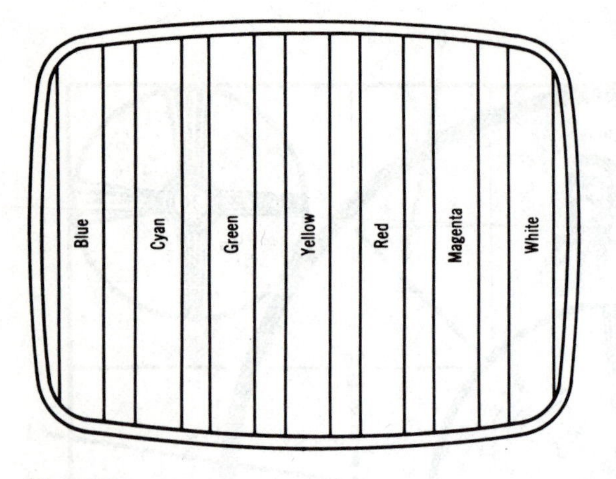

Fig. 4-2

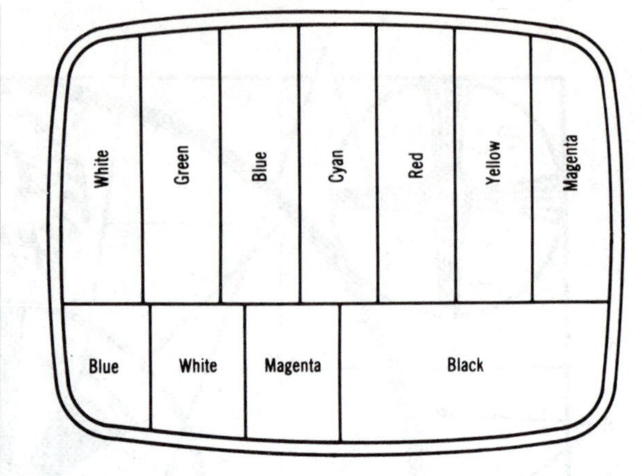

Fig. 4-3

The test pattern of Fig. 4-3 is a hybrid in which the test pattern is a set of color bars of different widths. The test pattern is also a part of the information for overall setup of the station transmitter or of the color receiver.

MINIATURE LAMP DATA

Table 4-2 lists the most common miniature lamps and their characteristics. The outline drawings for each lamp are shown in Fig. 4-4.

TABLE 4-2
Miniature Lamp Data

Lamp no.	Volts	Amps	Bead color	Base	Bulb type	Outline fig.
PR2	2.4	0.50	blue	flange	B-3½	A
PR3	3.6	0.50	green	flange	B-3½	A
PR4	2.3	0.27	yellow	flange	B-3½	A
PR6	2.5	0.30	brown	flange	B-3½	A
PR7	3.8	0.30		flange	B-3½	A
PR12	5.95	0.50	white	flange	B-3½	A
PR13	4.75	0.50		flange	B-3½	A
PR18	7.2	0.55		flange	B-3½	A
PR20	8.63	0.50		flange	B-3½	A
12	6.3	0.15		2-pin	G-3½	H
13	3.8	0.30	green	screw	G-3½	B
14	2.5	0.30	blue	screw	G-3½	B
19	14.4	0.10		2-pin	G-3½	H
27	4.9	0.30		bayonet	G-4½	F
31	6.15	0.30		bayonet	G-4½	F
41	2.5	0.50	white	screw	T-3¼	C
43	2.5	0.50	white	bayonet	T-3¼	D
44	6.3	0.25	blue	bayonet	T-3¼	D
45	3.2	0.35†	green†	bayonet	T-3¼	D
46	6.3	0.25	blue	screw	T-3¼*	C
47	6.3	0.15	brown	bayonet	T-3¼	D
48	2.0	0.06	pink	screw	T-3¼	C

TABLE 4-2 Cont.
Miniature Lamp Data

Lamp no.	Volts	Amps	Bead color	Base	Bulb type	Outline fig.
49	2.0	0.06	pink	bayonet	T-3¼	D
50	6.3	0.20	white	screw	G-3½	B
51	6.3	0.20	white	bayonet	G-3½	E
53	14.4	0.12		bayonet	G-3½	E
55	6.3	0.40	white	bayonet	G-4½	G
57	14.0	0.24	white	bayonet	G-4½	G
63	7.0	0.63		bayonet	G-5	P
67	13.5	0.59		bayonet	G-6	P
81	6.5	1.02		bayonet	G-5	P
82	6.5	1.02		bayonet	G-5	I
87	6.8	1.91		bayonet	S-8	K
88	6.8	1.91		bayonet	S-8	Q
89	13.0	0.58		bayonet	G-5	P
93	12.8	1.04		bayonet	S-8	K
112	1.2	0.22		screw	TL-3	DD
123	1.25	0.30		screw	G-3½	B
136	1.25	0.60		bayonet	G-4½	F
158	14.0	0.24		wedge	T-3¼	S
161	14.0	0.19		wedge	T-3¼	S
168	14.0	0.35		wedge	T-3¼	S
194	14.0	0.27		wedge	T-3¼	S
222	2.2	0.25	white	screw	TL-3	DD
301	28.0	0.17		bayonet	G-5	R
302	28.0	0.17		bayonet	G-5	T
303	28.0	0.30		bayonet	G-6	P
305	28.0	0.51		bayonet	S-8	K
307	28.0	0.67		bayonet	S-8	K
308	28.0	0.67		bayonet	S-8	Q
309	28.0	0.90		bayonet	S-11	U
313	28.0	0.17		bayonet	T-3¼	D
327	28.0	0.04		flanged	T-1¾	W
328	6.0	0.20		flanged	T-1¾	W
330	14.0	0.08		flanged	T-1¾	W
331	1.35	0.06		flanged	T-1¾	W
334	28.0	0.04		grooved	T-1¾	X
344	10.0	0.014		flanged	T-1¾	W
382	14.0	0.08		flanged	T-1¾	W
387	28.0	0.04		flanged	T-1¾	W
388	28.0	0.04		grooved	T-1¾	X
680	5.0	0.060		wire terminal	T-1	M
682	5.0	0.060		wire terminal	T-1	Y
683	5.0	0.060		wire terminal	T-1	M
685	5.0	0.060		wire terminal	T-1	Y
713	5.0	0.075		wire terminal	T-1	M
714	5.0	0.075		wire terminal	T-1	Y
715	5.0	0.115		wire terminal	T-1	M
718	5.0	0.115		wire terminal	T-1	Y
1003	12.8	0.94		bayonet	B-6	AA
1004	12.8	0.94		bayonet	B-6	BB
1034	12.8	1.80		bayonet	S-8	Q
1076	12.8	1.80		bayonet	S-8	Q
1133	6.2	3.91		bayonet	RP-11	CC

TABLE 4-2 Cont.
Miniature Lamp Data

Lamp no.	Volts	Amps	Bead color	Base	Bulb type	Outline fig.
1156	12.8	2.10		bayonet	S-8	K
1157	12.8	2.10		bayonet	S-8	Q
1176	12.8	1.34		bayonet	S-8	Q
1183	5.5	6.25		bayonet	RP-11	CC
1195	12.5	3.0		bayonet	RP-11	CC
1445	14.4	0.135		bayonet	G-3½	E
1447	18.0	0.15		screw	G-3½	B
1490	3.2	0.16	white	bayonet	T-3¼	D
1495	28.0	0.30		bayonet	T-4½	Z
1728	1.35	0.06		wire terminal	T-1¾	N
1738	2.7	0.06		wire terminal	T-1¾	N
1764	28.0	0.04		wire terminal	T-1¾	N
1784	6.0	0.20		wire terminal	T-1¾	N
1813	14.4	0.10		bayonet	T-3¼	D
1815	14.0	0.20		bayonet	T-3¼	D
1816	13.0	0.33		bayonet	T-3¼	D
1819	28.0	0.40	white	bayonet	T-3¼	D
1820	28.0	0.10		bayonet	T-3¼	D
1829	28.0	0.07		bayonet	T-3¼	D
1847	6.3	0.15	white	bayonet	T-3¼	D
1850	5.0	0.09		bayonet	T-3¼	D
1864	28.0	0.17		bayonet	T-3¼	D
1866	6.3	0.25		bayonet	T-3¼	D
1869	10.0	0.014		wire terminal	T-1¾	N
1888	6.3	0.46	white	bayonet	T-3¼	D
1889	14.0	0.27		bayonet	T-3¼	D
1891	14.0	0.24	pink	bayonet	T-3¼	D
1893	14.0	0.33		bayonet	T-3¼	D
1895	14.0	0.27		bayonet	G-4½	F
2162	14.0	0.10		wire terminal	T-1¾	N
2182	14.0	0.08		wire terminal	T-1¾	N
2187	28.0	0.04		wire terminal	T-1¾	N
6832	5.0	0.06		wire terminal	T-1‡	O
6833	5.0	0.06		wire terminal	T-¾	O
6838	28.0	0.24		wire terminal	T-1	M
6839	28.0	0.24		wire terminal	T-1	Y
7152	5.0	0.115		wire terminal	T-1‡	L
7327	28.0	0.04		wire terminal	T-1¾	V
7328	6.0	0.20		wire terminal	T-1¾	V
7344	10.0	0.014		wire terminal	T-1¾	V
7382	14.0	0.08		wire terminal	T-1¾	V
7387	28.0	0.04		wire terminal	T-1¾	V
7931	1.35	0.06		wire terminal	T-1¾	V

*Frosted.
†Some brands are 0.5 A and white bead.
‡Short.

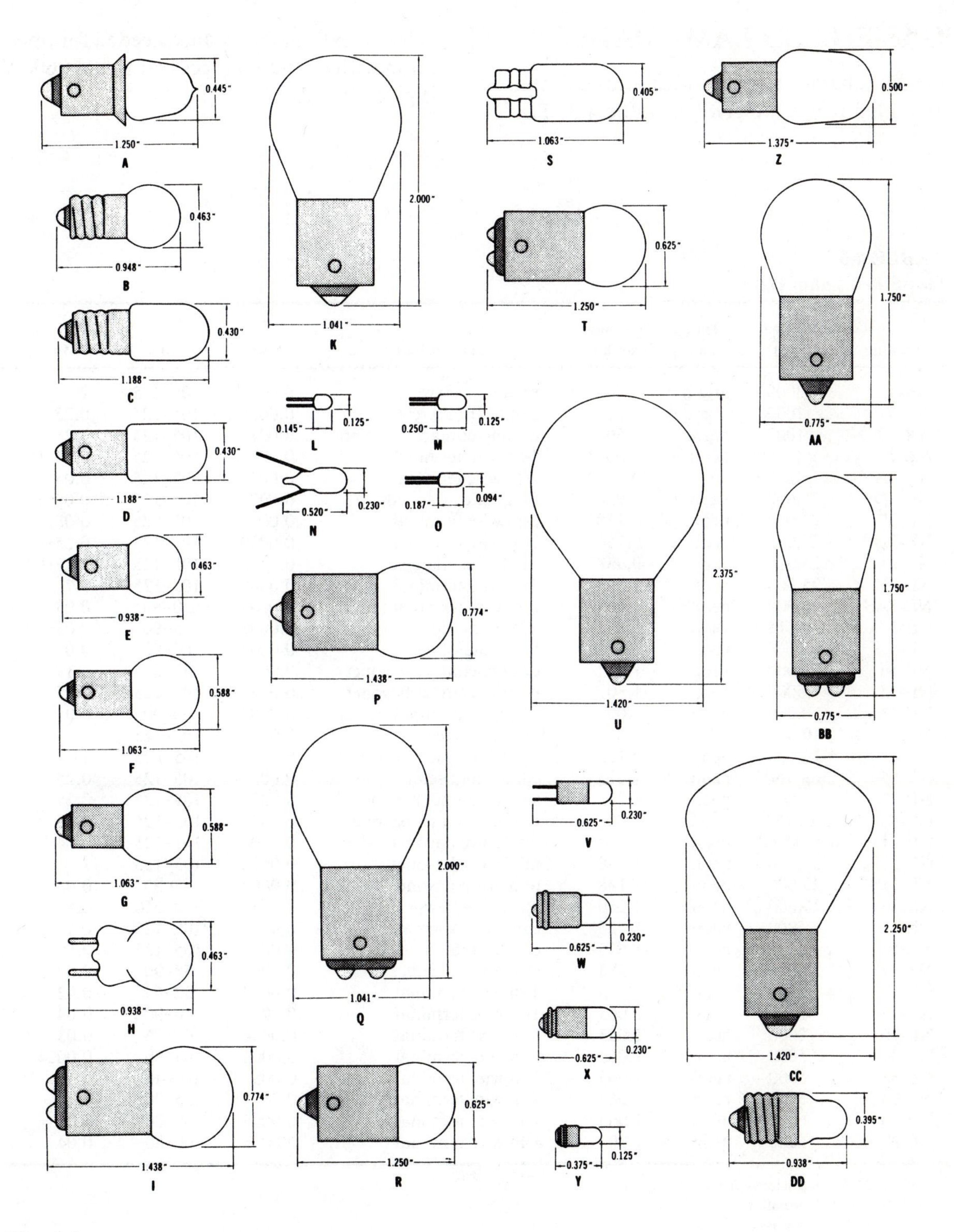
0.445"
1.250"
A
0.463"
0.948"
B
0.430"
1.188"
C
0.430"
1.188"
D
0.463"
0.938"
E
0.588"
1.063"
F
0.588"
1.063"
G
0.463"
0.938"
H
0.774"
1.438"
I
2.000"
1.041"
K
0.145"
0.125"
L
0.250"
0.125"
M
0.520"
0.230"
N
0.187"
0.094"
O
0.774"
1.438"
P
2.000"
1.041"
Q
0.625"
1.250"
R
0.405"
1.063"
S
0.625"
1.250"
T
2.375"
1.420"
U
0.625"
0.230"
V
0.625"
0.230"
W
0.625"
0.230"
X
0.375"
0.125"
Y
0.500"
1.375"
Z
1.750"
0.775"
AA
1.750"
0.775"
BB
2.250"
1.420"
CC
0.395"
0.938"
DD

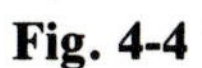
Fig. 4-4

GAS-FILLED LAMP DATA

The characteristics of the more common gas-filled lamps are given in Table 4-3. The value of external resistance needed for operation with circuit voltages from 110 to 600 V is given in Table 4-4.

TABLE 4-3
Gas-Filled Lamp Data

Number	Average life (h)*	Type gas	Maximum length (in)	Type of base	Amps	Volts	Watts†
AR-1	3000	argon	3.50	medium screw	0.018	105–125	2
AR-3	1000	argon	1.625	candelabra screw	0.0035	105–125	0.25
AR-4	1000	argon	1.50	double-contact bayonet	0.0035	105–125	0.25
NE-2	over 25,000	neon	1.063‡	1-in wire terminal	0.003	105–125	0.04
NE-2A	over 25,000	neon	0.844‡	2-in wire terminal	0.003	105–125	0.04
NE-2D	25,000	neon	0.938	S. C. mid: flanged	0.007	105–125	0.08
NE-2E	25,000	neon	0.750	2-in wire terminal	0.007	105–125	0.08
NE-2H	25,000	neon	0.750	2-in wire terminal	0.0019	105–125	0.25H-B
NE-2J	25,000	neon	0.938	S. C. mid. flanged	0.0019	105–125	0.25H-B
NE-2V	25,000	neon	1.063	2-in wire terminal	0.0065	105–125	0.7
NE-2AS	25,000	neon	1.063	2-in wire terminal	0.0003	60–90	0.03
NE-3	over 5000	neon	0.875	telephone slide	0.0003	55–90	0.03
NE-4	over 5000	neon	0.875	telephone slide	0.0003	60–90	0.03
NE-16	1000	neon	1.50	double-contact bayonet	0.0015	53–65	0.1
NE-17	5000	neon	1.50	double-contact bayonet	0.002	105–125	0.25
NE-23	6000	neon	1.00	1-in wire terminal	0.0003	60–90	0.03
NE-30	10,000	neon	2.250	medium screw	0.012	105–125	1
NE-32	10,000	neon	2.125	double-contact bayonet§	0.012	105–125	1
NE-45	over 7500	neon	1.625	candelabra screw	0.002	105–125	0.25
NE-47	5000	neon	1.375	single-contact bayonet	0.002	105–125	0.25
NE-48	over 7500	neon	1.50	double-contact bayonet	0.002	105–125	0.25
NE-51	over 15,000	neon	1.188	miniature bayonet	0.0003	105–125	0.04
NE-51H	25,000	neon	1.188	miniature bayonet	0.0012	105–125	1
NE-51S	25,000	neon	1.188	miniature bayonet	0.0002	55–90	0.02
NE-56	10,000	neon	2.250	medium screw§	0.005	210–250	0.5
NE-57	5000	neon	1.625	candelabra screw§	0.002	105–125	0.25
NE-58	over 7500	neon	1.625	candelabra screw	0.002	105–125	0.50
NE-67	25,000	neon	1.188	miniature bayonet	0.0002	55–90	0.02
NE-68	5000	neon	1.063	2-in wire terminal	0.0003	52–65	0.02
NE-75	2000	neon	1.063	1-in wire terminal	0.0004	60–90	0.04
NE-76	2000	neon	1.063	1-in wire terminal	0.0004	68–76	0.03
NE-81	5000	neon	1.063	1-in wire terminal	0.0003	64–80	0.0024
NE-83	5000	neon	1.063	1-in wire terminal	0.005	60–100	0.5
NE-86	5000	neon	1.063	1-in wire terminal	0.0015	55–90	0.14
NE-96	6000	neon	1.063	1-in wire terminal	0.0005	60–80	0.04
NE-97	6000	neon	1.00	1-in wire terminal	0.0005	60–80	0.04

*Life on DC is approximately 60% of AC values.
†For 105- to 125-V operation.
‡The dimension is for glass only.
§In DC circuits, the base should be negative.

TABLE 4-4
External Resistances Needed for Gas-Filled Lamps

Type	105–125 V	220–300 V	300–375 V	375–450 V	450–600 V
AR-1	included in base	10,000	18,000	24,000	30,000
AR-3	included in base	68,000	90,000	150,000	160,000
AR-4	15,000	82,000	100,000	160,000	180,000
NE-2	200,000	750,000	1,000,000	1,200,000	1,600,000
NE-2A	200,000	750,000	1,000,000	1,200,000	1,600,000
NE-2D	100,000	—	—	—	—
NE-2E	100,000	—	—	—	—
NE-2H	30,000	—	—	—	—
NE-2H	30,000	—	—	—	—
NE-2V	100,000	—	—	—	—
NE-17	30,000	110,000	150,000	180,000	240,000
NE-30	included in base	10,000	20,000	24,000	36,000
NE-32	7,500	18,000	27,000	33,000	43,000
NE-45	included in base	82,000	120,000	150,000	200,000
NE-47	30,000	—	—	—	—
NE-48	30,000	—	—	—	—
NE-51	200,000	750,000	1,000,000	1,200,000	1,600,000
NE-51H	47,000	—	—	—	—
NE-56	included in base	—	—	—	—
NE-57	included in base	82,000	120,000	150,000	200,000
NE-58	included in base	—	—	—	—

RECEIVER AUDIOPOWER AND FREQUENCY RESPONSE CHECK

Normally the first receiver check performed is the audiopower check. This determines whether the receiver is dead. If not, it will show whether it can deliver appropriate audiopower. If it is found that audiopower output is as specified, then the frequency response can be quickly and easily checked by comparing the audio output power at 400 and 2500 Hz to a 1000-Hz reference. This check shows the overall ability of the receiver to pass all audiosignals in the voice communications range (Fig. 2-11). Figures 4-5 and 4-6 provide conversion charts of audiovoltage to audiopower for 0.1–1.0 W and 1.0–10 W, respectively.

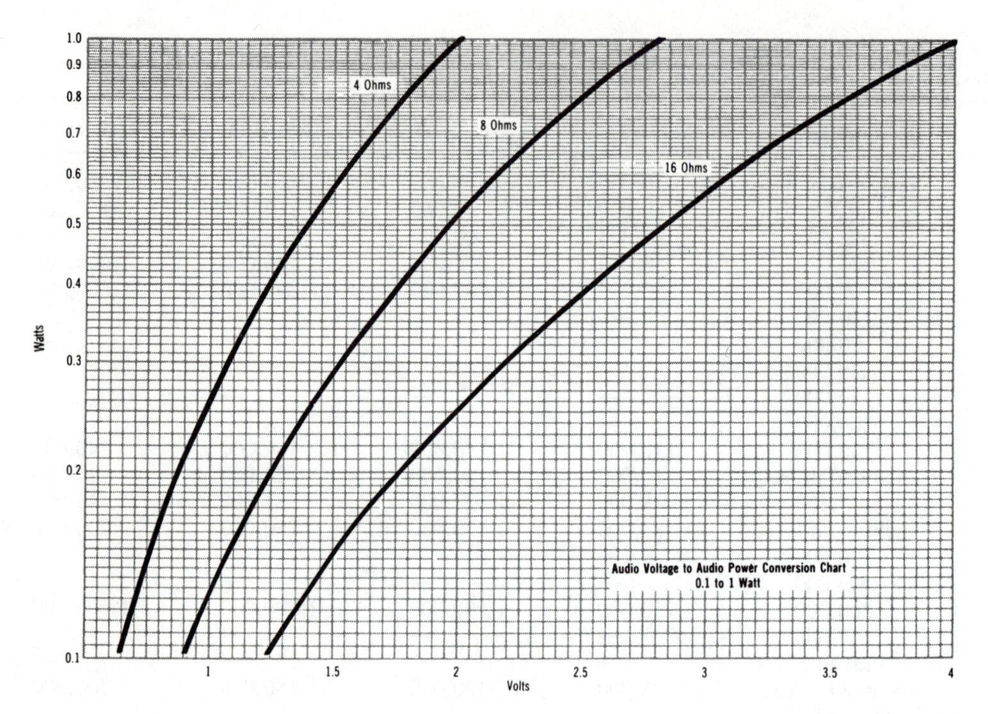

Audio Voltage to Audio Power Conversion Chart
0.1 to 1 Watt
4 Ohms
8 Ohms
16 Ohms
Watts
Volts
1.0
0.9
0.8
0.7
0.6
0.5
0.4
0.3
0.2
0.1
1
1.5
2
2.5
3
3.5
4

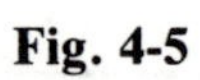

Fig. 4-5

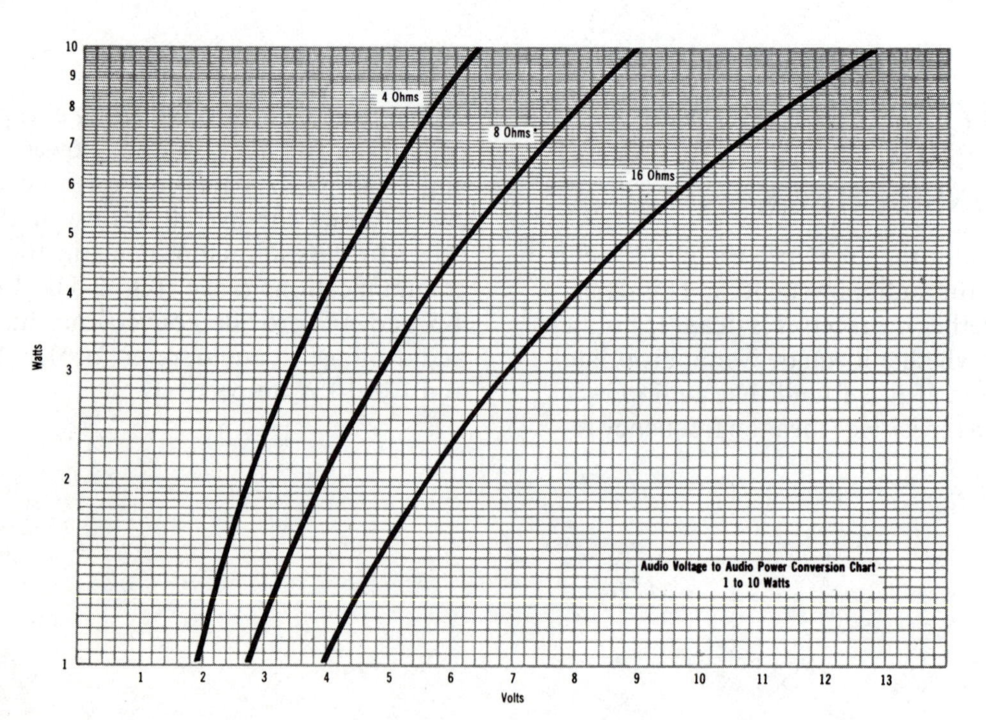

Audio Voltage to Audio Power Conversion Chart
1 to 10 Watts
4 Ohms
8 Ohms
16 Ohms
Watts
Volts
10
9
8
7
6
5
4
3
2
1
1
2
3
4
5
6
7
8
9
10
11
12
13

Fig. 4-6

SPEAKER CONNECTIONS

Figures 4-7 through 4-10 show the proper connection methods for single- or multiple-speaker operation.

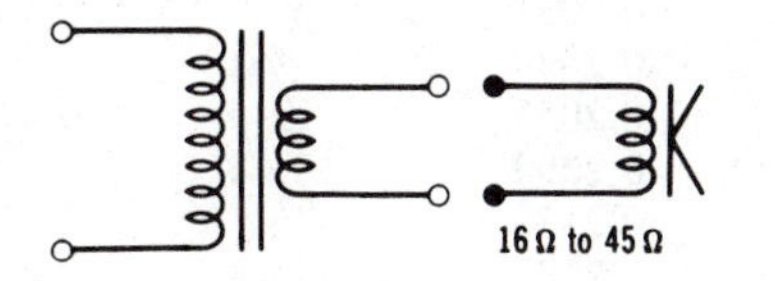

Fig. 4-7. Single speaker.

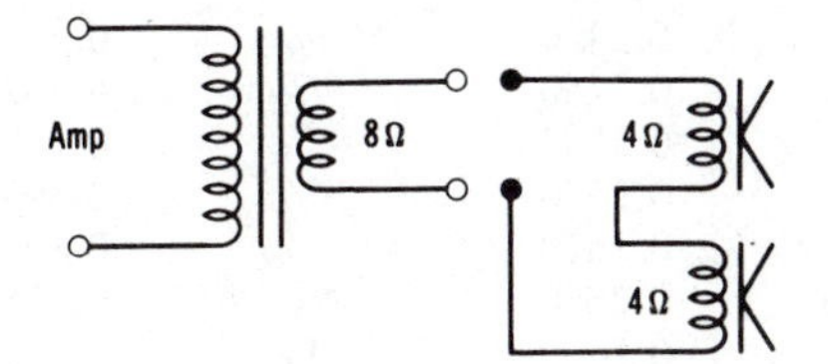

Fig. 4-8. Two speakers in series.

MACHINE SCREW AND DRILL SIZES

The decimal equivalents of No. 80 to 1-in drills are in Table 4-5.

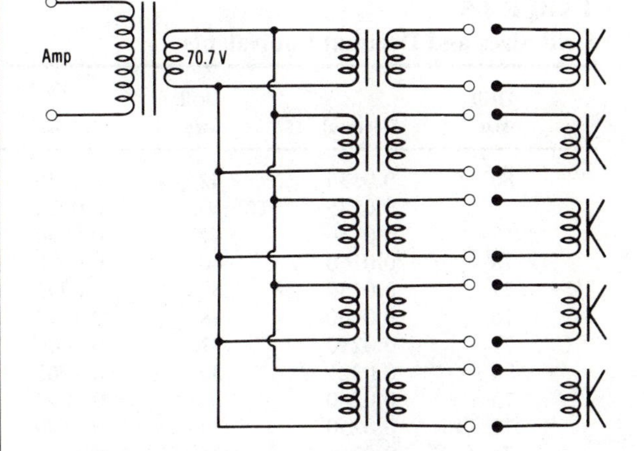

Fig. 4-9. 70.7-V hook-up using matching transformers.

The most common screw sizes and threads, together with the tap and clearance drill sizes, are given in Table 4-6. The number listed under the "Type" column is actually a combination of the screw size and the number of threads per inch. For example, a No. 6-32 screw denotes a size no. 6 screw with 32 threads per inch.

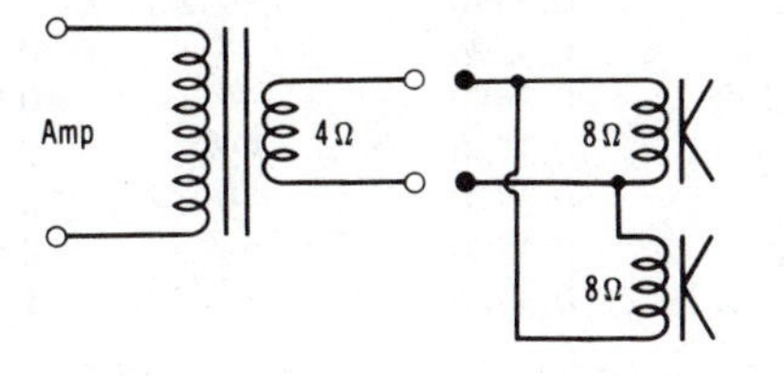

Fig. 4-10. Speakers in parallel.

TABLE 4-5
Drill Sizes and Decimal Equivalents

Drill size	Decimal	Drill size	Decimal	Drill size	Decimal	Drill size	Decimal
80	0.0135	42	0.0935	7	0.2010	X	0.3970
79	0.0145	3/32	0.0938	13/64	0.2031	Y	0.4040
1/64	0.0156	41	0.0960	6	0.2040	13/32	0.4062
78	0.0160	40	0.0980	5	0.2055	Z	0.4130
77	0.0180	39	0.0995	4	0.2090	27/64	0.4219
76	0.0200	38	0.1015	3	0.2130	7/16	0.4375
75	0.0210	37	0.1040	7/32	0.2188	29/64	0.4531
74	0.0225	36	0.1065	2	0.2210	15/32	0.4688
73	0.0240	7/64	0.1094	1	0.2280	31/64	0.4844
72	0.0250	35	0.1100	A	0.2340	1/2	0.5000
71	0.0260	34	0.1110	15/64	0.2344	33/64	0.5156
70	0.0280	33	0.1130	B	0.2380	17/32	0.5313
69	0.0292	32	0.1160	C	0.2420	35/64	0.5469
68	0.0310	31	0.1200	D	0.2460	9/16	0.5625
1/32	0.0313	1/8	0.1250	1/4	0.2500	37/64	0.5781
67	0.0320	30	0.1285	F	0.2570	19/32	0.5938
66	0.0330	29	0.1360	G	0.2610	39/64	0.6094
65	0.0350	28	0.1405	17/64	0.2656	5/8	0.6250
64	0.0360	9/64	0.1406	H	0.2660	41/64	0.6406
63	0.0370	27	0.1440	I	0.2720	21/32	0.6562
62	0.0380	26	0.1470	J	0.2770	43/64	0.6719
61	0.0390	25	0.1495	K	0.2810	11/16	0.6875
60	0.0400	24	0.1520	9/32	0.2812	45/64	0.7031
59	0.0410	23	0.1540	L	0.2900	23/32	0.7188
58	0.0420	5/32	0.1562	M	0.2950	47/64	0.7344
57	0.0430	22	0.1570	19/64	0.2969	3/4	0.7500
56	0.0465	21	0.1590	N	0.3020	49/64	0.7656
3/64	0.0469	20	0.1610	5/16	0.3125	25/32	0.7812
55	0.0520	19	0.1660	O	0.3160	51/64	0.7969
54	0.0550	18	0.1695	P	0.3230	13/16	0.8125
53	0.0595	11/64	0.1709	21/64	0.3281	53/64	0.8281
1/16	0.0625	17	0.1730	Q	0.3320	27/32	0.8438
52	0.0635	16	0.1770	R	0.3390	55/64	0.8594
51	0.0670	15	0.1800	11/32	0.3438	7/8	0.8750
50	0.0700	14	0.1820	S	0.3480	57/64	0.8906
49	0.0730	13	0.1850	T	0.3580	29/32	0.9062
48	0.0760	3/16	0.1875	23/64	0.3594	59/64	0.9219
5/64	0.0781	12	0.1890	U	0.3680	15/16	0.9375
47	0.0785	11	0.1910	3/8	0.3750	61/64	0.9531
46	0.0810	10	0.1935	V	0.3770	31/32	0.9688
45	0.0820	9	0.1960	W	0.3860	63/64	0.9844
44	0.0860	8	0.1990	25/64	0.3906	1	1.000
43	0.0890						

TYPES OF SCREW HEADS

The most common types of screw heads are listed and illustrated in Fig. 4-11.

TABLE 4-6
Machine Screw Tap and Clearance Drill Sizes

Type	Tap drill	Clearance drill
0-80	3/64	50
1-64	53	47
1-72	53	47
2-56	50	42
2-64	50	42
3-48	47	36
3-56	45	36
4-40	43	31
4-48	42	31
5-40	38	29
5-44	37	29
6-32	36	25
6-40	33	25
8-32	29	16
8-36	29	16
10-24	25	13/64
10-32	21	13/64
12-24	16	7/32
12-28	14	7/32
1/4-20	7	17/64
1/4-28	3	17/64
5/16-18	F	21/64
5/16-24	1	21/64
3/8-16	5/16	25/64
3/8-24	Q	25/64
7/16-14	U	29/64
7/16-20	25/64	29/64
1/2-12	27/64	33/64
1/2-13	27/64	33/64
1/2-20	29/64	33/64

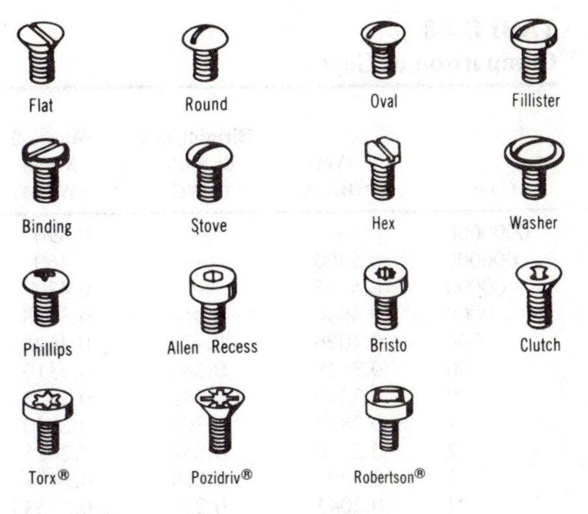

Fig. 4-11

SHEET-METAL GAGES

Materials are customarily made to certain gage systems. While materials can usually be had specially in any system, some usual practices are shown in Tables 4-7 and 4-8.

TABLE 4-7
Common Gage Practices

Material	Sheet	Wire
aluminum	B&S	AWG (B&S)
brass, bronze, sheet	B&S	—
copper	B&S	AWG (B&S)
iron, steel, band, and hoop	BWG	—
iron, steel, telephone, and telegraph wire	—	BWG
steel wire, except telephone and telegraph	—	W&M
steel sheet	US	—
tank steel	BWG	—
zinc sheet	"zinc gage" proprietary	—

TABLE 4-8
Comparison of Gages

Gage	AWG (B&S)	Birmingham or Stubs (BWG)	Wash. & Moen (W&M)	British Standard (NBS SWG)	London or Old English	United States Standard (US)	American Standard preferred thickness*
0000000	—	—	0.490	0.500	—	0.50000	—
000000	0.5800	—	0.460	0.464	—	0.46875	—
00000	0.5165	—	0.430	0.432	—	0.43750	—
0000	0.4600	0.454	0.3938	0.400	0.454	0.40625	—
000	0.4096	0.425	0.3625	0.372	0.425	0.37500	—
00	0.3648	0.380	0.3310	0.348	0.380	0.34375	—
0	0.3249	0.340	0.3065	0.324	0.340	0.31250	—
1	0.2893	0.300	0.2830	0.300	0.300	0.2815	—
2	0.2576	0.284	0.2625	0.276	0.284	0.265625	—
3	0.2294	0.259	0.2437	0.252	0.259	0.250000	0.224
4	0.2043	0.238	0.2253	0.232	0.238	0.234375	0.200
5	0.1819	0.220	0.2070	0.212	0.220	0.218750	0.180
6	0.1620	0.203	0.1920	0.192	0.203	0.203125	0.160
7	0.1443	0.180	0.1770	0.176	0.180	0.187500	0.140
8	0.1285	0.165	0.1620	0.160	0.165	0.171875	0.125
9	0.1144	0.148	0.1483	0.144	0.148	0.156250	0.112
10	0.1019	0.134	0.1350	0.128	0.134	0.140625	0.100
11	0.09074	0.120	0.1205	0.116	0.120	0.125000	0.090
12	0.08081	0.109	0.1055	0.104	0.109	0.109375	0.080
13	0.07196	0.095	0.0915	0.092	0.095	0.093750	0.071
14	0.06408	0.083	0.0800	0.080	0.083	0.078125	0.063
15	0.05707	0.072	0.0720	0.072	0.072	0.0703125	0.056
16	0.05082	0.065	0.0625	0.064	0.065	0.0625000	0.050
17	0.04526	0.058	0.0540	0.056	0.058	0.0562500	0.045
18	0.04030	0.049	0.0475	0.048	0.049	0.0500000	0.040
19	0.03589	0.042	0.0410	0.040	0.040	0.0437500	0.036
20	0.03196	0.035	0.0348	0.036	0.035	0.0375000	0.032
21	0.02846	0.032	0.03175	0.032	0.0315	0.0343750	0.028
22	0.02535	0.028	0.02860	0.028	0.0295	0.0312500	0.025
23	0.02257	0.025	0.02580	0.024	0.0270	0.0281250	0.022
24	0.02010	0.022	0.02300	0.022	0.0250	0.0250000	0.020
25	0.01790	0.020	0.02040	0.020	0.0230	0.0218750	0.018
26	0.01594	0.018	0.01810	0.018	0.0205	0.0187500	0.016
27	0.01420	0.016	0.01730	0.0164	0.0187	0.0171875	0.014
28	0.01264	0.014	0.01620	0.0148	0.0165	0.0156250	0.012
29	0.01126	0.013	0.01500	0.0136	0.0155	0.0140625	0.011
30	0.01003	0.012	0.01400	0.0124	0.01372	0.0125000	0.010
31	0.008928	0.010	0.01320	0.0116	0.01220	0.01093750	0.009
32	0.007950	0.009	0.01280	0.0108	0.01120	0.01015625	0.008
33	0.007080	0.008	0.01180	0.0100	0.01020	0.00937500	0.007
34	0.006305	0.007	0.01040	0.0092	0.00950	0.00859375	0.006
35	0.005615	0.005	0.00950	0.0084	0.00900	0.00781250	—
36	0.005000	0.004	0.00900	0.0076	0.00750	0.007031250	—
37	0.004453	—	0.00850	0.0068	0.00650	0.006640625	—
38	0.003965	—	0.00800	0.0060	0.00570	0.006250000	—
39	0.003531	—	0.00750	0.0052	0.00500	—	—
40	0.003145	—	0.00700	0.0048	0.00450	—	—

*These thicknesses are intended to express the desired thickness in decimal fractions of an inch. They have no relation to gage numbers; they are approximately related to AWG sizes 3–34.

Courtesy Whitehead Metal Products Co., Inc.

RESISTANCE OF METALS AND ALLOYS

The resistance for a given length of wire is determined by:

$$R = \frac{KL}{d^2}$$

where

R is the resistance of the length of wire, in ohms,
K is the resistance of the material, in ohms per circular mil foot,
L is the length of the wire, in feet,
d is the diameter of the wire, in mils.

The resistance, in ohms per circular mil foot, of many of the materials used for conductors or heating elements is given in Table 4-9. The resistance shown is for 20 °C (68 °F), unless otherwise stated.

TABLE 4-9
Resistance of Metals and Alloys

Material	Symbol	Resistance (Ω/cir mil foot)
nichrome	Ni-Fe-Cr	675
tophet A	Ni-Cr	659
nichrome V	Ni-Cr	650
chromax	Cr-Ni-Fe	610
steel, stainless	C-Cr-Ni-Fe	549
chromel	Ni-Cr	427
steel, manganese	Mn-C-Fe	427
kovar A	Ni-Co-Mn-Fe	1732
titanium	Ti	292
constantan	Cu-Ni	270
manganin	Cu-Mn-Ni	268
monel	Ni-Cu-Fe-Mn	256
arsenic	As	214
alumel	Ni-Al-Mn-Si	203
nickel-silver	Cu-Zn-Ni	171
lead	Pb	134
steel	C-Fe	103
manganese-nickel	Ni-Mn	85
tantalum	Ta	79.9
tin	Sn	69.5
palladium	Pd	65.9
platinum	Pt	63.8
iron	Fe	60.14
nickel, pure	Ni	60
phosphor-bronze	Sn-P-Cu	57.38
high-brass	Cu-Zn	50
potassium	K	42.7
molybdenum	Mo	34.27
tungsten	W	33.22
rhodium	Rh	31
aluminum	Al	16.06
chromium	Cr	15.87
gold	Au	14.55
copper	Cu	10.37
silver	Ag	9.706
selenium	Se	7.3

COPPER-WIRE CHARACTERISTICS

Copper-wire sizes ranging from American wire gage (B&S) 0000 to 60 are listed in Table 4-10. The turns per linear inch, diameter, area in circular mils, current-carrying capacity, feet per pound, and resistance per 1000 ft are included in the table.

TABLE 4-10
Copper-Wire Characteristics

AWG	Nominal bare diameter (in)	Nominal circular mils	Nominal feet per pound (bare)	Nominal ohms per 1000 ft @20 °C	Current-carrying capacity @700 CM/A	Turns per linear inch	
						Single film coated	*Heavy film coated*
0000	0.4600	211,600	1.561	0.04901	302.3		
000	0.4096	167,800	1.969	0.06182	239.7		
00	0.3648	133,100	2.482	0.07793	190.1		
0	0.3249	105,600	3.130	0.09825	150.9		
1	0.2893	83,690	3.947	0.1239	119.6		
2	0.2576	66,360	4.978	0.1563	94.8		
3	0.2294	52,620	6.278	0.1971	75.2		
4	0.2043	41,740	7.915	0.2485	59.6		4.80
5	0.1819	33,090	9.984	0.3134	47.3		5.38
6	0.1620	26,240	12.59	0.3952	37.5		6.03
7	0.1443	20,820	15.87	0.4981	29.7		6.75
8	0.1285	16,510	20.01	0.6281	23.6		7.57
9	0.1144	13,090	25.24	0.7925	18.7		8.48
10	0.1019	10,380	31.82	0.9988	14.8		9.50
11	0.0907	8230	40.2	1.26	11.8		10.6
12	0.0808	6530	50.6	1.59	9.33		11.9
13	0.0720	5180	63.7	2.00	7.40		13.3
14	0.0641	4110	80.4	2.52	5.87	15.2	14.8
15	0.0571	3260	101	3.18	4.66	17.0	16.6
16	0.0508	2580	128	4.02	3.69	19.0	18.5
17	0.0453	2050	161	5.05	2.93	21.3	20.7
18	0.0403	1620	203	6.39	2.31	23.9	23.1
19	0.0359	1290	256	8.05	1.84	26.7	25.9
20	0.0320	1020	323	10.1	1.46	29.9	28.9
21	0.0285	812	407	12.8	1.16	33.4	32.3
22	0.0253	640	516	16.2	0.914	37.5	36.1
23	0.0226	511	647	20.3	0.730	41.8	40.2
24	0.0201	404	818	25.7	0.577	46.8	44.8
25	0.0179	320	1030	32.4	0.457	52.5	50.1
26	0.0159	253	1310	41.0	0.361	58.8	56.0
27	0.0142	202	1640	51.4	0.289	65.6	62.3
28	0.0126	159	2080	65.3	0.227	73.3	69.4
29	0.0113	123	2590	81.2	0.183	81.6	76.9
30	0.0100	100	3300	104.0	0.143	91.7	86.2
31	0.0089	79.2	4170	131	0.113	103	96
32	0.0080	64.0	5160	162	0.091	114	106
33	0.0071	50.4	6550	206	0.072	128	118
34	0.0063	39.7	8320	261	0.057	145	133
35	0.0056	31.4	10,500	331	0.045	163	149
36	0.0050	25.0	13,200	415	0.036	182	167

TABLE 4-10 Cont.
Copper-Wire Characteristics

AWG	Nominal bare diameter (in)	Nominal circular mils	Nominal feet per pound (bare)	Nominal ohms per 1000 ft @20 °C	Current-carrying capacity @700 CM/A	Turns per linear inch	
						Single film coated	*Heavy film coated*
37	0.0045	20.2	16,300	512	0.029	202	183
38	0.0040	16.0	20,600	648	0.023	225	206
39	0.0035	12.2	27,000	847	0.017	260	235
40	0.0031	9.61	34,400	1080	0.014	290	263
41	0.0028	7.84	42,100	1320	0.011	323	294
42	0.0025	6.25	52,900	1660	0.0089	357	328
43	0.0022	4.84	68,300	2140	0.0069	408	370
44	0.0020	4.00	82,600	2590	0.0057	444	400
45	0.00176	3.10	107,000	3350	0.0044	520	465
46	0.00157	2.46	134,000	4210	0.0035	580	510
47	0.00140	1.96	169,000	5290	0.0028	630	560
48	0.00124	1.54	215,000	6750	0.0022	710	645
49	0.00111	1.23	268,000	8420	0.0018	800	720
50	0.00099	0.980	337,000	10,600	0.0014	880	780
51	0.00088	0.774	427,000	13,400	0.0011	970	855
52	0.00078	0.608	543,000	17,000	0.00087	1080	935
53	0.00070	0.490	674,000	21,200	0.00070	1270	1110
54	0.00062	0.384	859,000	27,000	0.00055	1430	1220
55	0.00055	0.302	1,090,000	34,300	0.00043	1560	1330
56	0.00049	0.240	1,380,000	43,200	0.00034	1690	1450
57	0.000438	0.192	1,722,000	54,100	0.00027	1960	
58	0.000390	0.152	2,166,000	68,000	0.00022	2160	
59	0.000347	0.121	2,737,000	85,900	0.00017	2450	
60	0.000309	0.090	3,453,000	108,400	0.00014	2740	

Chapter 5

DESIGN DATA

VACUUM-TUBE FORMULAS

The following formulas can be used to calculate the vacuum-tube properties listed.

Amplification Factor:

$$\mu = \frac{\Delta E_b}{\Delta E_c} \text{ (with } I_b \text{ constant)}$$

AC (Dynamic) Plate Resistance:

$$r_p = \frac{\Delta E_b}{\Delta I_b} \text{ (with } E_c \text{ constant)}$$

Mutual Conductance (Transconductance):

$$g_m = \frac{\Delta I_b}{\Delta E_c} \text{ (with } E_b \text{ constant)}$$

Gain of an Amplifier Stage:

$$\text{Gain} = \mu \frac{R_L}{R_L + r_p}$$

where

μ is the amplification factor,
Δ is the variation or change in value,
E_b is the plate voltage, in volts,
E_c is the grid voltage, in volts,
I_b is the plate current, in amperes,
R_L is the plate-load resistance, in ohms,
r_p is the AC plate resistance, in ohms,
g_m is the mutual conductance, in siemens.

TRANSISTOR FORMULAS

The following formulas can be used to calculate the transistor properties listed.

Input Resistance:

$$R_i = \frac{\Delta V_i}{\Delta I_i}$$

Current Gain (Fig. 5-1):

$$A_i = \frac{\Delta I_c}{\Delta I_b} \text{ (with } V_c \text{ constant)}$$

The current gain of the common-base configuration is alpha (α):

$$\alpha = \frac{\Delta I_c}{\Delta I_e} \text{ (with } V_c \text{ constant)}$$

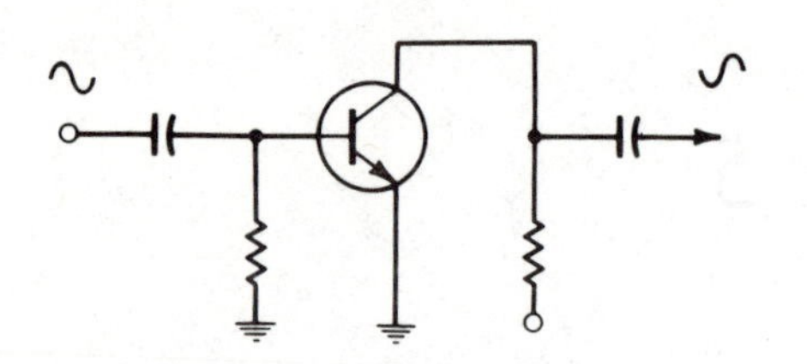

Fig. 5-1A. Common emitter.

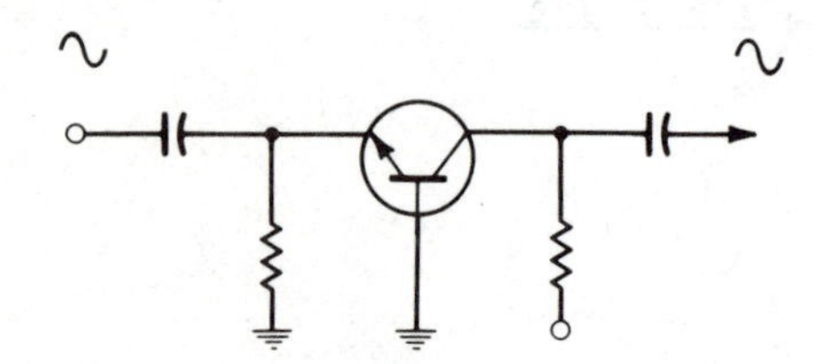

Fig. 5-1B. Common base.

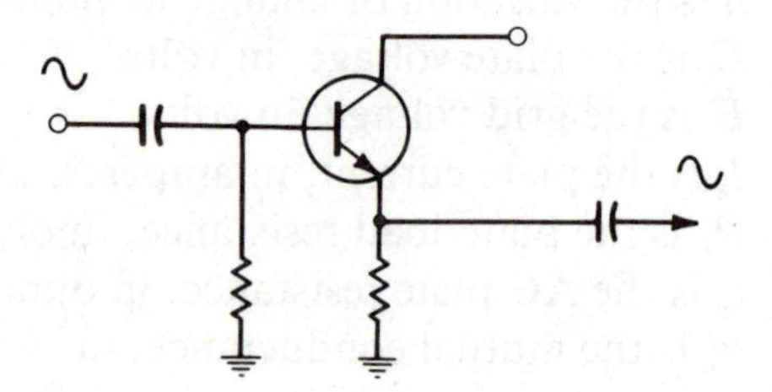

Fig. 5-1C. Common collector.

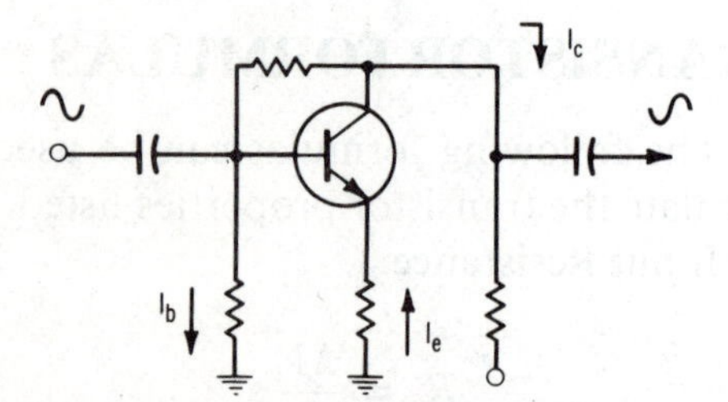

Fig. 5-1D. Basic current paths.

The current gain of the common emitter is beta (β):

$$\beta = \frac{\Delta I_c}{\Delta I_b} \text{ (with } V_c \text{ constant)}$$

A direct relationship exists between the *alpha* and *beta* of a transistor:

$$\alpha = \frac{\beta}{1 + \beta}$$

$$\beta = \frac{\alpha}{1 - \alpha}$$

Voltage Gain:

$$A_v = \frac{\Delta V_c}{\Delta V_b} \text{ (with } I_c \text{ constant)}$$

Output Resistance:

$$R_o = \frac{\Delta V_o}{\Delta I_o}$$

Power Gain:

$$A_p = \frac{\Delta P_o}{\Delta P_i}$$

Base Current:

$$I_b = I_e - I_c = \frac{I_c}{h_{fe}}$$

or

$$= \frac{I_e}{h_{fe}} - I_{co}$$

Collector Current:

$$I_c = I_e - I_b = \alpha I_e = h_{fe} I_b$$

Collector Power:

$$P_c = V_{ce} I_c$$

Emitter Current:

$$I_e = I_b + I_c \text{ (Total Current)}$$

Small-Signal Emitter Resistance:

$$r_e = \frac{26}{I_e}$$

where

I_e is emitter current, in milliamperes.

Transconductance:

$$g_m = \frac{I_e}{26}$$

where

I_e is emitter current, in milliamperes.

Input Capacitance:

$$C_{in} = \frac{g_m}{6.28 f_{hfb}}$$

Upper Frequency Limit:

$$f_u = \frac{g_m}{6.28\, C_t}$$

Bandwidth:

$$f_{hfe} = \frac{F_{hfb}}{h_{fe}}$$

or

$$f_{hfb} = h_{fe} f_{hfe}$$

where

α is the current gain of a common-base configuration, in amperes,
A_v is the voltage gain, in volts,
A_i is the current gain, in amperes,
A_p is the power gain, in watts,
β is the current gain in a common-emitter configuration, in amperes,
C_t is total capacitance, in picofarads,
C_{in} is input capacitance, in farads,
f_u is upper frequency limit (unity gain), in megahertz,
f_{hfe} is the beta cutoff frequency (3-dB point),
f_{hfb} is the alpha cutoff frequency (3-dB point),
g_m is transconductance, in microsiemens,
h_{fe} is $\beta = \alpha/(1 - \alpha)$,
I_b is the base current, in amperes,
I_c is the collector current, in amperes,
I_e is the emitter current, in amperes,
I_i is the input current, in amperes,
I_o is the output current, in amperes,
P_c is collector power, in watts,
P_i is the input power, in watts,
P_o is the output power, in watts,
R_i is the input resistance, in ohms,
R_o is the output resistance, in ohms,
r_e is small-signal emitter resistance, in ohms,
V_b is the base voltage, in volts,
V_c is the collector voltage, in volts,
V_{ce} is the collector-emitter voltage, in volts,
V_i is the input voltage, in volts,
V_o is the output voltage, in volts.

Terminology:

h_{ie} or h_{11e} = input impedance with output short-circuited (CE mode)

h_{oe} or h_{22e} = output admittance with open-circuit input

h_{re} or h_{12e} = reverse open-circuit voltage amplification factor

h_{fe} or h_{21e} = forward short-circuit current amplification factor

Conventionally, capital letters denote DC relations, and lowercase letters denote AC small-signal relations.

OPERATIONAL AMPLIFIERS (OP AMPS)

An op amp is a DC-coupled high-gain differential amplifier in integrated-circuit (IC) form. Typical op amps have very high input impedance and very low output impedance. Open-loop gain refers to the voltage gain of the op amp (without feedback) and working into its rated load value. An op amp has very high open-loop gain at zero frequency (DC), with progressively falling frequency response. The op amp is almost always operated with a large amount of negative feedback so that it has an essentially flat frequency response. A typical general-purpose op amp provides full power output to 10 kHz and decreases to unity gain at 1 MHz.

A general-purpose op amp typically develops an open-loop gain of 100,000 times, with an input impedance of 5 MΩ and a low output impedance rated for working into a 1000-Ω load. This type of op amp can provide an output-voltage swing of ± 10 V. A basic rating is its slew rate, or its maximum possible rate of change in output voltage (transient response). A general-purpose op amp may have a rated slew rate of 0.5 V/μs at unity gain. Although originally designed for analog computers, op amps are now used in radio and TV receivers, magnetic recording pickup head amplifiers, and a wide range of instrumentation applications.

Op amps are used as electronic integrators and differentiators to obtain characteristics that approximate mathematical integration and differentiation. Although many op amps require a positive as well as a negative power-supply source, various op amps can operate from a single power-supply source, and some operate from the same power supply as the digital ICs in a computer. As shown in Figs. 5-2 through 5-6, op amps are also used as basic amplifiers, mixers, limiters, comparators, and filters with optimized characteristics.

Since a typical op amp has very low input impedance due to substantial negative feedback, the input terminal is regarded as a virtual ground that is almost at ground potential. In turn, the voltage gain of the am-

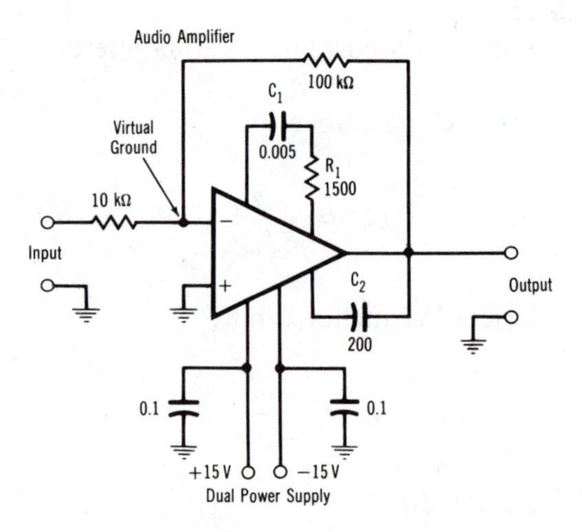

Fig. 5-2. Audio amplifier.

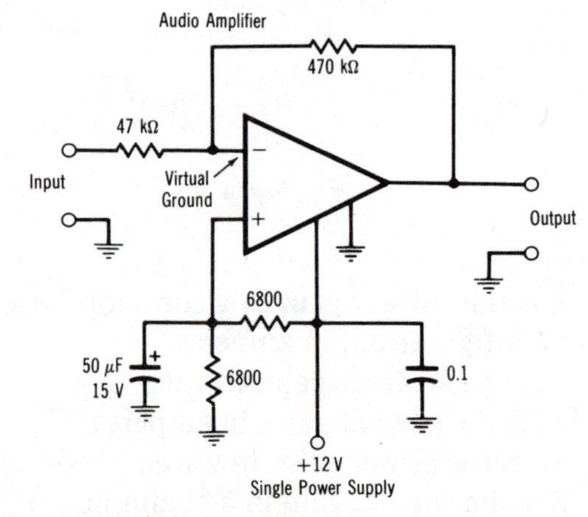

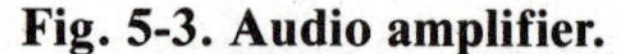
Fig. 5-3. Audio amplifier.

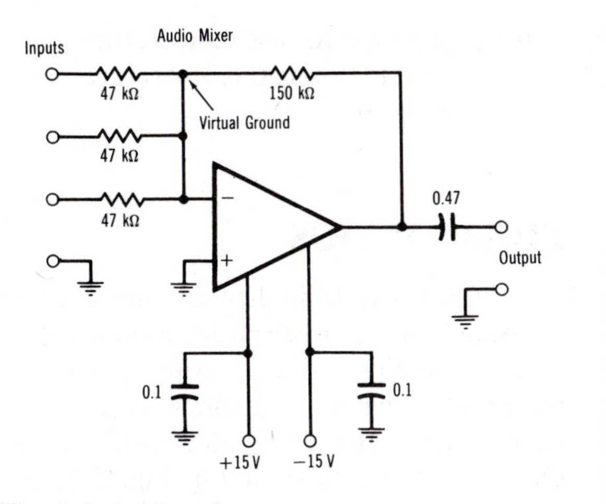

Fig. 5-4. Audio mixer.

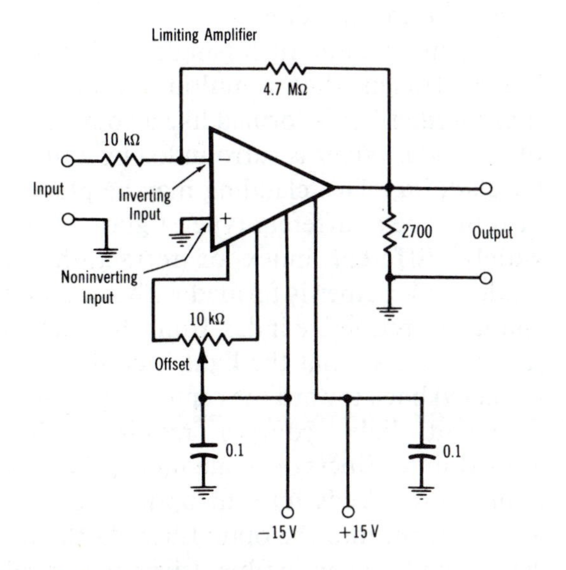

Fig. 5-5. Limiting amplifier.

plifier is equal to the sum of the series input resistance and feedback resistance, divided by the input resistance. For example, if the series input resistance is 1000 Ω and the feedback resistance is 100,000 Ω, the voltage gain of the op amp is essentially 101,000/1000, or 101 times. Note that the inverting input of an op amp is indicated by a minus sign; the noninverting input is indicated by a plus sign.

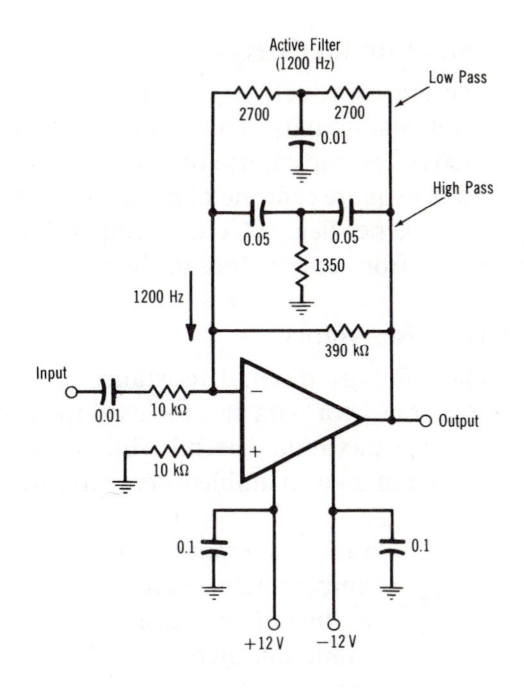

Fig. 5-6. Active filter (1200 Hz).

HEAT

When working with power transistors, integrated circuits, and heat sinks, the following equations will be useful:

Joule: The unit of energy required to move one coulomb between two points having a potential-energy difference of 1 V.

Electrical Equivalent of Heat

Calorie: 1 cal is equal to 4.18605 J.
1 cal will heat 1 g of water by 1°C.
252 cal will heat 1 lb of water 1°F.
1 kWh equals 860,000 cal.
1 cal equals 1/860 Wh.
0.293 Wh will heat 1 lb of water 1°F.

Thermal Conductivity

Thermal conductivity is analogous to electrical conductivity. The unit of thermal conductivity is one calorie of heat flow per second per square centimeter per centimeter of thickness per degree Celsius temperature difference from one surface to the next.

Thermal Resistance

The value of thermal resistance specified in a transistor data sheet is used to calculate the maximum permissible power dissipation at a given ambient temperature.

T_j = junction temperature
T_a = ambient temperature
θ_{ja} = ambient temperature junction thermal resistance
P_d = power dissipated
$\theta_{ja} - P_d$ = temperature rise

Example. A small transistor has a maximum rated junction temperature of 85°C and a thermal resistance of 0.5°C/mW. If the ambient temperature is 25°C, the transistor may dissipate an amount of power to raise the junction temperature to 85°C. This increase in temperature of 60°C will result, in this case, from a power of 120 mW in the transistor:

$$
\begin{aligned}
T_j &= T_a + \theta_{ja}(P_d) \\
&= 60°\text{C}/(0.5°\text{C/mW}) \\
&= 120 \text{ mW}
\end{aligned}
$$

If a heat sink is correctly installed with thermally conducting washers, silicone grease, and proper bolting pressure, then the thermal resistance from transistor case to heat sink can be neglected. The junction-to-case thermal resistance is added to the sink-to-ambient thermal resistance to obtain the total thermal resistance.

θ_{ja} = junction to ambient temperature
θ_{jc} = junction to case temperature
θ_{sa} = sink to ambient temperature
$\theta_{ja} = \theta_{jc} + \theta_{sa}$

FIBER OPTICS

Glass fibers (thin strands) are used to transmit light in curved paths. Bundles of fibers are analogous to coaxial cables and waveguides, where electromagnetic radiation is transmitted along an arbitrary path. Circular fibers, as shown in Fig. 5-7, are generally used, and fiber optics is based on principles of reflection, refraction, absorption, and transmission.

A simple round fiber (single mode fiber) is less efficient than a multimode fiber. A multimode fiber is formed like a coaxial cable; the glass fiber is surrounded by a tubular cladding. This cladding may be plastic, or it may be a different type of glass with a widely different index of refraction. A graded index fiber is formed with a gradual change in refractive index from the core to the surface, so that the light is continually refracted back toward the center of the core. Coherent light sources require either graded-index fibers or single-mode fibers. A light source feeds into an optical coupler and, in turn, into the optic fiber. At the receiving end, the optic fiber drives an optical coupler, which in turn feeds into a photodetector. Although fiber-optics transmission is very costly at present, it has the advantages of unusually high information capacity, miniature construction, and immunity to various types of electromagnetic interference. An optic-fiber cable is composed of hundreds of fibers, each of which may have a diameter of 50 or 100 mm. Half of the fibers transmit data in one direction, while the other half transmit data in the opposite direction.

Fig. 5-7A. Single-mode fiber.

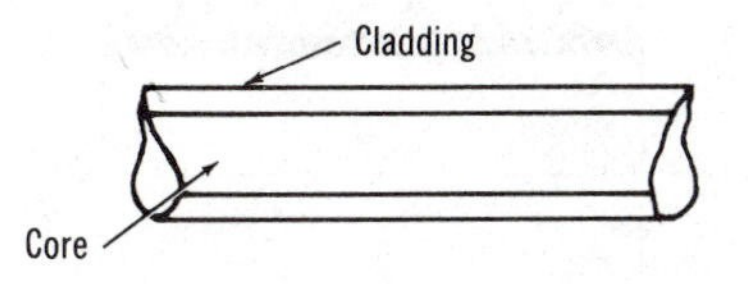

Fig. 5-7B. Multimode fiber.

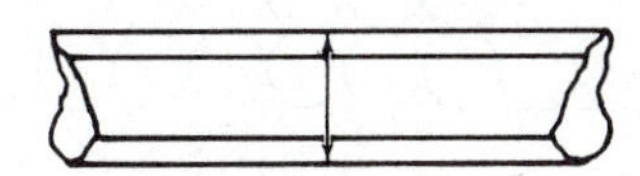

Fig. 5-7C. Graded-index fiber.

THREE-PHASE POWER FORMULAS

In a three-phase system, there are three voltages, each separated by a phase difference of 120°. The power-supply input transformers may be connected in either a delta or a Y (star). Figure 5-8 shows how the terminals are placed in relationship to the coils. In the delta connection, there is one coil between each pair of terminals; in the Y connection, there are two. The voltage between two terminals of the Y-connected coil is equal to $\sqrt{3}$ times the voltage across one winding.

The formulas for determining the voltage across the secondary winding for each of the four possible connections are as follows:

Δ to Y:

$$E_s = E_p \times N \times \sqrt{3}$$

Y to Δ:

$$E_s = \frac{E_p \times N}{\sqrt{3}}$$

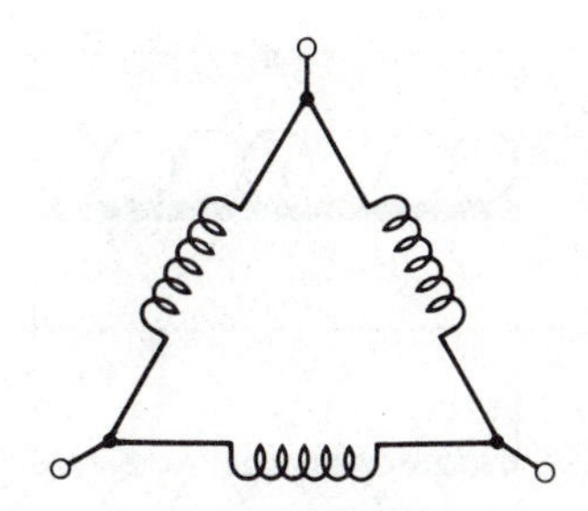

Fig. 5-8A. Delta connection.

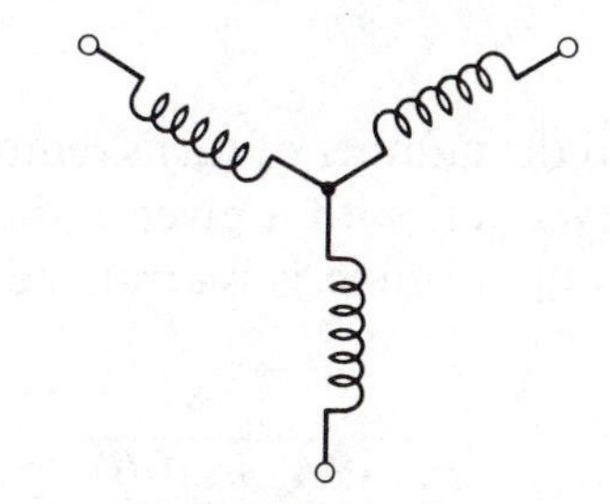

Fig. 5-8B. Y connection.

Δ to Δ:

$$E_s = E_p \times N$$

Y to Y:

$$E_s = E_p \times N$$

where

E_s is the secondary voltage, in volts,
E_p is the primary voltage, in volts,
N is the turns ratio.

COIL WINDINGS

Single-Layer Coils (Fig. 5-9)

The inductance of single-layer coils can be calculated to an accuracy of approximately 1% with the formula:

$$L = \frac{(N \times A)^2}{9A + 10B}$$

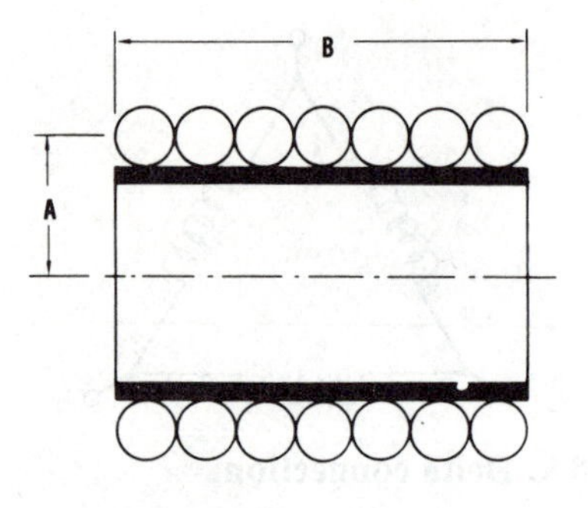

Fig. 5-9

To find the number of turns required for a single-layer coil with a given inductance, the foregoing formula is rearranged as follows:

$$N = \frac{\sqrt{L(9A + 10B)}}{A}$$

where

L is the inductance, in microhenrys,
N is the number of turns,
A is the mean radius, in inches,
B is the length of the coil, in inches.

Multilayer Coils (Fig. 5-10)

The inductance of a multilayer coil of rectangular cross section can be computed from the formula:

$$L = \frac{0.8(N \times A)^2}{6A + 9B + 10C}$$

where

L is the inductance, in microhenrys,
N is the number of turns,
A is the mean radius, in inches,
B is the length of the coil, in inches,
C is the depth of the coil, in inches.

Single-Layer Air-Core Coil Chart

The chart in Fig. 5-11 provides an easy method for determining either the inductance or the number of turns for single-layer coils. When the length of the winding, the diameter, and the number of turns of the coil are known, the inductance can be found by placing a straightedge from the "Turns" scale to the "Ratio" (diameter ÷ length) scale and noting the point where the straightedge intersects the "Axis" scale. Then lay the straightedge from the point of intersection of the "Axis" scale to the "Diameter" scale. The point at which this line intersects the "Inductance" scale indicates the inductance, in microhenrys, of the coil. The number of turns can be determined by reversing the procedure.

After finding the number of turns, consult Table 4-8 to determine the size of wire to be used.

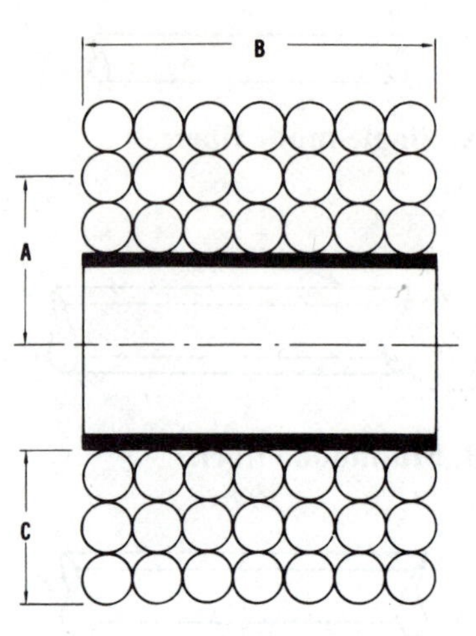

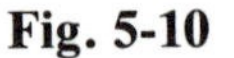

Fig. 5-10

Example. What is the inductance of a single-layer air-core coil having 80 turns wound to 4 in in length on a coil form 2 in in diameter?

Answer. 130 μH. (First lay the straightedge as indicated by the line labeled "Example 1A." Then lay the straightedge as indicated by the line labeled "Example 1B.")

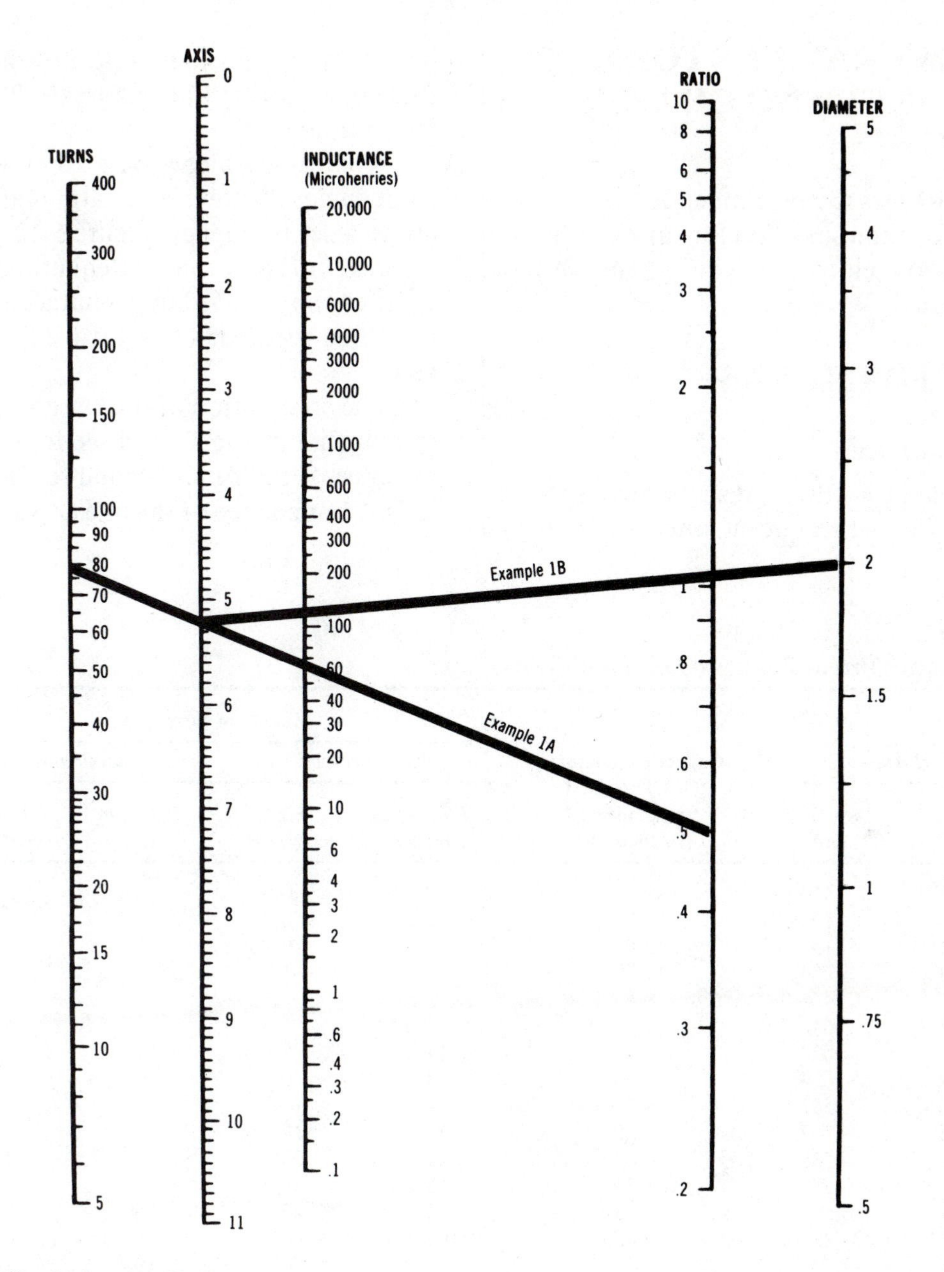

Fig. 5-11. Single-layer coil chart.

CURRENT RATINGS FOR EQUIPMENT AND CHASSIS WIRING

Table 5-1 lists the recommended current rating (for continuous duty) for various wire sizes used on electronic equipment and chassis wiring.

FILTER FORMULAS

Constant-k Filters

A constant-k filter presents an impedance match to the line at only one frequency and a mismatch at all others. The three basic configurations are the T, L (half-section), and pi.

A constant-k low-pass filter will pass frequencies below and attenuate those above a set frequency. Figure 5-12 gives the circuit configurations, attenuation characteristics, and impedance characteristics of the three types of constant-k low-pass filters.

The attenuation of the L section is equal to half that of the T or pi sections. The impedance of the filter is equal to the characteristic impedance of the line (Z_0) at the zero

TABLE 5-1
Recommended Current Ratings (Continuous Duty)

Wire size		Copper conductor (100°C) nominal resistance	Maximum current (A)			
			Copper wire		*Aluminum wire*	
AWG	*Circular mils*	(Ω/1000 ft)	*Wiring in free air*	*Wiring confined**	*Wiring in free air*	*Wiring confined**
32	63.2	188.0	0.53	0.32		
30	100.5	116.0	0.86	0.52		
28	159.8	72.0	1.4	0.83		
26	254.1	45.2	2.2	1.3		
24	404.0	28.4	3.5	2.1		
22	642.4	22.0	7.0	5.0		
20	1022	13.7	11.0	7.5		
18	1624	6.50	16	10		
16	2583	5.15	22	13		
14	4107	3.20	32	17		
12	6530	2.02	41	23		
10	10,380	1.31	55	33		
8	16,510	0.734	73	46	60	36
6	26,250	0.459	101	60	83	50
4	41,740	0.290	135	80	108	66
2	66,370	0.185	181	100	152	82
1	83,690	0.151	211	125	174	105
0	105,500	0.117	245	150	202	123
00	133,100	0.092	283	175	235	145
000	167,800	0.074	328	200	266	162
0000	211,600	0.059	380	225	303	190

*"Wiring confined" ratings are based on 15 or more wires in a bundle, with the sum of all the actual load currents of the bundled wires not exceeding 20% of the permitted "Wiring confined" sum total carrying capacity of the bundled wires. These ratings approximate 60% of the free-air ratings (with some variations due to rounding). They should be used for wire in harnesses, cable, conduit, and general chassis conditions. Bundles of fewer than 15 wires may have the allowable sum of the load currents increased as the bundle approaches the single-wire condition.

frequency only. For all other frequencies, the input and output impedance of the filter are equal to Z_I or Z_I', as shown in Fig. 5-12.

The values for L_1, C_2, Z_0, and f_c can be computed from the following formulas:

$$L_1 = \frac{Z_0}{\pi f_c}$$

$$C_2 = \frac{1}{\pi f_c Z_0}$$

$$Z_0 = \sqrt{\frac{L_1}{C_2}}$$

$$f_c = \frac{1}{\pi\sqrt{L_1 C_2}}$$

The values computed for L_1 and C_2 must be divided in half, where specified in Fig. 5-12. That is, the coils in the T and L sections, and the capacitors in the L and pi sections, are equal to one-half the computed value.

A high-pass filter will pass all frequencies above and attenuate all those below a set frequency.

The circuit configurations, attenuation characteristics, and impedance characteristics of constant-k high-pass filters are given in Fig. 5-13. The formulas for computing L_2, C_1, Z_0, and f_c are as follows:

$$L_2 = \frac{Z_0}{4\pi f_c}$$

$$C_1 = \frac{1}{4\pi f_c Z_0}$$

$$Z_0 = \sqrt{\frac{L_2}{C_1}}$$

$$f_c = \frac{1}{4\pi\sqrt{L_2 C_1}}$$

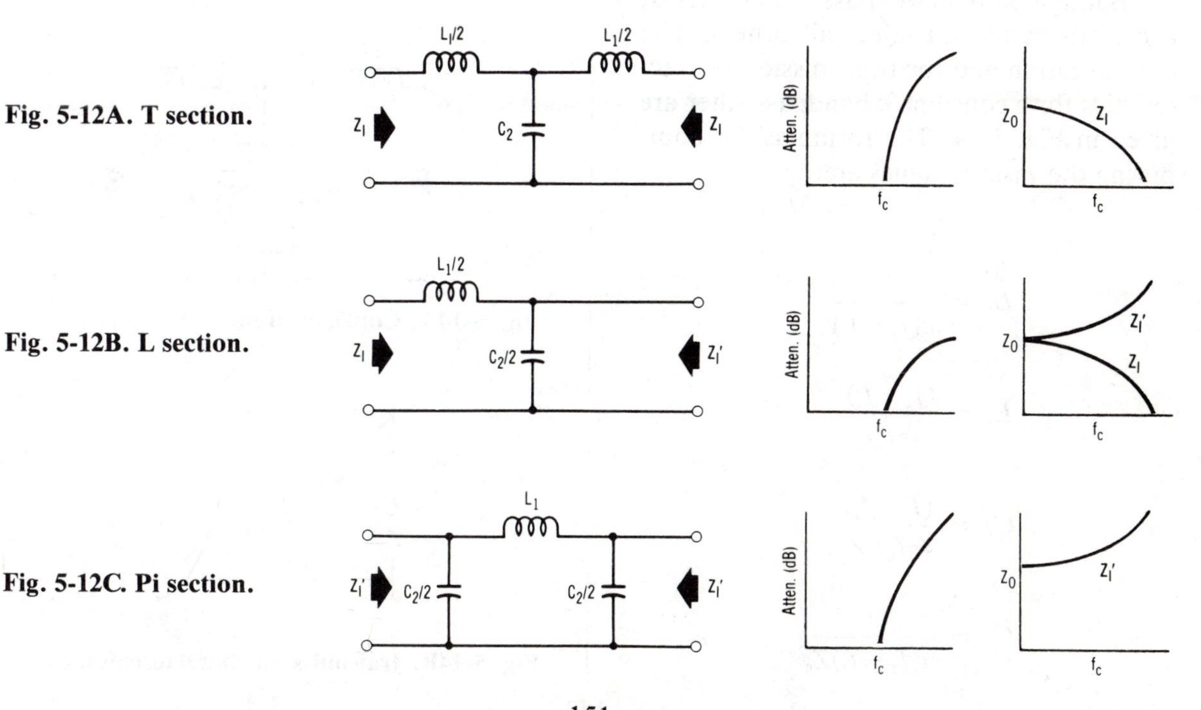

Fig. 5-12A. T section.

Fig. 5-12B. L section.

Fig. 5-12C. Pi section.

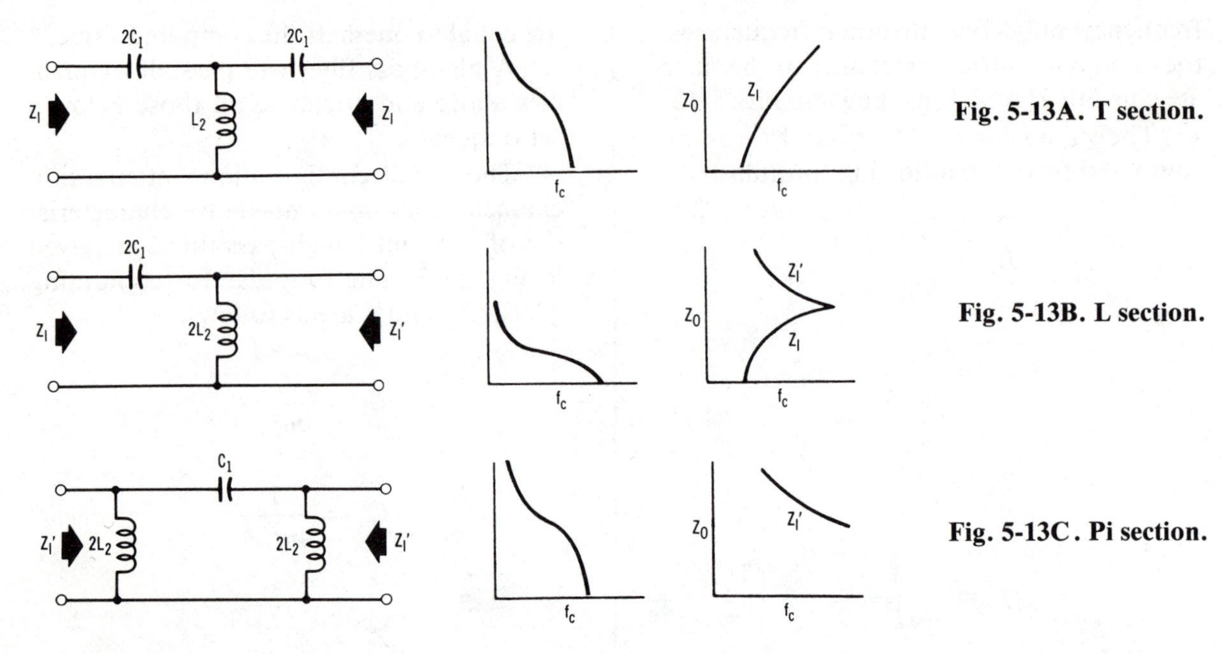

Fig. 5-13A. T section.

Fig. 5-13B. L section.

Fig. 5-13C. Pi section.

Notice that the values computed for C in the foregoing formulas must be doubled in the T and L sections. Likewise, the value computed for L must be doubled in the L and pi sections.

Bandpass filters will pass frequencies of a certain band and reject all others. The configuration and the transmission characteristics for a constant-k bandpass filter are given in Fig. 5-14. The formulas for computing the various values are:

$$L_1 = \frac{Z_0}{\pi(f_2 - f_1)}$$

$$L_2 = \frac{(f_2 - f_1)}{4\pi f_1 f_2}$$

$$C_1 = \frac{(f_2 - f_1)}{4\pi f_1 f_2 Z_0}$$

$$C_2 = \frac{1}{\pi(f_2 - f_1)Z_0}$$

$$f_m = \sqrt{f_1 f_2} = \frac{1}{2\pi\sqrt{L_1 C_1}} = \frac{1}{2\pi\sqrt{L_2 C_2}}$$

$$Z_0 = \sqrt{\frac{L_1}{C_2}} = \sqrt{\frac{L_2}{C_1}}$$

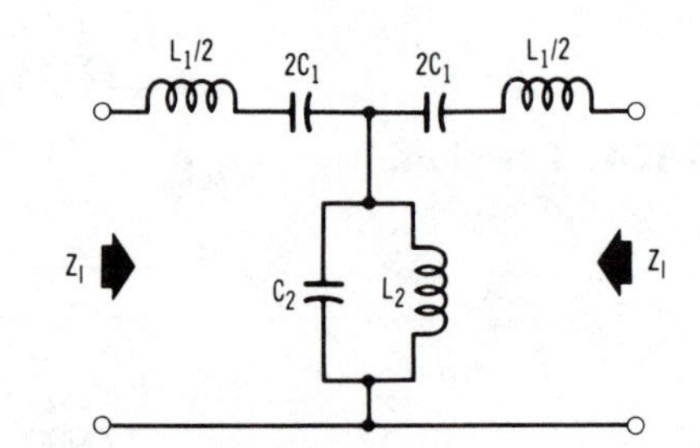

Fig. 5-14A. Configuration.

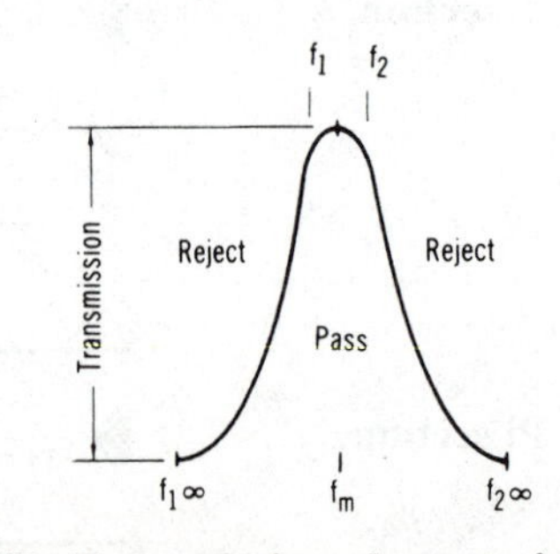

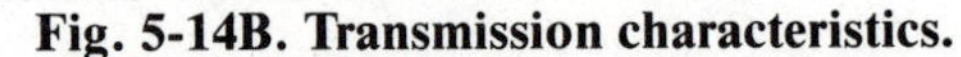

Fig. 5-14B. Transmission characteristics.

As before, some values must be doubled or halved, as shown in Fig. 5-14.

A band-rejection filter will reject a certain band of frequencies and pass all others. The configuration and the transmission characteristics of a constant-k band-rejection filter are given in Fig. 5-15. The formulas for computing the component values, frequencies, and line impedance are:

$$L_1 = \frac{(f_2 - f_1)Z_0}{\pi f_1 f_2}$$

$$L_2 = \frac{Z_0}{4\pi(f_2 - f_1)}$$

$$C_1 = \frac{1}{4\pi(f_2 - f_1)Z_0}$$

$$C_2 = \frac{(f_2 - f_1)}{\pi f_1 f_2 Z_0}$$

$$f_m = \sqrt{f_1 f_2} = \frac{1}{2\pi\sqrt{L_1 C_1}} = \frac{1}{2\pi\sqrt{L_2 C_2}}$$

$$Z_0 = \sqrt{\frac{L_1}{C_2}} = \sqrt{\frac{L_2}{C_1}}$$

where

L_1 and L_2 are the inductances of the coils, in henrys,

C_1 and C_2 are the capacitances of the capacitors, in farads,

f_1 and f_2 are the frequencies at the edge of the passband, in hertz,

f_m is the frequency at the center of the passband, in hertz,

$f_{1\infty}$ and $f_{2\infty}$ are the frequencies of infinite attenuation, in hertz,

Z_0 is the line impedance, in ohms.

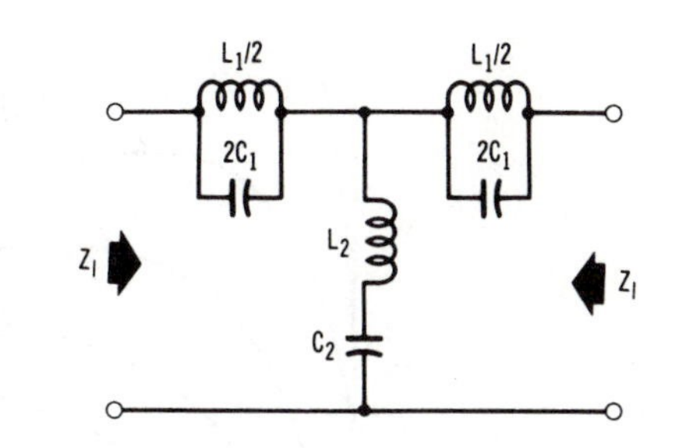

Fig. 5-15A. Configuration.

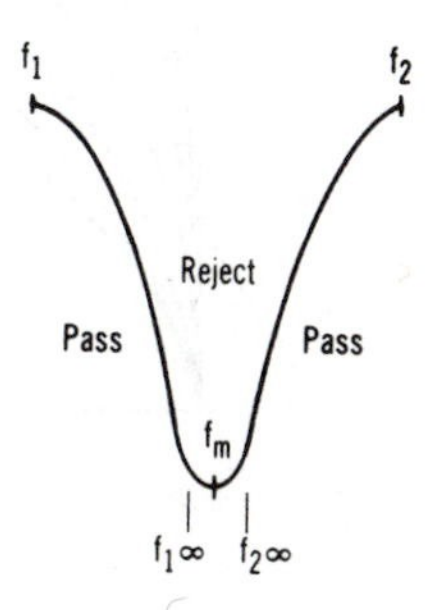

Fig. 5-15B. Transmission characteristics.

m-Derived Filters

In an *m*-derived filter, the designer can control either the impedance or the attenuation characteristics. The values are first computed as for a constant-k filter and then modified by an algebraic expression containing the constant *m*. The term *m* is a positive number between zero and one. Its value governs the characteristics of the filter.

Two frequencies — the cutoff and the frequency of infinite attenuation — are involved in the design of *m*-derived filters. By selecting the proper value for *m*, it is possible to control the spacing between the two frequencies. Figure 5-16 shows the effect different values of *m* have on the impedance characteristics. The best impedance match is obtained when *m* is equal to 0.6; hence, this value is usually employed.

The attenuation characteristics for the various values of *m* are given in Fig. 5-17.

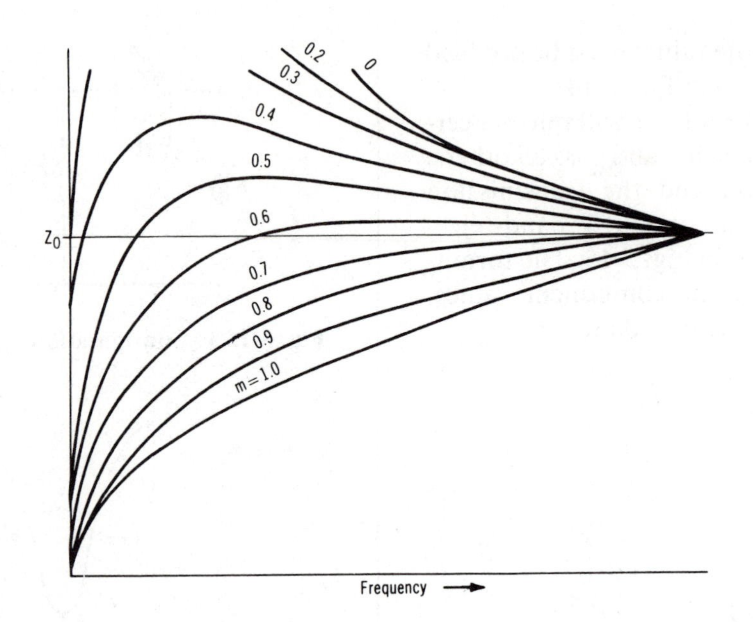

Fig. 5-16

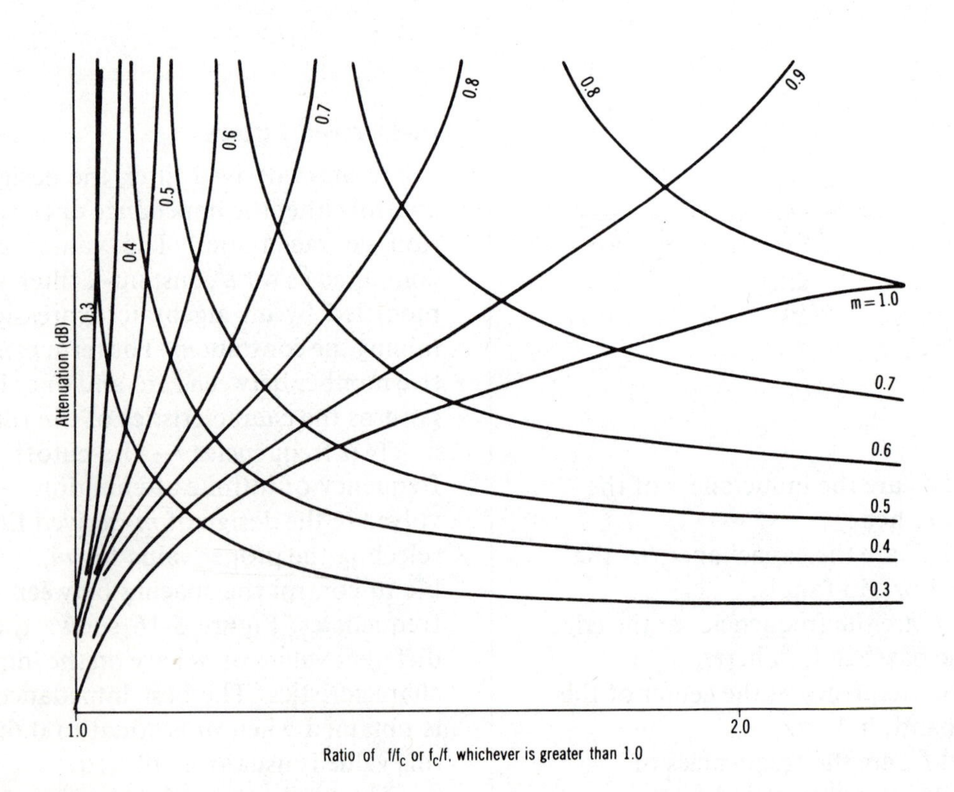

Fig. 5-17

The attenuation rises to maximum and then drops on all curves. This graph applies to both low- and high-pass filters.

The value of m is determined from the formulas:

$$m = \sqrt{1 - \left(\frac{f_c}{f_\infty}\right)^2}$$

or

$$m = \sqrt{1 - \left(\frac{f_\infty}{f_c}\right)^2}$$

Select the formula that will give a positive number.

The configurations for m-derived filters are classified as either series or shunt. Those for the series m-derived low-pass filters are given in Fig. 5-18. The formulas are as follows:

$$L_1 = m\left(\frac{Z_0}{2\pi f_C}\right)$$

$$L_2 = \left(\frac{1 - m^2}{4m}\right)\left(\frac{Z_0}{2\pi f_C}\right)$$

$$C_2 = \left(\frac{1}{\pi f_C Z_0}\right)$$

For a series m-derived high-pass filter (Fig. 5-19), the formulas are:

$$L_2 = \frac{\left(\frac{Z_0}{4\pi f_C}\right)}{m}$$

$$C_1 = \frac{\left(\frac{1}{4\pi f_C Z_0}\right)}{m}$$

$$C_2 = \left(\frac{4m}{1 - m^2}\right)\left(\frac{1}{4\pi f_C Z_0}\right)$$

The configurations for shunt m-derived low-pass filters are given in Fig. 5-20. The formulas for computing the component values are:

$$L_1 = m\left(\frac{Z_0}{\pi f_C}\right)$$

$$C_1 = \left(\frac{1 - m^2}{4m}\right)\left(\frac{1}{\pi f_C Z_0}\right)$$

$$C_2 = m\left(\frac{1}{\pi f_C Z_0}\right)$$

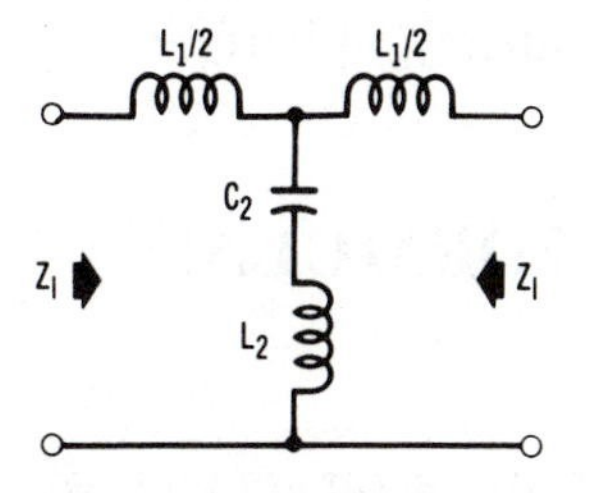

Fig. 5-18A. T section.

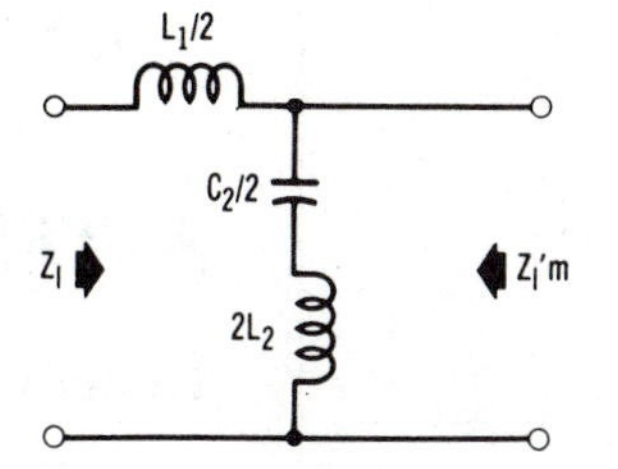

Fig. 5-18B. L section.

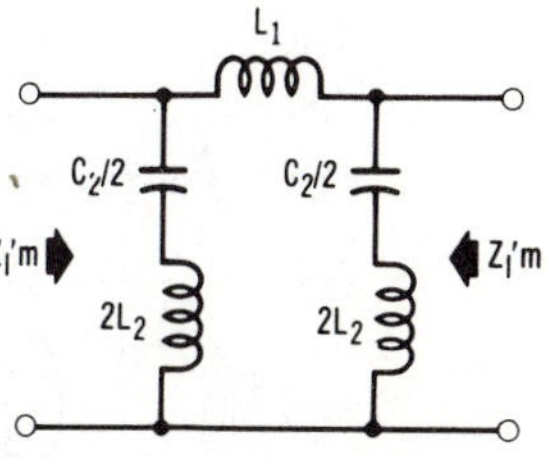

Fig. 5-18C. Pi section.

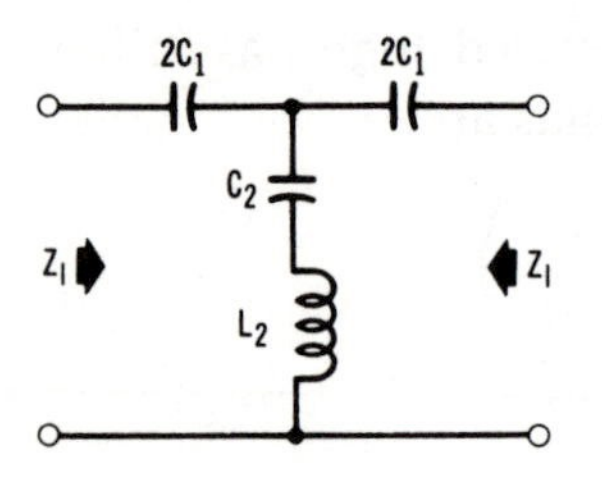

Fig. 5-19A. T section.

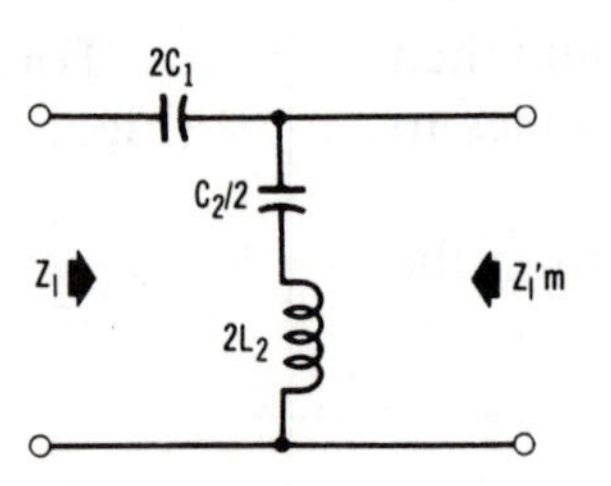

Fig. 5-19B. L section.

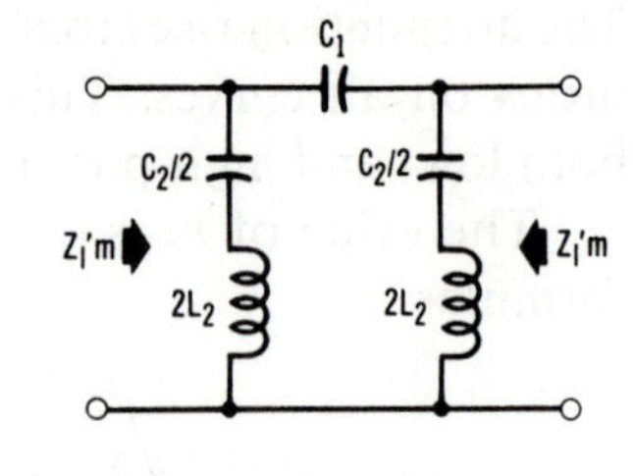

Fig. 5-19C. Pi section.

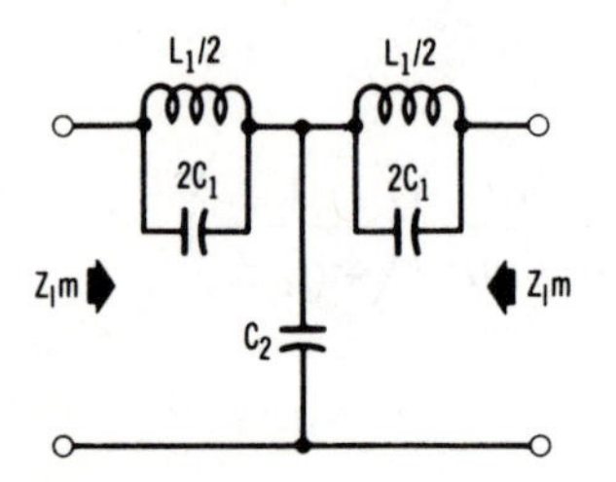

Fig. 5-20A. T section.

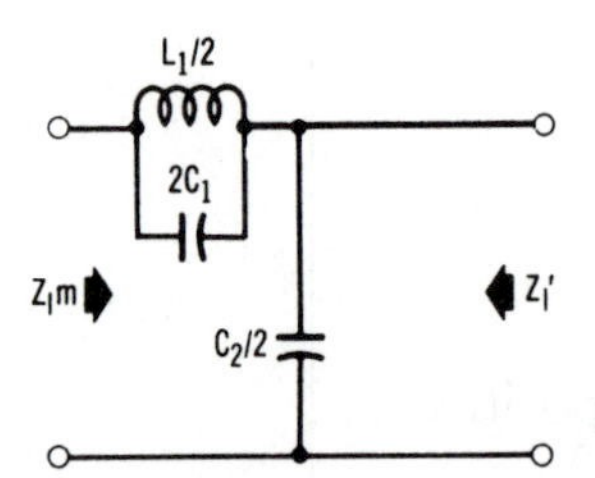

Fig. 5-20B. L section.

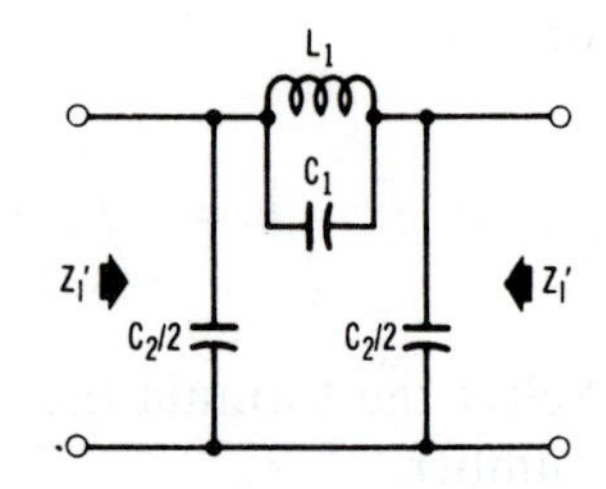

Fig. 5-20C. Pi section.

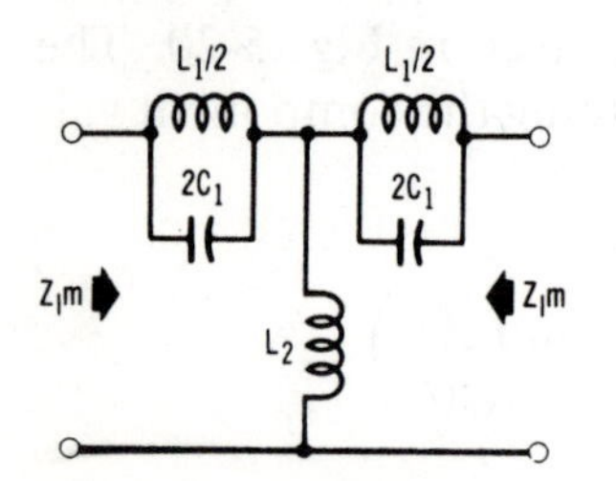

Fig. 5-21A. T section.

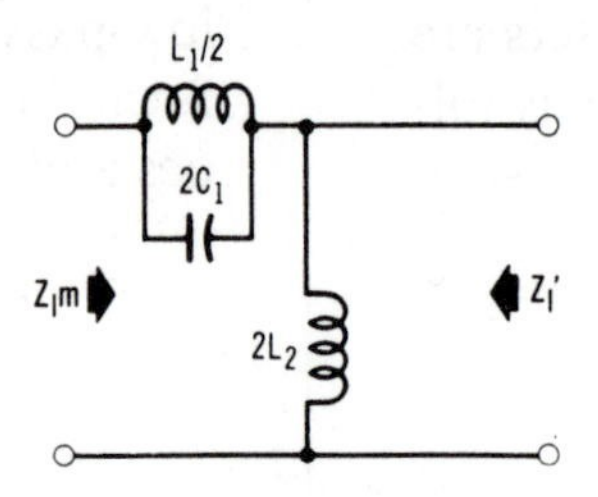

Fig. 5-21B. L section.

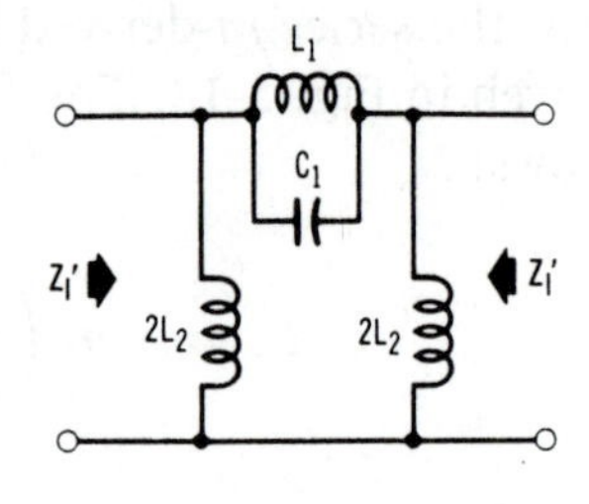

Fig. 5-21C. Pi section.

For shunt *m*-derived high-pass filters (Fig. 5-21), the formulas are:

$$L_1 = \left(\frac{4m}{1 - m^2}\right)\left(\frac{Z_0}{4\pi f_C}\right)$$

$$L_2 = \frac{\left(\frac{Z_0}{4\pi f_C}\right)}{m}$$

$$C_1 = \frac{\left(\frac{1}{4\pi f_C Z_0}\right)}{m}$$

where

L_1 and L_2 are the inductances of the coils, in henrys,

C_1 and C_2 are the capacitances of the capacitors, in farads,

m is a constant between 0 and 1,

Z_0 is the line impedance, in ohms,

f_C is the cutoff frequency, in hertz.

ATTENUATOR FORMULAS

General

An attenuator is an arrangement of noninductive resistors used in an electrical circuit to reduce the audio- or radiosignal strength without introducing distortion. The resistors may be fixed or variable. Attenuators can be designed to work between equal or unequal impedances; hence, they are often used as impedance-matching networks.

Any attenuator working between unequal impedances must introduce a certain minimum loss. These values are given in the graph of Fig. 5-22. The impedance ratio is the input impedance divided by the output impedance, or vice versa—whichever gives a value of more than one.

A factor is used in the calculation of resistor values in attenuator networks. Called K, it is the ratio of current, voltage, or power corresponding to a given value of attenuation in decibels. Table 5-2 gives the value of K for the more common loss values.

The four steps in the design of a pad are: (1) Determine the type of network required. (2) If impedances are unequal, calculate the ratio of input to output impedance (or output to input impedance) and refer to Fig. 5-22 for the minimum loss value. (3) From Table 5-2, find the value of K for the desired loss. (4) Calculate the resistor values using the following formulas.

Combining or Dividing Network (Fig. 5-23)

$$R_B = \left(\frac{N-1}{N+1}\right) Z$$

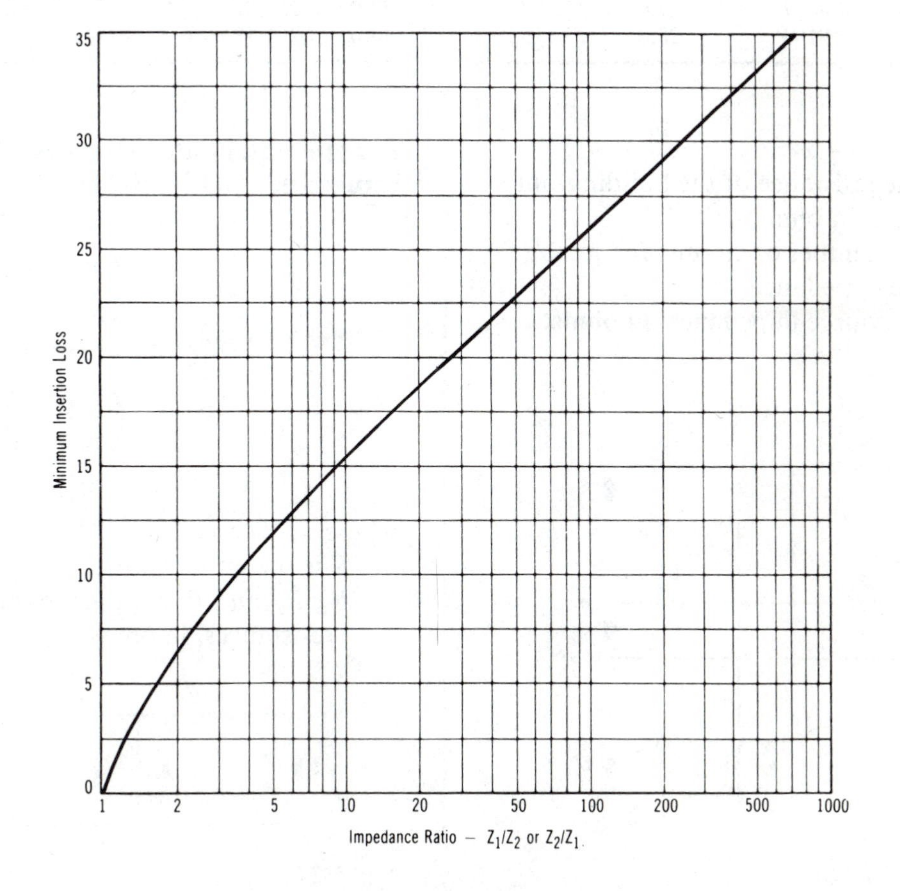

Fig. 5-22

TABLE 5-2
***K* Factors for Calculating Attenuator Loss**

dB	K	dB	K	dB	K	dB	K
0.05	1.0058	9.5	2.9854	29.0	28.184	49.0	281.84
0.1	1.0116	10.0	3.1623	30.0	31.623	50.0	316.23
0.5	1.0593	11.0	3.5481	31.0	35.481	51.0	354.81
1.0	1.1220	12.0	3.9811	32.0	39.811	52.0	398.11
1.5	1.1885	13.0	4.4668	33.0	44.668	54.0	501.19
2.0	1.2589	14.0	5.0119	34.0	50.119	55.0	562.34
2.5	1.3335	15.0	5.6234	35.0	56.234	56.0	630.96
3.0	1.4125	16.0	6.3096	36.0	63.096	57.0	707.95
3.5	1.4962	17.0	7.0795	37.0	70.795	58.0	794.33
4.0	1.5849	18.0	7.9433	38.0	79.433	60.0	1000.0
4.5	1.6788	19.0	8.9125	39.0	89.125	65.0	1778.3
5.0	1.7783	20.0	10.0000	40.0	100.000	70.0	3162.3
5.5	1.8837	21.0	11.2202	41.0	112.202	75.0	5623.4
6.0	1.9953	22.0	12.589	42.0	125.89	80.0	10,000
6.5	2.1135	23.0	14.125	43.0	141.25	85.0	17,783
7.0	2.2387	24.0	15.849	44.0	158.49	90.0	31,623
7.5	2.3714	25.0	17.783	45.0	177.83	95.0	56,234
8.0	2.5119	26.0	19.953	46.0	199.53	100.0	10^5
8.5	2.6607	27.0	22.387	47.0	223.87		
9.0	2.8184	28.0	25.119	48.0	251.19		

where
- R_B is the resistance of the building-out resistors, in ohms,
- N is the number of circuits fed by the source impedance,
- Z is the source impedance, in ohms.

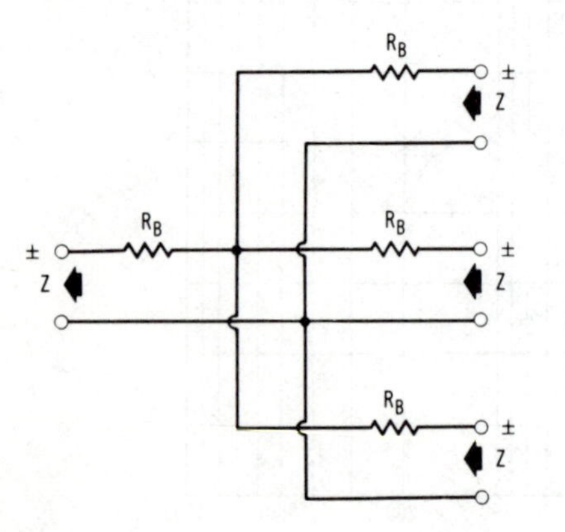

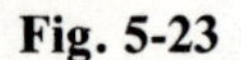
Fig. 5-23

T-Type Attenuator (Between Equal Impedances) (Fig. 5-24)

$$R_1 \text{ and } R_2 = \left(\frac{K-1}{K+1}\right) Z$$

$$R_3 = \left(\frac{K}{K^2-1}\right) 2Z$$

where
- K is the impedance factor,
- R_1, R_2, and R_3 are the measured resistances, in ohms.

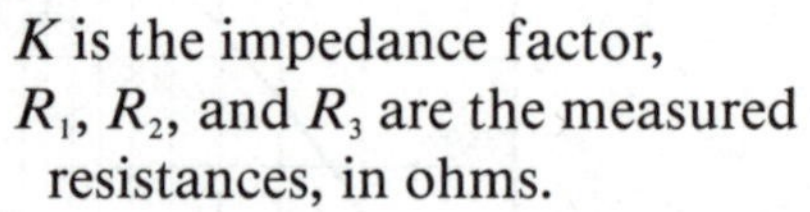

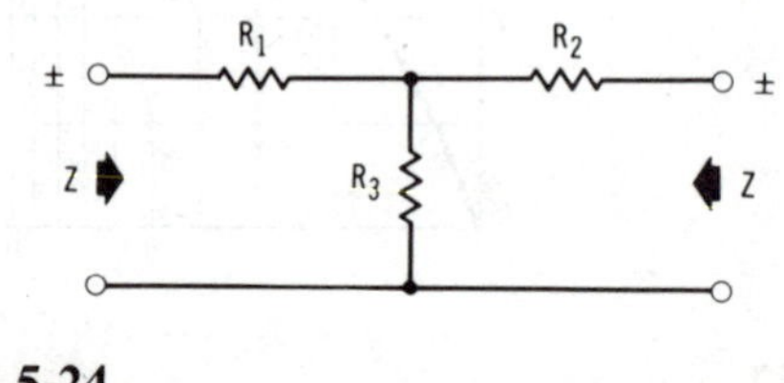

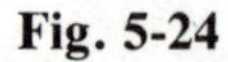
Fig. 5-24

H-Type Attenuator (Balanced-T Attenuator)

Calculate the values for R_1, R_2, and R_3 as for an unbalanced T-attenuator (Fig. 5-24). Then halve the values of R_1 and R_2, as shown in Fig. 5-25. The tap on R_3 is exactly in the center.

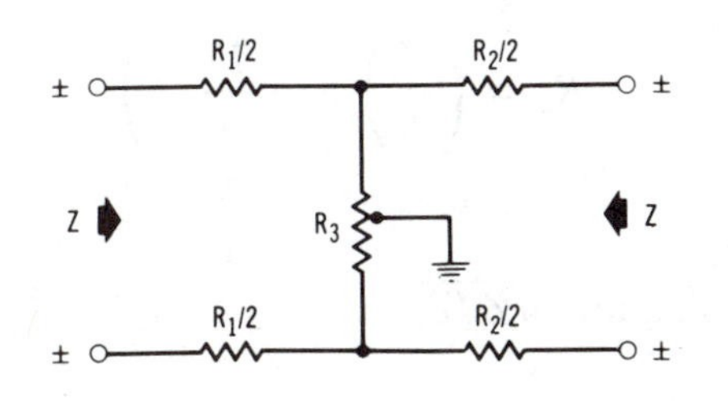

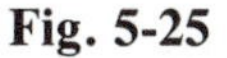

Fig. 5-25

Taper Pad (T-Type Attenuator Between Unequal Impedances) (Fig. 5-26)

$$R_1 = Z_1\left(\frac{K^2+1}{K^2-1}\right) - 2\sqrt{Z_1Z_2}\left(\frac{K}{K^2-1}\right)$$

$$R_2 = Z_2\left(\frac{K^2+1}{K^2-1}\right) - 2\sqrt{Z_1Z_2}\left(\frac{K}{K^2-1}\right)$$

$$R_3 = 2\sqrt{Z_1Z_2}\left(\frac{K}{K^2-1}\right)$$

where

K is the impedance factor,

R_1, R_2, and R_3 are the measured resistances, in ohms,

Z_1 is the larger impedance, in ohms,

Z_2 is the smaller impedance, in ohms.

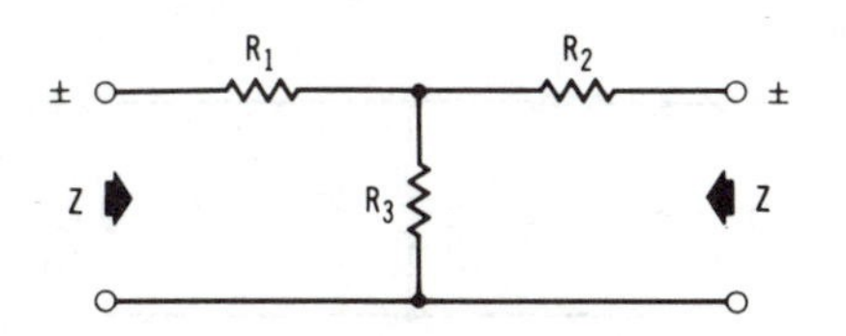

Fig. 5-26

Bridged-T Attenuator (Unbalanced) (Fig. 5-27)

$$R_1 = Z$$

$$R_5 = (K-1)Z$$

$$R_6 = \left(\frac{1}{K-1}\right)Z$$

R_5 and R_6 are connected to a common shaft, and each varies inversely in value with respect to the other.

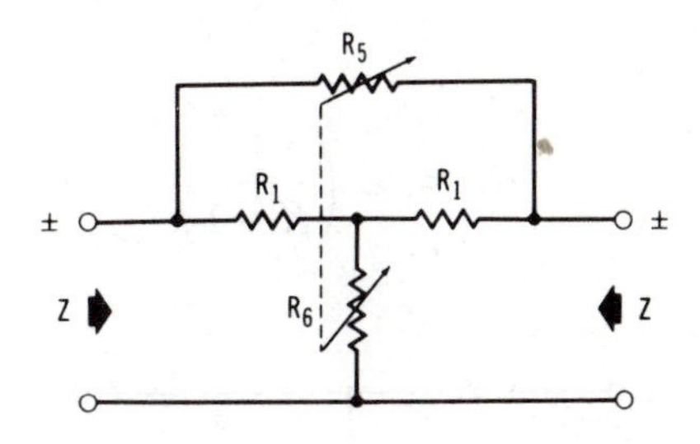

Fig. 5-27

Balanced Bridged-T Attenuator

Calculate the values for R_1, R_5, and R_6 as for an unbalanced bridged-T attenuator (Fig. 5-27). Then halve the values as shown in Fig. 5-28.

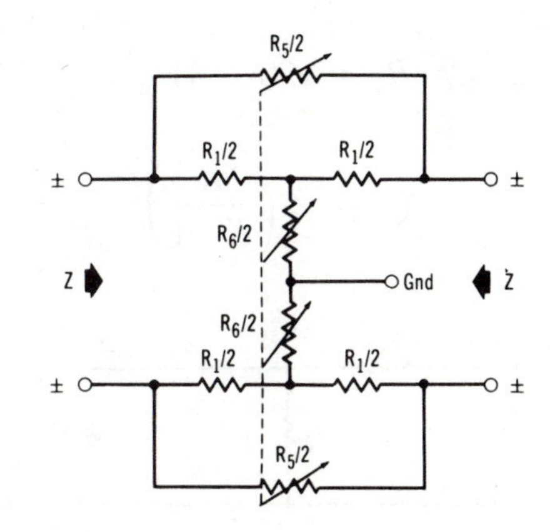

Fig. 5-28

L-Type Attenuators

An L-type attenuator can supply an impedance match in only one direction. If the impedances it works out of and into are unequal, it can be made to match either—but not both—impedances. The arrows in Figs. 5-29 through 5-32 indicate the direction of impedance match.

Between equal impedances and with the impedance match in the direction of the series arm:

$$R_1 = Z\left(\frac{K-1}{K}\right)$$

$$R_2 = Z\left(\frac{1}{K-1}\right)$$

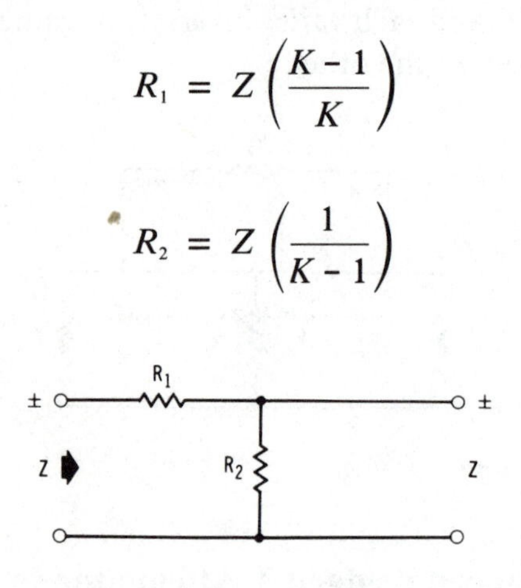

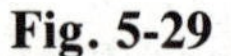
Fig. 5-29

Between equal impedances and with the impedance match in the direction of the shunt arm:

$$R_1 = Z(K-1)$$

$$R_2 = Z\left(\frac{K}{K-1}\right)$$

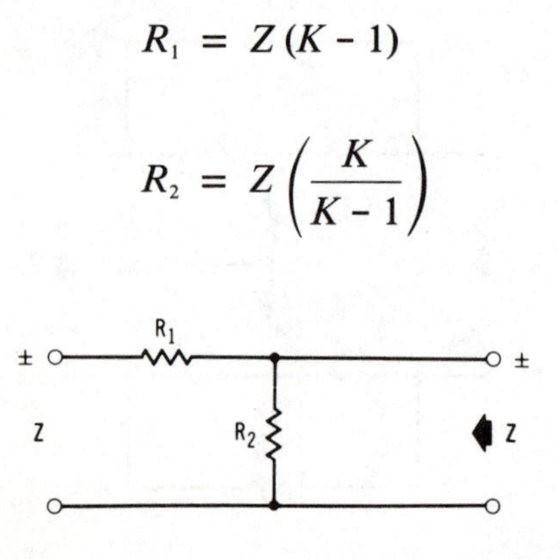

Fig. 5-30

Between unequal impedances and with the impedance match toward the larger value:

$$R_1 = \left(\frac{Z_1}{S}\right)\left(\frac{KS-1}{K}\right)$$

$$R_2 = \left(\frac{Z_1}{S}\right)\left(\frac{1}{K-S}\right)$$

where

S equals $\sqrt{Z_1/Z_2}$.

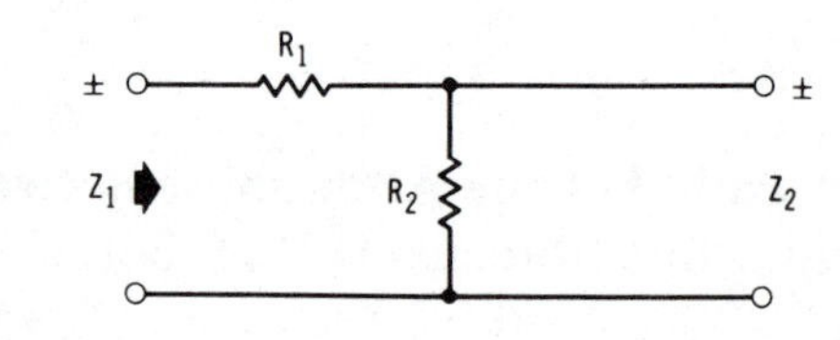

Fig. 5-31

Between unequal impedances and with the impedance match toward the smaller value:

$$R_1 = \left(\frac{Z_1}{S}\right)(K-S)$$

$$R_2 = \frac{Z_1}{S}\left(\frac{K}{KS-1}\right)$$

where

S equals $\sqrt{Z_1/Z_2}$.

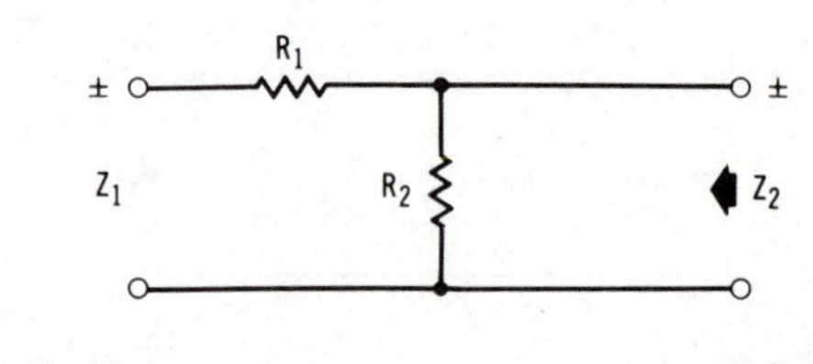

Fig. 5-32

Pi-Type Attenuator (Between Equal Impedances) (Fig. 5-33)

$$R_1 = Z\left(\frac{K + 1}{K - 1}\right)$$

$$R_2 = \left(\frac{Z}{2}\right)\left(\frac{K^2 - 1}{K}\right)$$

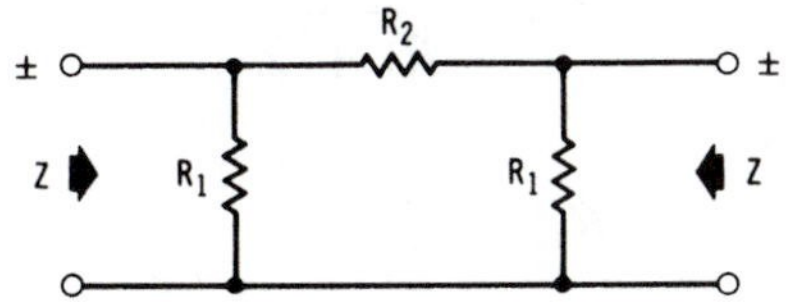

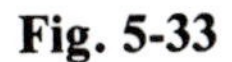
Fig. 5-33

Pi-Type Attenuator (Between Unequal Impedances) (Fig. 5-34)

$$R_1 = Z_1\left(\frac{K^2 - 1}{K^2 - 2KS + 1}\right)$$

$$R_2 = \left(\frac{\sqrt{Z_1Z_2}}{2}\right)\left(\frac{K^2 - 1}{K}\right)$$

$$R_3 = Z_2\left(\frac{K^2 - 1}{K_2 - 2\frac{K}{S} + 1}\right)$$

where
  S equals $\sqrt{Z_1/Z_2}$.

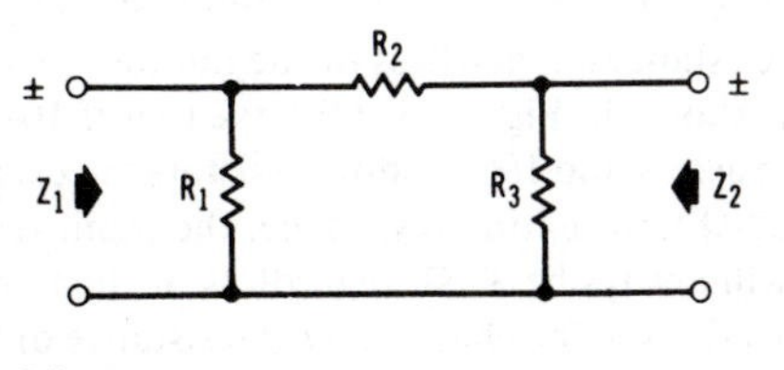

Fig. 5-34

O-Type Attenuators

Calculate the values for a pi-type attenuator (Figs. 5-33 and 5-34), then halve the values for the series resistors as shown in Figs. 5-35 (balanced) and 5-36 (unbalanced).

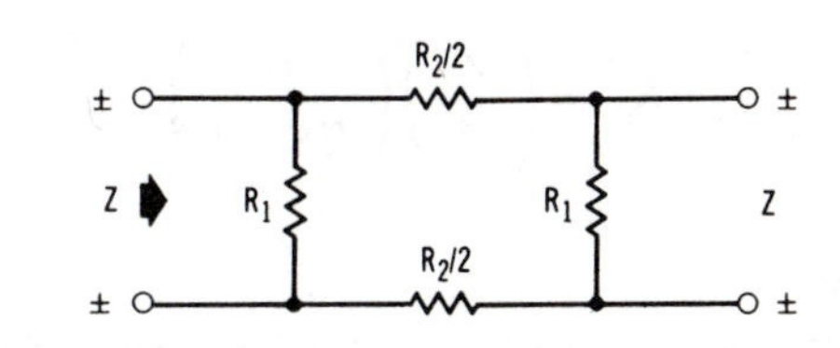

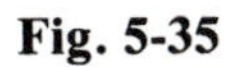
Fig. 5-35

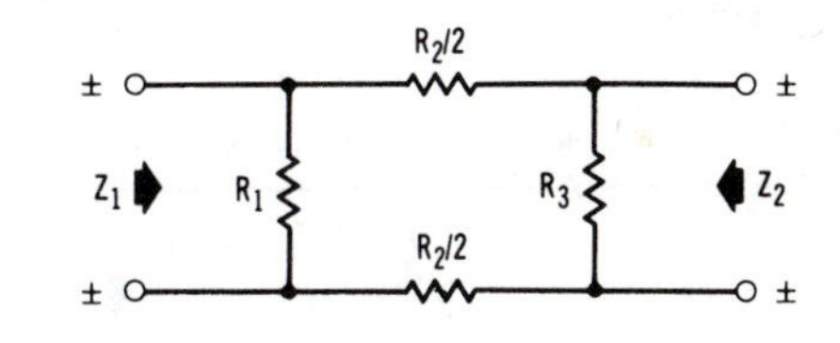

Fig. 5-36

U-Type Attenuator (Figs. 5-37 and 5-38)

For impedance match in the direction of the series arms:

$$R_1 = \left(\frac{Z_1}{2S}\right)\left(\frac{KS - 1}{K}\right)$$

$$R_2 = \left(\frac{Z_1}{S}\right)\left(\frac{1}{K - S}\right)$$

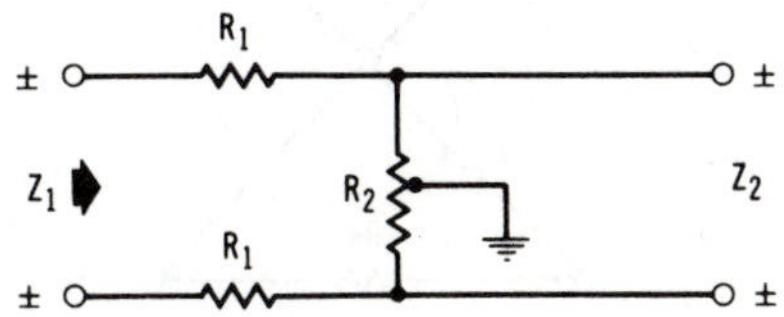

Fig. 5-37

For impedance match in the direction of the shunt arm:

$$R_1 = \left(\frac{Z_1}{2S}\right)(K - S)$$

$$R_2 = \left(\frac{Z_1}{S}\right)\left(\frac{K}{KS - 1}\right)$$

where
The arrows indicate the direction of the impedance match,
S equals $\sqrt{Z_1/Z_2}$.

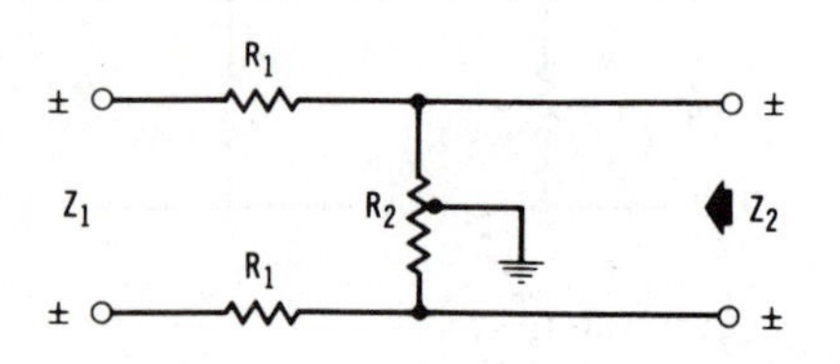

Fig. 5-38

Lattice-Type Attenuator (Fig. 5-39)

$$R_1 = \left(\frac{K - 1}{K + 1}\right) Z$$

$$R_2 = \left(\frac{K + 1}{K - 1}\right) Z$$

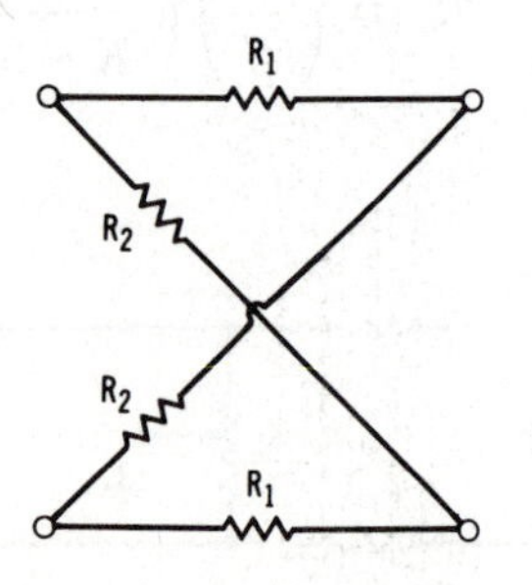

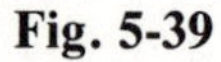

Fig. 5-39

Ladder-Type Attenuator (Fig. 5-40)

$$R_1 = \left(\frac{K^2 - 1}{2K}\right) Z$$

$$R_2 = \left(\frac{K + 1}{K - 1}\right) Z$$

$$R_3 = \frac{R_2 \times Z}{R_2 + Z}$$

$$R_4 = \frac{Z}{2}$$

$$Z_{in} = Z_{out}$$

where
K depends on the loss per step—not on the total loss.

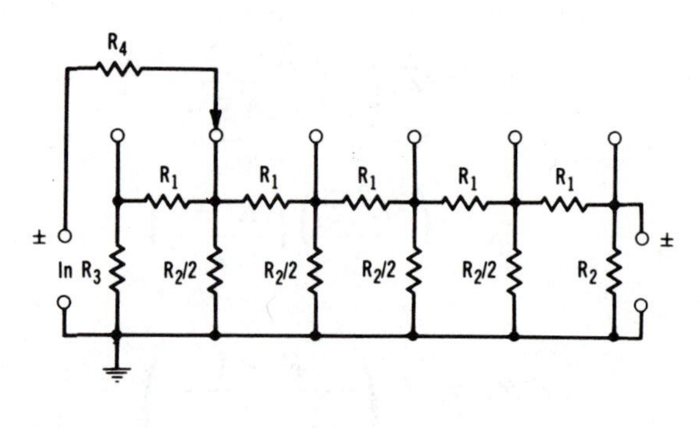

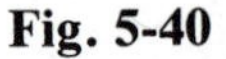

Fig. 5-40

Note. An instructive special case of an L-section resistive network with equal input and terminating resistances, regardless of the number of sections, is shown in Fig. 5-41. Observe that if 10-Ω series resistors and 100-Ω shunt resistors are used with a 27-Ω terminating resistance, the input resistance will always be 37 Ω, regardless of the number of sections. (The characteristic resistance of the network is 37 Ω.)

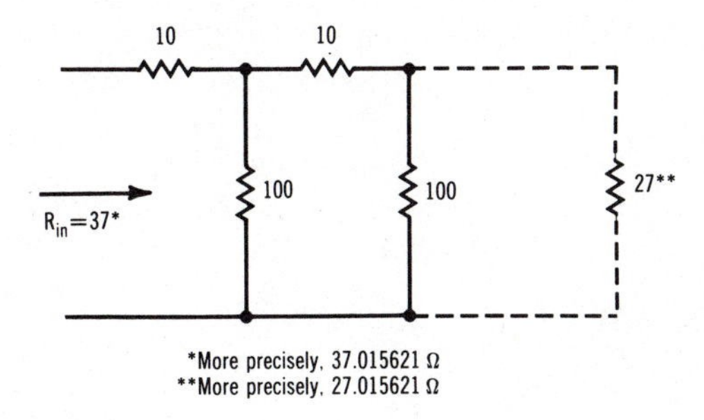

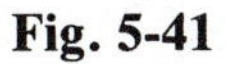
Fig. 5-41

STANDARD POTENTIOMETER TAPERS (Fig. 5-42)

Taper S	straight or uniform resistance change with rotation
Taper T	right-hand 30% resistance at 50% of counterclockwise rotation
Taper V	right-hand 20% resistance at 50% of counterclockwise rotation

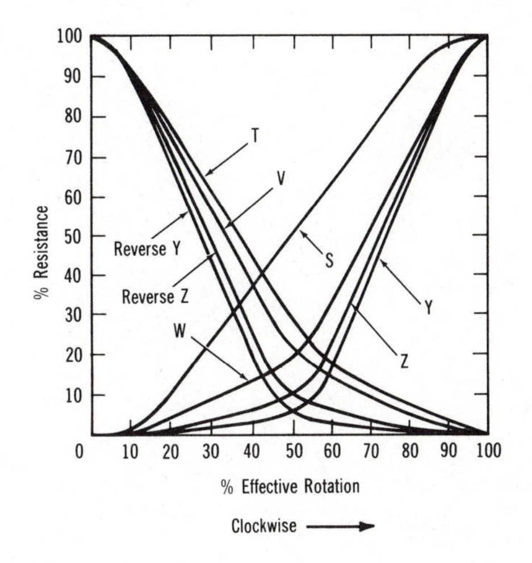

Fig. 5-42

Taper W	left-hand 20% resistance at 50% of clockwise rotation
Taper Z	left-hand (log. audio) 10% resistance at 50% clockwise rotation
Taper Y	left-hand 5% resistance at 50% of clockwise rotation

Chapter 6

Mathematical Tables and Formulas

MATHEMATICAL CONSTANTS

$\pi = 3.1416$

$\pi^2 = 9.8696$

$\pi^3 = 31.0063$

$\frac{1}{\pi} = 0.3183$

$\frac{1}{\pi^2} = 0.1013$

$\frac{1}{\pi^3} = 0.0323$

$\sqrt{\pi} = 1.7725$

$\frac{1}{\sqrt{\pi}} = 0.5642$

$\frac{1}{2\pi} = 0.1592$

$\left(\frac{1}{2\pi}\right)^2 = 0.0253$

$2\pi = 6.2832$

$(2\pi)^2 = 39.4786$

$4\pi = 12.5664$

$\frac{\pi}{2} = 1.5708$

$\sqrt{\frac{\pi}{2}} = 1.2533$

$\sqrt{2} = 1.4142$

$\sqrt{3} = 1.7321$

$\frac{1}{\sqrt{2}} = 0.7071$

$\frac{1}{\sqrt{3}} = 0.5773$

$\log \pi = 0.4971$

$\log \pi^2 = 0.9943$

$\log \sqrt{\pi} = 0.2486$

$\log \frac{\pi}{2} = 0.1961$

MATHEMATICAL SYMBOLS

$\times$ or $\cdot$ multiplied by

$\div$ divided by

$=$ equals

$\neq$ does not equal

$<$ is less than

$\pm$ plus or minus

≡ identical with

+ positive, add, and plus

- negative, subtract, and minus

> is greater than

≥ equal to or greater than

≤ equal to or less than

∴ therefore

|| parallel to

∠ angle

≪ is much less than

≫ is much greater than

⊥ perpendicular to

|n| absolute value of *n*

≅ is approximately equal to

√ square root

FRACTIONAL INCH, DECIMAL, AND MILLIMETER EQUIVALENTS

Table 6-1 gives the decimal inch and millimeter equivalents of fractional parts of an inch by 64ths to four significant figures.

POWERS OF 10

Exponent Determination

Large numbers can be simplified by using powers of 10. For example, some of the multiples of 10 from 1 to 1,000,000, with their equivalents in powers of 10, are:

*Any number to the zero power is 1.

TABLE 6-1
Fractional Inch, Decimal, and Millimeter Equivalents

Fractional inch	Decimal inch	Millimeter equivalent	Fractional inch	Decimal inch	Millimeter equivalent
1/64	0.0156	0.397	33/64	0.5156	13.097
1/32	0.0313	0.794	17/32	0.5313	13.494
3/64	0.0469	1.191	9/16	0.5469	13.891
1/16	0.0625	1.588	35/64	0.5625	14.288
5/64	0.0781	1.934	37/64	0.5781	14.684
3/32	0.0938	2.381	19/32	0.5938	15.081
7/64	0.1094	2.778	39/64	0.6094	15.478
1/8	0.1250	3.175	5/8	0.6250	15.875
9/64	0.1406	3.572	41/64	0.6406	16.272
5/32	0.1563	3.969	21/32	0.6563	16.669
11/64	0.1719	4.366	43/64	0.6719	17.066
3/16	0.1875	4.763	11/16	0.6875	17.463
13/64	0.2031	5.159	45/64	0.7031	17.859
7/32	0.2188	5.556	23/32	0.7188	18.256
15/64	0.2344	5.953	47/64	0.7344	18.653
1/4	0.2500	6.350	3/4	0.7500	19.050
17/64	0.2656	6.747	49/64	0.7656	19.447
9/32	0.2813	7.144	25/32	0.7813	19.844
19/64	0.2969	7.541	51/64	0.7969	20.241
5/16	0.3125	7.938	13/16	0.8125	20.638
21/64	0.3281	8.334	53/64	0.8281	21.034
11/32	0.3438	8.731	27/32	0.8438	21.431
23/64	0.3594	9.128	55/64	0.8594	21.828
3/8	0.3750	9.525	7/8	0.8750	22.225
25/64	0.3906	9.922	57/64	0.8906	22.622
13/32	0.4063	10.319	29/32	0.9063	23.019
27/64	0.4219	10.716	59/64	0.9219	23.416
7/16	0.4375	11.113	15/16	0.9375	23.813
29/64	0.4531	11.509	61/64	0.9531	24.209
15/32	0.4688	11.906	31/32	0.9688	24.606
31/64	0.4844	12.303	63/64	0.9844	25.003
1/2	0.5000	12.700	1	1.000	25.400

$$1 = 10^0\text{*}$$
$$10 = 10^1$$
$$100 = 10^2$$
$$1000 = 10^3$$
$$10{,}000 = 10^4$$
$$100{,}000 = 10^5$$
$$1{,}000{,}000 = 10^6$$

Likewise, powers of 10 can be used to simplify decimal expressions. Some of the submultiples of 10 from 0.1 to 0.000001, with their equivalents in powers of 10, are:

$$0.1 = 10^{-1}$$
$$0.01 = 10^{-2}$$
$$0.001 = 10^{-3}$$
$$0.0001 = 10^{-4}$$
$$0.00001 = 10^{-5}$$
$$0.000001 = 10^{-6}$$

Any whole number can be expressed as a smaller whole number, and any decimal can be expressed as a whole number, by moving the decimal point to the left or right and expressing the number as a power of 10. If the decimal point is moved to the left, the power is positive and is equal to the number of places the decimal point was moved. If the decimal point is moved to the right, the power is negative and is equal to the number of places the decimal point was moved.

Example.

$$1.23 = 0.0123 \times 10^2$$
$$456.7 = 4.567 \times 10^2$$
$$78{,}900 = 78.9 \times 10^3$$
$$0.00012 = 1.2 \times 10^{-4}$$
$$0.0345 = 34.5 \times 10^{-3}$$
$$0.678 = 67.8 \times 10^{-2}$$

Addition and Subtraction

To add or subtract using powers of 10, first convert all numbers to the same power of 10. The numbers can then be added or subtracted, and the answer will be in the same power of 10.

Example.

$$9.32 \times 10^2 + 17.63 \times 10^3 + 297 = ?$$

$$\begin{aligned} 9.32 \times 10^2 &= 0.932 \times 10^3 \\ 17.63 \times 10^3 &= 17.630 \times 10^3 \\ 297 &= \underline{0.297 \times 10^3} \\ & 18.859 \times 10^3 = 18{,}859 \end{aligned}$$

Example.

$$18.47 \times 10^2 - 1.59 \times 10^3 = ?$$

$$\begin{aligned} 18.47 \times 10^2 &= 1.847 \times 10^3 \\ 1.59 \times 10^3 &= \underline{1.590 \times 10^3} \\ & 0.257 \times 10^3 = 257 \end{aligned}$$

Multiplication

To multiply using powers of 10, add the exponents.

Example.

$$\begin{aligned} 1000 \times 3721 &= 10^3 \times 37.21 \times 10^2 \\ &= 37.21 \times 10^{3+2} \\ &= 37.21 \times 10^5 \\ &= 3{,}721{,}000 \end{aligned}$$

Example.

$$\begin{aligned} 225 \times 0.00723 &= 2.25 \times 10^2 \times 7.23 \times 10^{-3} \\ &= 2.25 \times 7.23 \times 10^{2+(-3)} \\ &= 2.25 \times 7.23 \times 10^{-1} \\ &= 16.2675 \times 10^{-1} \\ &= 1.62675 \end{aligned}$$

Division

To divide using powers of 10, subtract the exponent of the denominator from the exponent of the numerator.

Example.

$$\begin{aligned} \frac{10^5}{10^3} &= 10^{5-3} \\ &= 10^2 \\ &= 100 \end{aligned}$$

Example.

$$\begin{aligned} \frac{72{,}600}{0.002} &= \frac{72.6 \times 10^3}{2 \times 10^{-3}} \\ &= \frac{72.6 \times 10^{3+3}}{2} \\ &= 36.3 \times 10^6 \\ &= 36{,}300{,}000 \end{aligned}$$

Combination Multiplication and Division

Problems involving a combination of multiplication and division can be solved using powers of 10 by multiplying and dividing, as called for, until the problem is completed.

Example.

$$\frac{3900 \times 0.007 \times 420}{142{,}000 \times 0.00005} = \frac{3.9 \times 10^3 \times 7 \times 10^{-3} \times 4.2 \times 10^2}{1.42 \times 10^5 \times 5 \times 10^{-5}}$$
$$= \frac{3.9 \times 7 \times 4.2 \times 10^2}{1.42 \times 5}$$
$$= \frac{114.66 \times 10^2}{7.1}$$
$$= 16.1493 \times 10^2$$
$$= 1614.93$$

Reciprocal

To take the reciprocal of a number using powers of 10, first (if necessary) state the number so the decimal point precedes the first significant figure of the number. Then divide this number into 1. The power of 10 in the answer will be the same value as in the original number, but it will have the opposite sign.

Example.

$$\text{Reciprocal of } 400 = \frac{1}{400}$$

$$\frac{1}{400} = \frac{1}{0.4 \times 10^3}$$
$$= 2.5 \times 10^{-3}$$
$$= 0.0025$$

Example.

$$\text{Reciprocal of } 0.0025 = \frac{1}{0.0025}$$

$$\frac{1}{0.0025} = \frac{1}{0.25 \times 10^{-2}}$$
$$= 4 \times 10^2$$
$$= 400$$

Square and Square Root

To square a number using powers of 10, multiply the number by itself, and double the exponent.

Example.

$$(7 \times 10^3)^2 = 49 \times 10^6$$
$$= 49{,}000{,}000$$

Example.

$$(9.2 \times 10^{-4})^2 = 84.64 \times 10^{-8}$$
$$= 0.0000008464$$

To extract the square root of a number using powers of 10, do the opposite. (If the number is an odd power of 10, first convert it to an even power of 10.) Extract the square root of the number, and divide the power of 10 by 2.

Example.

$$\sqrt{36 \times 10^{10}} = 6 \times 10^5$$
$$= 600{,}000$$

Example.

$$\sqrt{5.72 \times 10^3} = \sqrt{57.2 \times 10^2}$$
$$= 7.56 \times 10$$
$$= 75.6$$

ALGEBRAIC OPERATIONS

Transposition of Terms

The following rules apply to the transposition of terms in algebraic equations:

If $A = B/C$, then:

$$B = AC$$

$$C = \frac{B}{A}$$

If $A/B = C/D$, then:

$$A = \frac{BC}{D}$$

$$B = \frac{AD}{C}$$

$$C = \frac{AD}{B}$$

$$D = \frac{BC}{A}$$

If $A + B = C$, then:

$$A = C - B$$

$$A + B - C = 0$$

If $A^2 = 1/(D\sqrt{BC})$, then:

$$A^2 = \frac{1}{D^2 BC}$$

$$B = \frac{1}{D^2 A^2 C}$$

$$C = \frac{1}{D^2 A^2 B}$$

$$D = \frac{1}{A\sqrt{BC}}$$

If $A = \sqrt{B^2 + C^2}$, then:

$$A^2 = B^2 + C^2$$

$$B = \sqrt{A^2 - C^2}$$

$$C = \sqrt{A^2 - B^2}$$

Laws of Exponents

A power of a fraction is equal to that power of the numerator divided by the same power of the denominator:

$$\left(\frac{a}{b}\right)^x = \frac{a^x}{b^x}$$

The product of two powers of the same base is also a power of that base; the exponent of the product is equal to the sum of the exponents of the two factors:

$$a^x \cdot a^y = a^{x+y}$$

The quotient of two powers of the same base is also a power of that base; the exponent of the quotient is equal to the numerator exponent minus the denominator exponent:

$$\frac{a^x}{a^z} = a^{x-z}$$

The power of a power of a base is also a power of that base; the exponent of the product is equal to the product of the exponents:

$$(a^x)^y = a^{xy}$$

A negative exponent of a base is equal to the reciprocal of that base, with a positive exponent numerically equal to the original exponent:

$$a^{-x} = \frac{1}{a^x}$$

A fractional exponent indicates that the base should be raised to the power indicated by the numerator of the fraction; the root indicated by the denominator should then be extracted:

$$a^{\frac{x}{y}} = \sqrt[y]{a^x}$$

A root of a fraction is equal to the identical root of the numerator divided by the identical root of the denominator:

$$\sqrt[x]{\frac{a}{b}} = \frac{\sqrt[x]{a}}{\sqrt[x]{b}}$$

A root of a product is equal to the product of the roots of the individual factors:

$$\sqrt[x]{ab} = \sqrt[x]{a} \times \sqrt[x]{b}$$

Quadratic Equation

The general quadratic equation:

$$ax^2 + bx + c = 0$$

may be solved by:

$$x = \frac{-b \pm \sqrt{b^2 - 4ac}}{2a}$$

GEOMETRIC FORMULAS

Triangle (Fig. 6-1):

$$\text{area } (A) = \frac{bh}{2}$$

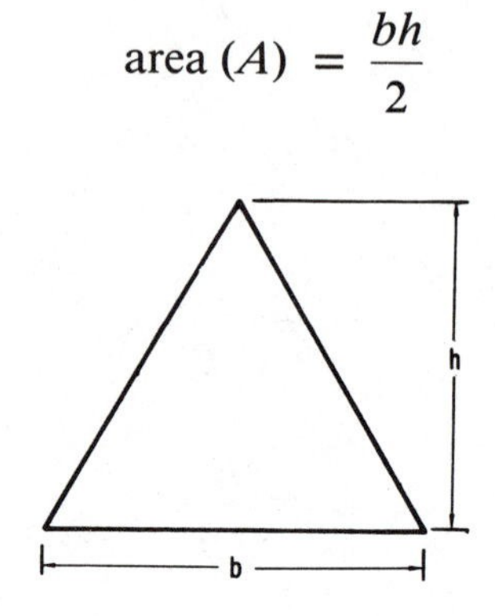

Fig. 6-1

Square (Fig. 6-2):

$$\text{area } (A) = b^2$$

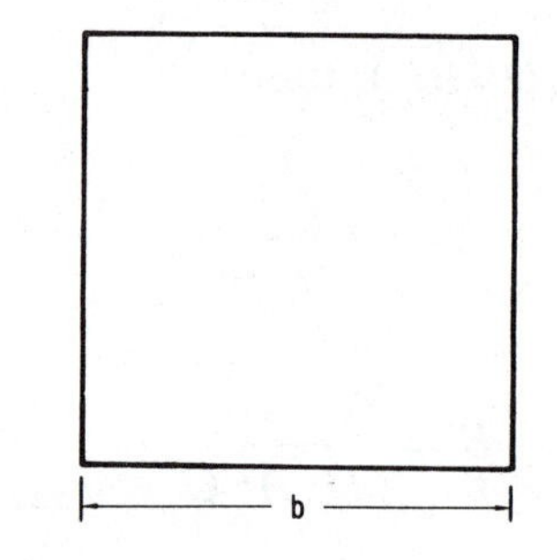

Fig. 6-2

Rectangle (Fig. 6-3):

$$\text{area } (A) = ab$$

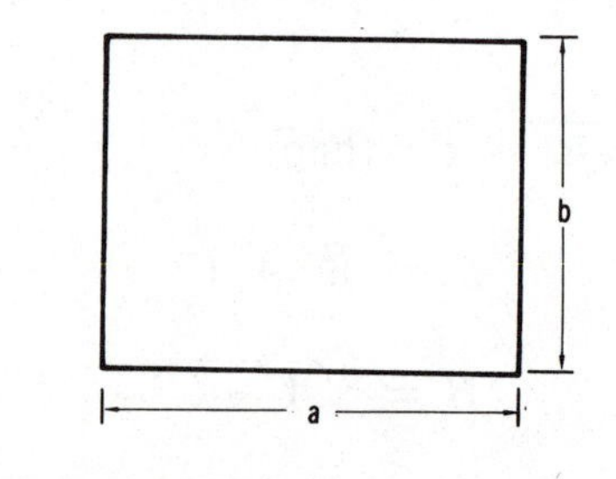

Fig. 6-3

Parallelogram (Fig. 6-4):

$$\text{area } (A) = ah$$

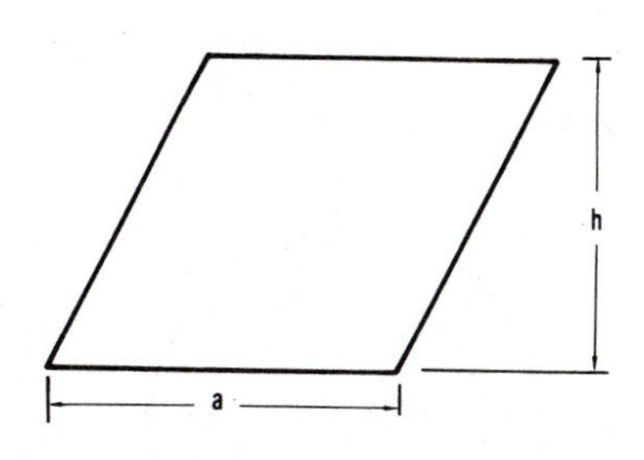

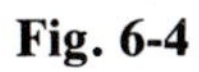
Fig. 6-4

Trapezoid (Fig. 6-5):

$$\text{area } (A) = \frac{h}{2}(a + b)$$

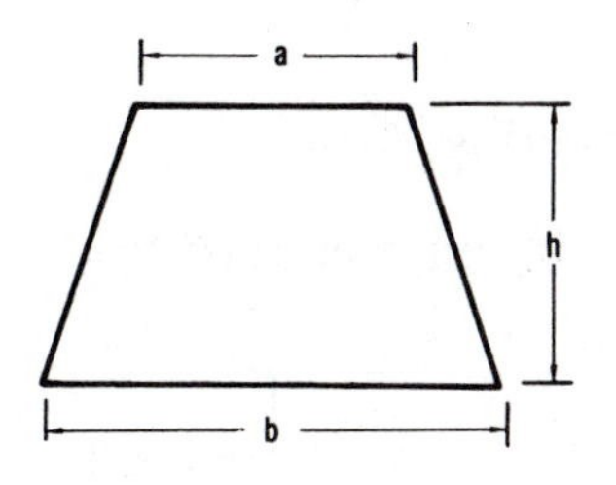

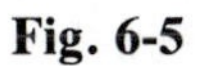
Fig. 6-5

Trapezium (Fig. 6-6):

$$\text{area } (A) = \frac{1}{2}[b(H + h) + ah + cH]$$

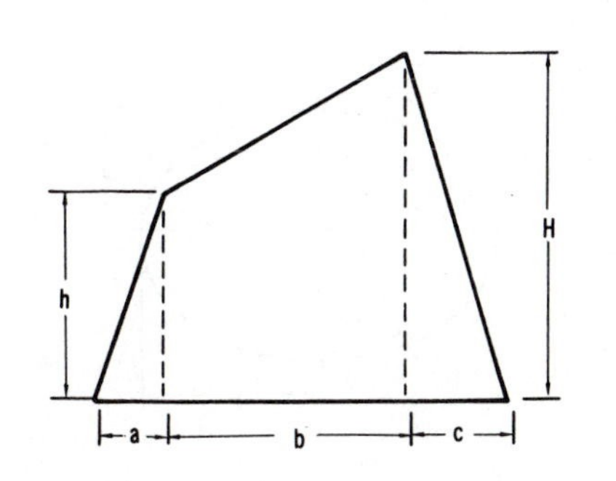

Fig. 6-6

Regular pentagon (Fig. 6-7):

$$\text{area } (A) = 1.720\, a^2$$

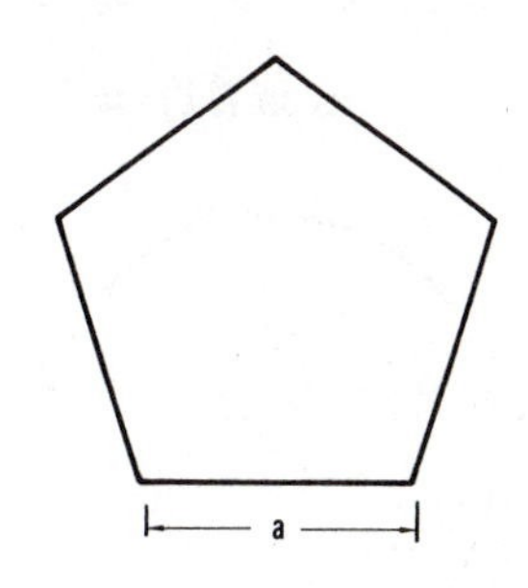

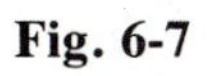
Fig. 6-7

Regular hexagon (Fig. 6-8):

$$\text{area } (A) = 2.598\, a^2$$

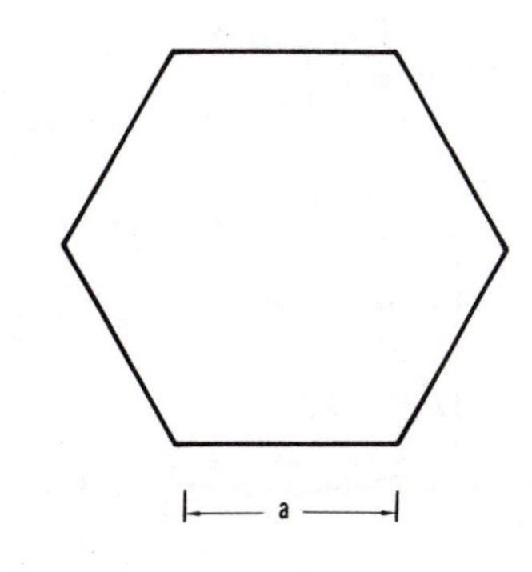

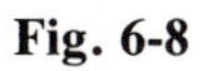
Fig. 6-8

Octagon (Fig. 6-9):

$$\text{area } (A) = 4.828\, a^2$$

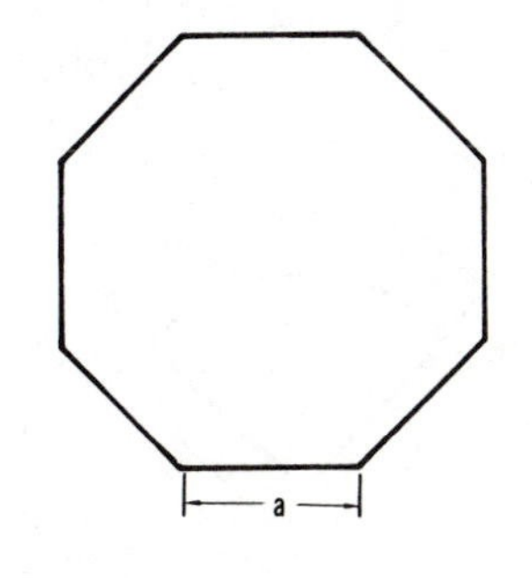

Fig. 6-9

Circle (Figs. 6-10 through 6-12):

$$\text{circumference } (C) = 2\pi R$$
$$= \pi D$$

$$\text{area } (A) = \pi R^2$$

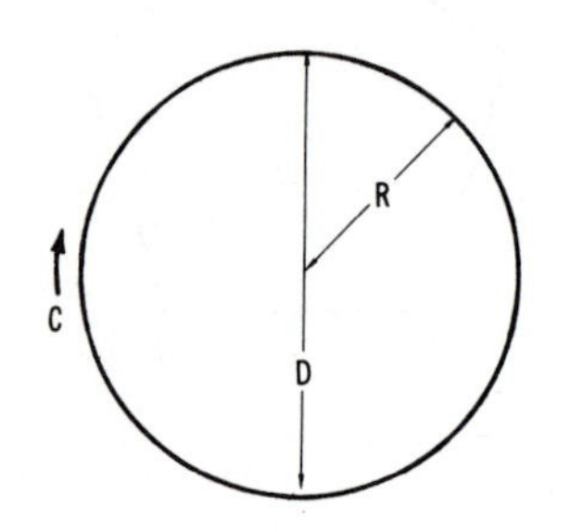

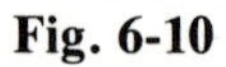

Fig. 6-10

$$\text{chord } (c) = \sqrt{4(2hR - h^2)}$$

$$\text{area } (A) = \pi R^2\left(\frac{\theta}{360}\right) - \left(\frac{c(R-h)}{2}\right)$$

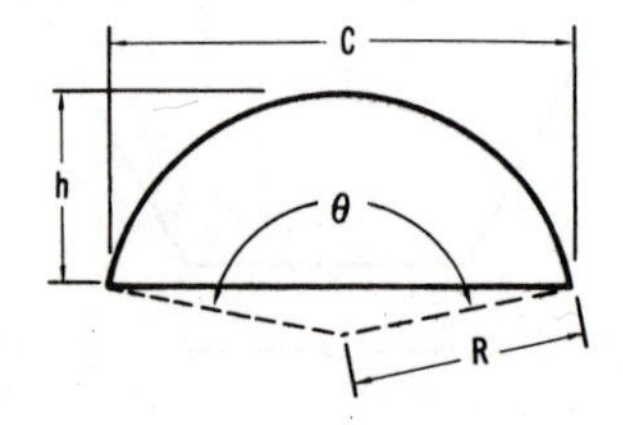

Fig. 6-11

$$\text{area } (A) = \frac{bR}{2}$$
$$= \pi R^2\left(\frac{\theta}{360}\right)$$

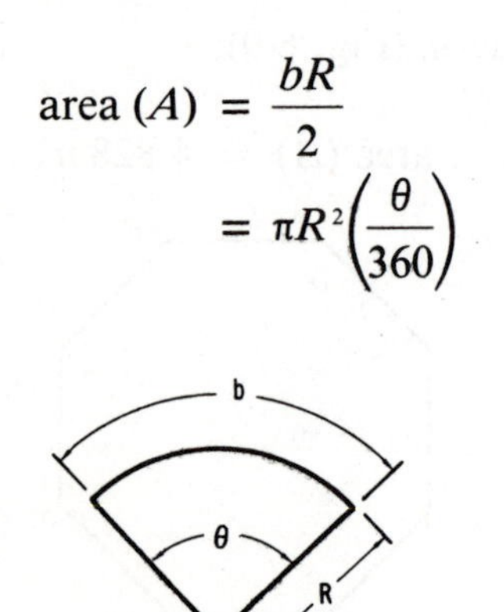

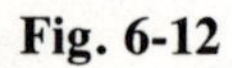

Fig. 6-12

Circular ring (Fig. 6-13):

$$\text{area } (A) = \pi(R^2 - r^2) = 0.7854\,(D^2 - d^2)$$

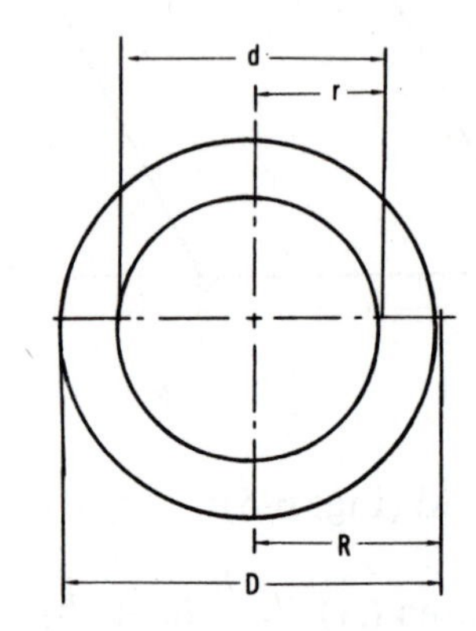

Fig. 6-13

Ellipse (Fig. 6-14):

$$\text{circumference } (C) =$$

$$\pi(a + b)\left[\frac{64 - 3\left(\dfrac{b - a^4}{b + a}\right)}{64 - 16\left(\dfrac{b - a^2}{b + a}\right)}\right]$$

$$\text{area } (A) = \pi ab$$

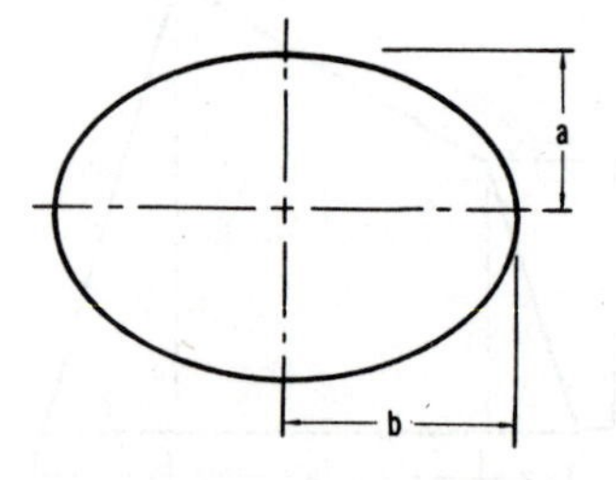

Fig. 6-14

Sphere (Fig. 6-15):

$$\text{area } (A) = 4R^2$$
$$= \pi D^2$$

$$\text{volume } (V) = \frac{4}{3}\pi R^3$$

$$= \frac{1}{6}\pi D^3$$

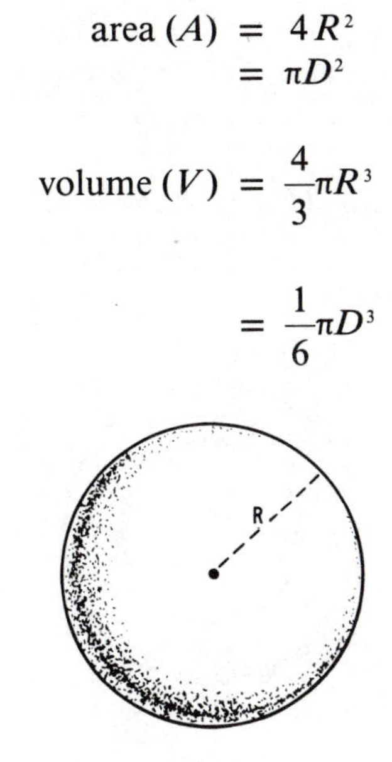

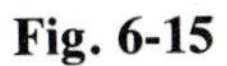
Fig. 6-15

Cube (Fig. 6-16):

$$\text{area } (A) = 6b^2$$

$$\text{volume } (V) = b^3$$

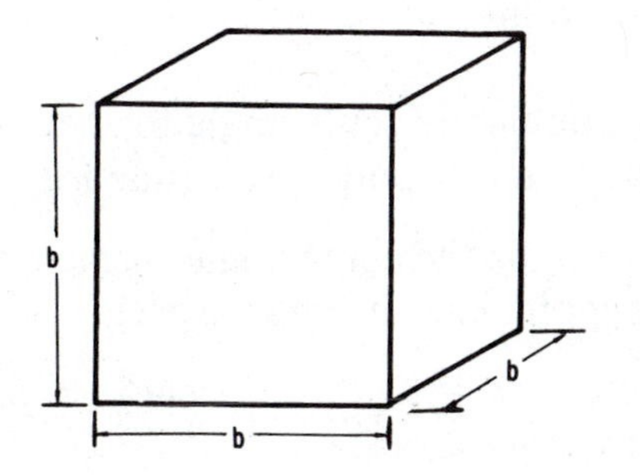

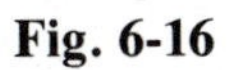
Fig. 6-16

Rectangular solid (Fig. 6-17):

$$\text{area } (A) = 2(ab + bc + ac)$$

$$\text{volume } (V) = abc$$

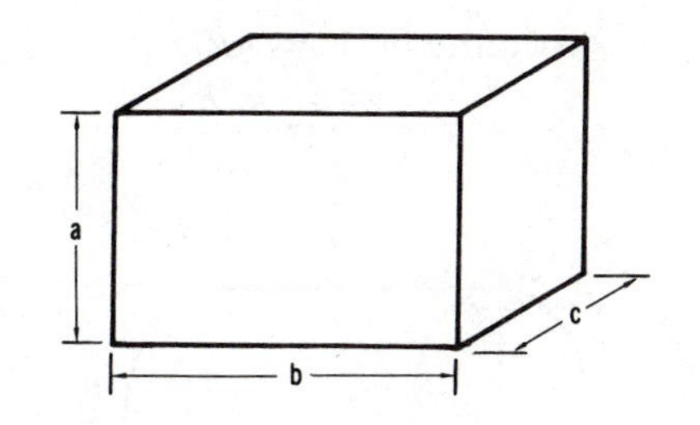

Fig. 6-17

Cone (Fig. 6-18):

$$\text{area } (A) = \pi RS$$
$$= \pi R\sqrt{R^2 + h^2}$$

$$\text{volume } (V) = \frac{\pi R^2 h}{3}$$
$$= 1.047R^2h$$
$$= 0.2618D^2h$$

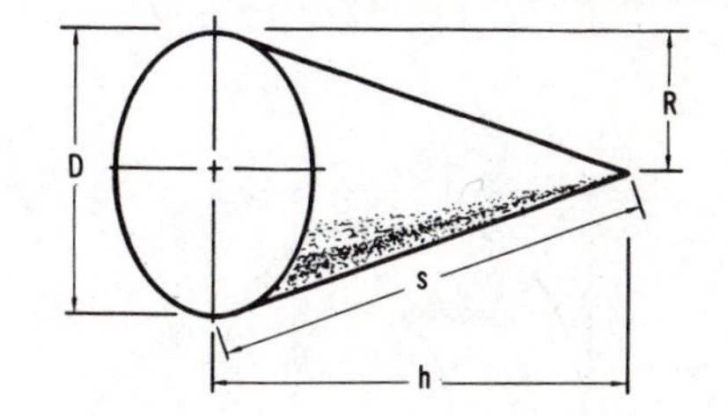

Fig. 6-18

Cylinder (Fig. 6-19):

$$\text{cylindrical surface} = \pi Dh$$

$$\text{total surface} = 2\pi R(R + h)$$

$$\text{volume } (V) = \pi R^2 h = \frac{c^2 h}{4\pi}$$

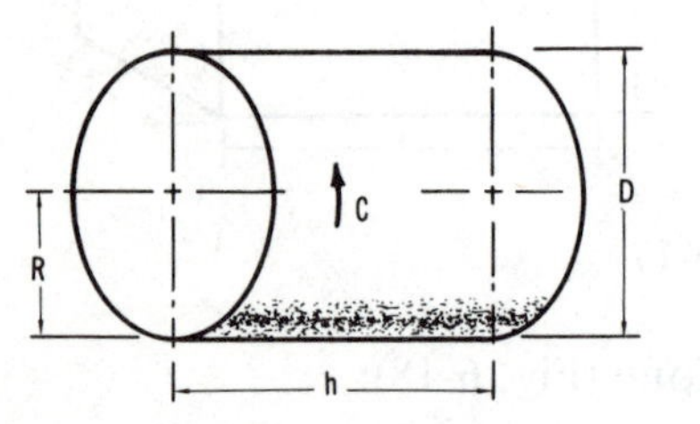

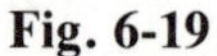
Fig. 6-19

Ring of rectangular cross section (Fig. 6-20):

$$\text{volume } (V) = \frac{\pi c}{4}(D^2 - d^2) = \left(\frac{D + d}{2}\right)\pi bc$$

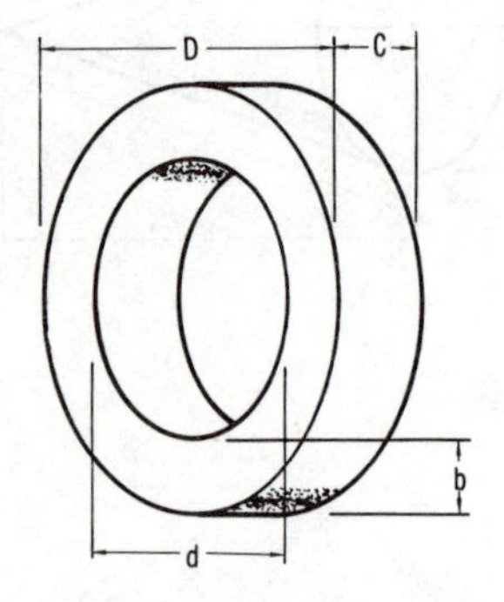

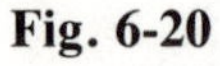
Fig. 6-20

Torus—ring of circular cross section (Fig. 6-21):

$$\text{total surface} = 4\pi^2 Rr = \pi^2 Dd$$

$$\text{volume } (V) = 2\pi R \times r^2 = 2.463D \times d^2$$

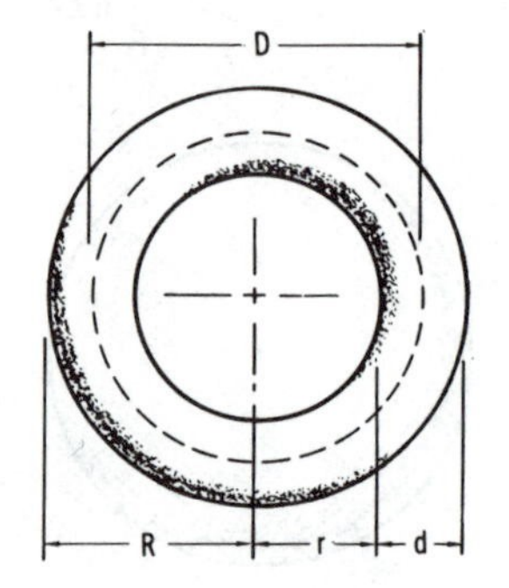

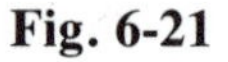
Fig. 6-21

TRIGONOMETRIC FUNCTIONS

Plane Trigonometry (Fig. 6-22)

In any right triangle, the values in Table 6-2 are valid if:

> a equals the acute angle formed by the hypotenuse and the altitude leg,
>
> b equals the acute angle formed by the hypotenuse and the base leg,
>
> A equals the side adjacent to $\angle b$ and opposite $\angle a$,
>
> B equals the side opposite $\angle b$ and adjacent to $\angle a$,
>
> C equals the hypotenuse.

TABLE 6-2
Trigonometric Formulas

Known values	Formulas for unknown values of				
	A	B	C	$\angle b$	$\angle a$
A & B	—	—	$\sqrt{A^2 + B^2}$	arc tan $\frac{B}{A}$	arc tan $\frac{A}{B}$
A & C	—	$\sqrt{C^2 - A^2}$	—	arc cos $\frac{A}{C}$	arc sin $\frac{A}{C}$
A & $\angle b$	—	$A \tan \angle b$	$\frac{A}{\cos \angle b}$	—	$90° - \angle b$
A & $\angle a$	—	$\frac{A}{\tan \angle a}$	$\frac{A}{\sin \angle a}$	$90° - \angle a$	—
B & C	$\sqrt{C^2 - B^2}$	—	—	arc sin $\frac{B}{C}$	arc cos $\frac{B}{C}$
B & $\angle b$	$\frac{B}{\tan \angle b}$	—	$\frac{B}{\sin \angle b}$	—	$90° - \angle b$
B & $\angle a$	$B \tan \angle a$	—	$\frac{B}{\cos \angle a}$	$90° - \angle a$	—
C & $\angle b$	$C \cos \angle b$	$C \sin \angle b$	—	—	$90° - \angle b$
C & $\angle a$	$C \sin \angle a$	$C \cos \angle a$	—	$90° - \angle a$	—

The expression "arc sin" or "$\sin^{-1}$" indicates "the angle whose sine is . . . ." Similarly, "arc tan" or "$\tan^{-1}$" indicates "the angle whose tangent is . . .".

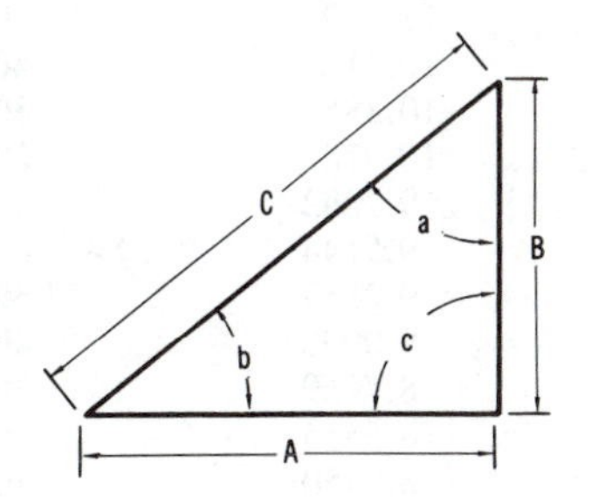

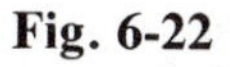
Fig. 6-22

Table of Trigonometric Functions

Table 6-3 gives the natural sines, cosines, tangents, and cotangents of angles. To find these values for angles from 0° to 45°, use the headings at the top of the table and the degree listings in the left-hand column. For angles from 45° to 90°, use the headings at the bottom of the table and the degree listings in the right-hand column.

Note. Read the degree listings in the right-hand column from bottom to top; thus, the 10′ listing directly above 89° signifies 89° 10′.

BINARY NUMBERS

Binary Digits

In the binary system of numbers, there are only two digits—0 and 1. All numbers

TABLE 6-3
Natural Trigonometric Functions

Degrees	sin	cos	tan	cot	
0° 00′	0.0000	1.0000	0.0000	∞	**90° 00′**
10	0.0029	1.0000	0.0029	343.77	**50**
20	0.0058	1.0000	0.0058	171.89	**40**
30	0.0087	1.0000	0.0087	114.59	**30**
40	0.0116	0.9999	0.0116	85.940	**20**
50	0.0145	0.9999	0.0145	68.750	**10**
1° 00′	0.0175	0.9998	0.0175	57.290	**89° 00′**
10	0.0204	0.9998	0.0204	49.104	**50**
20	0.0233	0.9997	0.0233	42.964	**40**
30	0.0262	0.9997	0.0262	38.188	**30**
40	0.0291	0.9996	0.0291	34.368	**20**
50	0.0320	0.9995	0.0320	31.242	**10**
2° 00′	0.0349	0.9994	0.0349	28.636	**88° 00′**
10	0.0378	0.9993	0.0378	26.432	**50**
20	0.0407	0.9992	0.0407	24.542	**40**
30	0.0436	0.9990	0.0437	22.904	**30**
40	0.0465	0.9989	0.0466	21.470	**20**
50	0.0494	0.9988	0.0495	20.206	**10**
3° 00′	0.0523	0.9986	0.0524	19.081	**87° 00′**
10	0.0552	0.9985	0.0553	18.075	**50**
20	0.0581	0.9983	0.0582	17.169	**40**
30	0.0610	0.9981	0.0612	16.350	**30**
40	0.0640	0.9980	0.0641	15.605	**20**
50	0.0669	0.9978	0.0670	14.924	**10**
4° 00′	0.0698	0.9976	0.0699	14.301	**86° 00′**
10	0.0727	0.9974	0.0729	13.727	**50**
20	0.0756	0.9971	0.0758	13.197	**40**
30	0.0785	0.9969	0.0787	12.706	**30**
40	0.0814	0.9967	0.0816	12.251	**20**
50	0.0843	0.9964	0.0846	11.826	**10**
5° 00′	0.0872	0.9962	0.0875	11.430	**85° 00′**
10	0.0901	0.9959	0.0904	11.059	**50**
20	0.0929	0.9957	0.0934	10.712	**40**
30	0.0958	0.9954	0.0963	10.385	**30**
40	0.0987	0.9951	0.0992	10.078	**20**
50	0.1016	0.9948	0.1022	9.7882	**10**
6° 00′	0.1045	0.9945	0.1051	9.5144	**84° 00′**
10	0.1074	0.9942	0.1080	9.2553	**50**
20	0.1103	0.9939	0.1110	9.0098	**40**
30	0.1132	0.9936	0.1139	8.7769	**30**
40	0.1161	0.9932	0.1169	8.5555	**20**
50	0.1190	0.9929	0.1198	8.3450	**10**
7° 00′	0.1219	0.9925	0.1228	8.1443	**83° 00′**
10	0.1248	0.9922	0.1257	7.9530	**50**
20	0.1276	0.9918	0.1287	7.7704	**40**
30	0.1305	0.9914	0.1317	7.5958	**30**
40	0.1334	0.9911	0.1346	7.4287	**20**
50	0.1363	0.9907	0.1376	7.2687	**10**
	cos	**sin**	**cot**	**tan**	**Degrees**

TABLE 6-3 Cont.
Natural Trigonometric Functions

Degrees	sin	cos	tan	cot	
8° 00′	0.1392	0.9903	0.1405	7.1154	82° 00′
10	0.1421	0.9899	0.1435	6.9682	50
20	0.1449	0.9894	0.1465	6.8269	40
30	0.1478	0.9890	0.1495	6.6912	30
40	0.1507	0.9886	0.1524	6.5606	20
50	0.1536	0.9881	0.1554	6.4348	10
9° 00′	0.1564	0.9877	0.1584	6.3138	81° 00′
10	0.1593	0.9872	0.1614	6.1970	50
20	0.1622	0.9868	0.1644	6.0844	40
30	0.1650	0.9863	0.1673	5.9758	30
40	0.1679	0.9858	0.1703	5.8708	20
50	0.1708	0.9853	0.1733	5.7694	10
10° 00′	0.1736	0.9848	0.1763	5.6713	80° 00′
10	0.1765	0.9843	0.1793	5.5764	50
20	0.1794	0.9838	0.1823	5.4845	40
30	0.1822	0.9833	0.1853	5.3955	30
40	0.1851	0.9827	0.1883	5.3093	20
50	0.1880	0.9822	0.1914	5.2257	10
11° 00′	0.1908	0.9816	0.1944	5.1446	79° 00′
10	0.1937	0.9811	0.1974	5.0658	50
20	0.1965	0.9805	0.2004	4.9894	40
30	0.1994	0.9799	0.2035	4.9152	30
40	0.2022	0.9793	0.2065	4.8430	20
50	0.2051	0.9787	0.2095	4.7729	10
12° 00′	0.2079	0.9781	0.2126	4.7046	78° 00′
10	0.2108	0.9775	0.2156	4.6382	50
20	0.2136	0.9769	0.2186	4.5736	40
30	0.2164	0.9763	0.2217	4.5107	30
40	0.2193	0.9757	0.2247	4.4494	20
50	0.2221	0.9750	0.2278	4.3897	10
13° 00′	0.2250	0.9744	0.2309	4.3315	77° 00′
10	0.2278	0.9737	0.2339	4.2747	50
20	0.2306	0.9730	0.2370	4.2193	40
30	0.2334	0.9724	0.2401	4.1653	30
40	0.2363	0.9717	0.2432	4.1126	20
50	0.2391	0.9710	0.2462	4.0611	10
14° 00′	0.2419	0.9703	0.2493	4.0108	76° 00′
10	0.2447	0.9696	0.2524	3.9617	50
20	0.2476	0.9689	0.2555	3.9136	40
30	0.2504	0.9681	0.2586	3.8667	30
40	0.2532	0.9674	0.2617	3.8208	20
50	0.2560	0.9667	0.2648	3.7760	10
15° 00′	0.2588	0.9659	0.2679	3.7321	75° 00′
10	0.2616	0.9652	0.2711	3.6891	50
20	0.2644	0.9644	0.2742	3.6470	40
30	0.2672	0.9636	0.2773	3.6059	30
40	0.2700	0.9628	0.2805	3.5656	20
50	0.2728	0.9621	0.2836	3.5261	10
	cos	sin	cot	tan	Degrees

TABLE 6-3 Cont.
Natural Trigonometric Functions

Degrees	sin	cos	tan	cot	
16° 00′	0.2756	0.9613	0.2867	3.4874	74° 00′
10	0.2784	0.9605	0.2899	3.4495	50
20	0.2812	0.9596	0.2931	3.4124	40
30	0.2840	0.9588	0.2962	3.3759	30
40	0.2868	0.9580	0.2994	3.3402	20
50	0.2896	0.9572	0.3026	3.3052	10
17° 00′	0.2924	0.9563	0.3057	3.2709	73° 00′
10	0.2952	0.9555	0.3089	3.2371	50
20	0.2979	0.9546	0.3121	3.2041	40
30	0.3007	0.9537	0.3153	3.1716	30
40	0.3035	0.9528	0.3185	3.1397	20
50	0.3062	0.9520	0.3217	3.1084	10
18° 00′	0.3090	0.9511	0.3249	3.0777	72° 00′
10	0.3118	0.9502	0.3281	3.0475	50
20	0.3145	0.9492	0.3314	3.0178	40
30	0.3173	0.9483	0.3346	2.9887	30
40	0.3201	0.9474	0.3378	2.9600	20
50	0.3228	0.9465	0.3411	2.9319	10
19° 00′	0.3256	0.9455	0.3443	2.9042	71° 00′
10	0.3283	0.9446	0.3476	2.8770	50
20	0.3311	0.9436	0.3508	2.8502	40
30	0.3338	0.9426	0.3541	2.8239	30
40	0.3365	0.9417	0.3574	2.7980	20
50	0.3393	0.9407	0.3607	2.7725	10
20° 00′	0.3420	0.9397	0.3640	2.7475	70° 00′
10	0.3448	0.9387	0.3673	2.7228	50
20	0.3475	0.9377	0.3706	2.6985	40
30	0.3502	0.9367	0.3739	2.6746	30
40	0.3529	0.9356	0.3772	2.6511	20
50	0.3557	0.9346	0.3805	2.6279	10
21° 00′	0.3584	0.9336	0.3839	2.6051	69° 00′
10	0.3611	0.9325	0.3872	2.5826	50
20	0.3638	0.9315	0.3906	2.5605	40
30	0.3665	0.9304	0.3939	2.5386	30
40	0.3692	0.9293	0.3973	2.5172	20
50	0.3719	0.9283	0.4006	2.4960	10
22° 00′	0.3746	0.9272	0.4040	2.4751	68° 00′
10	0.3773	0.9261	0.4074	2.4545	50
20	0.3800	0.9250	0.4108	2.4342	40
30	0.3827	0.9239	0.4142	2.4142	30
40	0.3854	0.9228	0.4176	2.3945	20
50	0.3881	0.9216	0.4210	2.3750	10
23° 00′	0.3907	0.9205	0.4245	2.3559	67° 00′
10	0.3934	0.9194	0.4279	2.3369	50
20	0.3961	0.9182	0.4314	2.3183	40
30	0.3987	0.9171	0.4348	2.2998	30
40	0.4014	0.9159	0.4383	2.2817	20
50	0.4041	0.9147	0.4417	2.2637	10
	cos	sin	cot	tan	Degrees

TABLE 6-3 Cont.
Natural Trigonometric Functions

Degrees	sin	cos	tan	cot	
24° 00′	0.4067	0.9135	0.4452	2.2460	**66° 00′**
10	0.4094	0.9124	0.4487	2.2286	**50**
20	0.4120	0.9112	0.4522	2.2113	**40**
30	0.4147	0.9100	0.4557	2.1943	**30**
40	0.4173	0.9088	0.4592	2.1775	**20**
50	0.4200	0.9075	0.4628	2.1609	**10**
25° 00′	0.4226	0.9063	0.4663	2.1445	**65° 00′**
10	0.4253	0.9051	0.4699	2.1283	**50**
20	0.4279	0.9038	0.4734	2.1123	**40**
30	0.4305	0.9026	0.4770	2.0965	**30**
40	0.4331	0.9013	0.4806	2.0809	**20**
50	0.4358	0.9001	0.4841	2.0655	**10**
26° 00′	0.4384	0.8988	0.4877	2.0503	**64° 00**
10	0.4410	0.8975	0.4913	2.0353	**50**
20	0.4436	0.8962	0.4950	2.0204	**40**
30	0.4462	0.8949	0.4986	2.0057	**30**
40	0.4488	0.8936	0.5022	1.9912	**20**
50	0.4514	0.8923	0.5059	1.9768	**10**
27° 00′	0.4540	0.8910	0.5095	1.9626	**63° 00′**
10	0.4566	0.8897	0.5132	1.9486	**50**
20	0.4592	0.8884	0.5169	1.9347	**40**
30	0.4617	0.8870	0.5206	1.9210	**30**
40	0.4643	0.8857	0.5243	1.9074	**20**
50	0.4669	0.8843	0.5280	1.8940	**10**
28° 00′	0.4695	0.8829	0.5317	1.8807	**62° 00′**
10	0.4720	0.8816	0.5354	1.8676	**50**
20	0.4746	0.8802	0.5392	1.8546	**40**
30	0.4772	0.8788	0.5430	1.8418	**30**
40	0.4797	0.8774	0.5467	1.8291	**20**
50	0.4823	0.8760	0.5505	1.8165	**10**
29° 00′	0.4848	0.8746	0.5543	1.8040	**61° 00′**
10	0.4874	0.8732	0.5581	1.7917	**50**
20	0.4899	0.8718	0.5619	1.7796	**40**
30	0.4924	0.8704	0.5658	1.7675	**30**
40	0.4950	0.8689	0.5696	1.7556	**20**
50	0.4975	0.8675	0.5735	1.7437	**10**
30° 00′	0.5000	0.8660	0.5774	1.7321	**60° 00′**
10	0.5025	0.8646	0.5812	1.7205	**50**
20	0.5050	0.8631	0.5851	1.7090	**40**
30	0.5075	0.8616	0.5890	1.6977	**30**
40	0.5100	0.8601	0.5930	1.6864	**20**
50	0.5125	0.8587	0.5969	1.6753	**10**
31° 00′	0.5150	0.8572	0.6009	1.6643	**59° 00′**
10	0.5175	0.8557	0.6048	1.6534	**50**
20	0.5200	0.8542	0.6088	1.6426	**40**
30	0.5225	0.8526	0.6128	1.6319	**30**
40	0.5250	0.8511	0.6168	1.6212	**20**
50	0.5275	0.8496	0.6208	1.6107	**10**
	cos	**sin**	**cot**	**tan**	**Degrees**

TABLE 6-3 Cont.
Natural Trigonometric Functions

Degrees	sin	cos	tan	cot	
32° 00′	0.5299	0.8480	0.6249	1.6003	**58° 00′**
10	0.5324	0.8465	0.6289	1.5900	**50**
20	0.5348	0.8450	0.6330	1.5798	**40**
30	0.5373	0.8434	0.6371	1.5697	**30**
40	0.5398	0.8418	0.6412	1.5597	**20**
50	0.5422	0.8403	0.6453	1.5497	**10**
33° 00′	0.5446	0.8387	0.6494	1.5399	**57° 00′**
10	0.5471	0.8371	0.6536	1.5301	**50**
20	0.5495	0.8355	0.6577	1.5204	**40**
30	0.5519	0.8339	0.6619	1.5108	**30**
40	0.5544	0.8323	0.6661	1.5013	**20**
50	0.5568	0.8307	0.6703	1.4919	**10**
34° 00′	0.5592	0.8290	0.6745	1.4826	**56° 00′**
10	0.5616	0.8274	0.6787	1.4733	**50**
20	0.5640	0.8258	0.6830	1.4641	**40**
30	0.5664	0.8241	0.6873	1.4550	**30**
40	0.5688	0.8225	0.6916	1.4460	**20**
50	0.5712	0.8208	0.6959	1.4370	**10**
35° 00′	0.5736	0.8192	0.7002	1.4281	**55° 00′**
10	0.5760	0.8175	0.7046	1.4193	**50**
20	0.5783	0.8158	0.7089	1.4106	**40**
30	0.5807	0.8141	0.7133	1.4019	**30**
40	0.5831	0.8124	0.7177	1.3934	**20**
50	0.5854	0.8107	0.7221	1.3848	**10**
36° 00′	0.5878	0.8090	0.7265	1.3764	**54° 00′**
10	0.5901	0.8073	0.7310	1.3680	**50**
20	0.5925	0.8056	0.7355	1.3597	**40**
30	0.5948	0.8039	0.7400	1.3514	**30**
40	0.5972	0.8021	0.7445	1.3432	**20**
50	0.5995	0.8004	0.7490	1.3351	**10**
37° 00′	0.6018	0.7986	0.7536	1.3270	**53° 00′**
10	0.6041	0.7969	0.7581	1.3190	**50**
20	0.6065	0.7951	0.7627	1.3111	**40**
30	0.6088	0.7934	0.7673	1.3032	**30**
40	0.6111	0.7916	0.7720	1.2954	**20**
50	0.6134	0.7898	0.7766	1.2876	**10**
38° 00′	0.6157	0.7880	0.7813	1.2799	**52° 00′**
10	0.6180	0.7862	0.7860	1.2723	**50**
20	0.6202	0.7844	0.7907	1.2647	**40**
30	0.6225	0.7826	0.7954	1.2572	**30**
40	0.6248	0.7808	0.8002	1.2497	**20**
50	0.6271	0.7790	0.8050	1.2423	**10**
39° 00′	0.6293	0.7771	0.8098	1.2349	**51° 00′**
10	0.6316	0.7753	0.8146	1.2276	**50**
20	0.6338	0.7735	0.8195	1.2203	**40**
30	0.6361	0.7716	0.8243	1.2131	**30**
40	0.6383	0.7698	0.8292	1.2059	**20**
50	0.6406	0.7679	0.8342	1.1988	**10**
	cos	**sin**	**cot**	**tan**	**Degrees**

TABLE 6-3 Cont.
Natural Trigonometric Functions

Degrees	sin	cos	tan	cot	
40° 00′	0.6428	0.7660	0.8391	1.1918	50° 00′
10	0.6450	0.7642	0.8441	1.1847	50
20	0.6472	0.7623	0.8491	1.1778	40
30	0.6494	0.7604	0.8541	1.1708	30
40	0.6517	0.7585	0.8591	1.1640	20
50	0.6539	0.7566	0.8642	1.1571	10
41° 00′	0.6561	0.7547	0.8693	1.1504	49° 00′
10	0.6583	0.7528	0.8744	1.1436	50
20	0.6604	0.7509	0.8796	1.1369	40
30	0.6626	0.7490	0.8847	1.1303	30
40	0.6648	0.7470	0.8899	1.1237	20
50	0.6670	0.7451	0.8952	1.1171	10
42° 00′	0.6691	0.7431	0.9004	1.1106	48° 00′
10	0.6713	0.7412	0.9057	1.1041	50
20	0.6734	0.7392	0.9110	1.0977	40
30	0.6756	0.7373	0.9163	1.0913	30
40	0.6777	0.7353	0.9217	1.0850	20
50	0.6799	0.7333	0.9271	1.0786	10
43° 00′	0.6820	0.7314	0.9325	1.0724	47° 00′
10	0.6841	0.7294	0.9380	1.0661	50
20	0.6862	0.7274	0.9435	1.0599	40
30	0.6884	0.7254	0.9490	1.0538	30
40	0.6905	0.7234	0.9545	1.0477	20
50	0.6926	0.7214	0.9601	1.0416	10
44° 00′	0.6947	0.7193	0.9657	1.0355	46° 00′
10	0.6967	0.7173	0.9713	1.0295	50
20	0.6988	0.7163	0.9770	1.0235	40
30	0.7009	0.7133	0.9827	1.0176	30
40	0.7030	0.7112	0.9884	1.0117	20
50	0.7050	0.7092	0.9942	1.0058	10
45° 00′	0.7071	0.7071	1.0000	1.0000	45° 00′
	cos	sin	cot	tan	Degrees

are written as successive powers of 2. Actually, in the decimal system, all numbers are written as successive powers of 10.

Example. Decimal 3487 is actually:

$$\begin{aligned} 3 \times 10^3 &= 3000 \\ 4 \times 10^2 &= 400 \\ 8 \times 10^1 &= 80 \\ 7 \times 10^0 &= \underline{\quad 7} \\ & \quad 3487 \end{aligned}$$

With binary numbers, a like system is used except the base (radix) is 2 instead of 10. For example, the binary numbers corresponding to decimal numbers 0 through 10 are 0, 1, 10, 11, 100, 101, 110, 111, 1000, 1001, 1010. Each number is written as a succession of powers of 2.

Example. Binary 1010 actually means:

$$\begin{aligned} 1 \times 2^3 &= 8 \\ + 1 \times 2^1 &= \underline{2} \\ & \quad 10 \end{aligned}$$

The powers of 2, from 0 to 20, are given in Table 6-4. Thus, to write a number above decimal 1,048,056 using binary numbers requires a minimum of 21 digits!

TABLE 6-4
Powers of 2

Power	Decimal	Power	Decimal	Power	Decimal
2^0	1	2^7	128	2^{14}	16,384
2^1	2	2^8	256	2^{15}	32,768
2^2	4	2^9	512	2^{16}	65,536
2^3	8	2^{10}	1024	2^{17}	131,072
2^4	16	2^{11}	2048	2^{18}	262,144
2^5	32	2^{12}	4096	2^{19}	524,288
2^6	64	2^{13}	8192	2^{20}	1,048,576

Binary numbers are also arranged into widely used codes such as the Excess-3 and Gray Codes shown in Tables 6-5 and 6-6, respectively. (See also the discussion of the ASCII Code in Chapter 7.)

TABLE 6-5
Excess-3 Code

Decimal	Binary code
0	0011
1	0100
2	0101
3	0110
4	0111
5	1000
6	1001
7	1010
8	1011
9	1100

Conversion

To convert from binary to decimal or from decimal to binary, you could use Table 6-4 and compute the equivalent in the other numbering system. However, there are simpler methods. To convert from decimal to binary, successively divide the decimal number by 2. Write down a 1 if there is a remainder and a 0 if not, until the division gives a 0.

TABLE 6-6
Gray Code

Decimal	Gray code $a_3a_2a_1a_0$
0	0000
1	0001
2	0011
3	0010
4	0110
5	0111
6	0101
7	0100
8	1100
9	1101
10	1111
11	1110
12	1010
13	1011
14	1001
15	1000

Example. To covert decimal 22 to binary.

2)22
2)11 R = 0
2)5 R = 1
2)2 R = 1
2)1 R = 0
0 R = 1

The least significant figure is at the top; thus, the binary number corresponding to decimal 22 is 10110.

To convert from binary to decimal, take the first binary digit, double it, and add your answer to the second digit. Write this sum under the second digit. Then double this number, add it to the third digit, and write the sum under the third digit. Continue this process up to and including the last digit, as follows:

```
1   0   1   1   0   1
    2   5   11  22  45
```

The number under the last digit (45) is the decimal equivalent of binary 101101.

Addition

Binary addition has only four rules:

```
0  0  1   1
0  0  1   1
-  -  -  --
0  1  1  10
```

Following these rules, any binary number can be added.

Example.

```
 1011
  110
-----
10001
```

To simplify the carry when $1 + 1 = 10$, place the carry under the next digit. Then add the partial total and the carries, as follows:

```
 111101
  10110
-------
 101011
 1 1
-------
1010011
```

Subtraction

Binary numbers can be subtracted directly, as follows:

```
 1111
 -111
-----
 1000
```

However, it is simpler to complement the subtracted number and add. In the binary system, a number is complemented by merely changing all 0's to 1's and all 1's to 0's and adding 1 to the final digit.

Example.

```
 1111                   1111
-0111   complemented   +1001
-----                  -----
                       11000
```

Answer. The first digit in the answer is disregarded. Hence, the answer is 1000 (decimal 8), the same as before.

Multiplication

Binary multiplication is similar to decimal multiplication. All products are the same as in decimal multiplication. That is:

$$0 \times 0 = 0$$

$$1 \times 0 = 0$$

$$1 \times 1 = 1$$

Example. To multiply 1011 by 101:

```
  1011
  0101
------
  1011
 0000
1011
------
110111
```

Division

Binary division is similar to decimal division.

Example. To divide 1101001 by 101:

```
       10101
101)1101001
    101
    ---
     110
     101
     ---
       101
       101
       ---
```

Handling Negative Remainders

In the preceding example of binary division, there was no 0 generated in the quotient. That is, there were no remainders smaller than the divisor. The following example shows what steps must be taken when a 0 is generated.

Example. Dividing 45 by 9:

```
         101
1001)101101
      1001
      ----
        1001
        1001
        ----
```

Bringing down the next digit resulted in a remainder smaller than the divisor. Therefore a 0 was placed in the quotient. When the next digit was transferred down, the remainder was larger than the quotient.

The successive-subtraction method used by a computer follows:

		Quotient
	101101	
Subtract	100100	
	001001	1
Shift	010010	
	100100	
	101110	0
Restore-add	100100	
	010010	
	100100	
Shift	100100	1

After the first subtraction, the number is positive. This positive remainder is shifted left but is still smaller than the divisor. Subtracting now would result in a negative number. A computer has no way of detecting that the remainder is larger than the divisor until the subtraction operation is performed. One indication is that the highest order will require a borrow, or what is termed an overdraw.

When a subtraction causes a negative number, the computer must restore the remainder to the original value before the next dividend is used. This is done by adding the divisor to the remainder prior to the next shift operation.

Binary Coded Decimal

Various codes based on the binary 1 and 0 concept have evolved to meet the needs of digital equipment operation. The binary coded decimal (BCD) is widely used. Here, four bits are used to represent each digit of the decimal number. Each digit position has a definite value or weight in the order 8, 4, 2, 1 and each four digit combination in the BCD represents one digit of a decimal number as follows:

Decimal	BCD
0	0000
1	0001
2	0010
3	0011
4	0100
5	0101
6	0110
7	0111
8	1000
9	1001

For larger decimal numbers, where two or more decimal digits are needed, additional four-bit BCD combinations are used. For example, decimal 25 in BCD is 0010 0101 and decimal 372 in BCD is 0011 0111 0010.

OTHER NUMBER SYSTEMS

Octal is a numbering system with a base of 8. Since 8 is a power of 2, conversion between the octal and binary systems is an easy operation. Thus, the digits 0 through 7 are used as follows:

Decimal	Octal equivalent
0	0
1	1
2	2
3	3
4	4
5	5
6	6
7	7
8	10

Thus, 21 in octal is the equivalent of decimal 17 and binary 10001.

Hexadecimal numbers (often abbreviated hex) also find wide usage. The hexadecimal system has 16 as its base. The conventional numbers are used from 0 through 9 and the letters A through F for decimal 10 through 15. Thus:

Decimal	Hexadecimal equivalent
0	0
1	1
2	2
3	3
4	4
5	5
6	6
7	7
8	8
9	9
10	A
11	B
12	C
13	D
14	E
15	F
16	10

Thus, 21 in hexadecimal is the equivalent of decimal 33 and binary 100001. Likewise 1A in hexadecimal is the equivalent of decimal 26.

FUNDAMENTALS OF BOOLEAN ALGEBRA

Boolean algebra is based on symbolic logic, which requires that a statement be either true or false—it can be nothing else. The symbols A, B, and C are used to designate various conditions, which may be characterized by statements. Two logical connectives—AND and OR—express relationships between two statements.

Two or more statements connected by the word "or" are considered to form a single true statement if at least one of the original statements is true. Similarly, if two parallel switches are connected in a circuit, the circuit is considered to be closed if at least one of the switches is closed. Thus, OR is the logical equivalent of a parallel switch circuit. It is symbolized by a plus sign.

Two or more statements connected by the word "and" are considered to form a single true statement if all of the original statements are true. Similarly, if two series switches are connected in a circuit, the circuit is considered to be closed if all of the switches are closed. Thus, AND is the logical equivalent of a series switch circuit. AND is symbolized by a multiplication sign (·) or no sign at all.

Given any statement, the opposite or contradiction of that statement can be formed. The contradiction of any statement A is called the negation of A. If A is true, then the negation of A is false, and vice versa. Similarly, if a switch has two positions, then the open position may be considered the opposite or negation of the closed

TABLE 6-7
Basic Rules of Symbolic Logic

Symbol	Logic	Switch	Meaning	Circuit
1	true	closed	The statement is true. The circuit is closed.	
0	false	open	The statement is false. The circuit is open.	
·	series	*A* and *B*	*A* is in series with *B*.	
+	parallel	*A* or *B*	*A* is in parallel with *B*.	
$\bar{A}$ or A'	not *A*		*Opposite of A* (if $A = 0$, $\bar{A} = 1$; if $A = 1$, $\bar{A} = 0$).	

position. Negation is indicated by a superior bar (¯) or prime (′).

A true statement, and hence a closed circuit, is generally said to have a truth value of 1. Conversely, a false statement, and hence an open circuit, is generally said to have a truth value of 0. Applying the AND and OR (· and +) relations to the truth values (0 and 1) yields the multiplication and addition tables of binary arithmetic.

The various symbols are given in Table 6-7. Table 6-8 summarizes the various logical statements, explains their meanings, and shows the equivalent switch circuits for the statements.

A further explanation of these Boolean algebra concepts is given by Figs. 6-23 through 6-26, which shown AND, OR, NAND, and NOR gates. At *A* in each figure, the symbol for the gate is given; at *B*, the gate is represented by an appropriate electrical circuit; and a truth table for each circuit is given at *C*.

The AND gate is shown as having two inputs *A* and *B* and an output *C*. The gate is not limited to two inputs; any number could be used. Regardless of the number, they would all be shown in series in the circuit of *B* (Fig. 6-23).

In the AND truth tables, a 0 represents a false statement, or an open switch in the circuit. A 1 represents a true statement, or a closed switch. The truth table for the AND

TABLE 6-8
Summary of Logical Statements

Logic	Meaning	Circuit
$0 \cdot 0 = 0$	An open in series with an open is open.	
$0 \cdot 1 = 0$	An open in series with a closed is open.	
$1 \cdot 1 = 1$	A closed in series with a closed is closed.	
$0 + 0 = 0$	An open in parallel with an open is open.	
$0 + 1 = 1$	An open in parallel with a closed is closed.	
$1 + 1 = 1$	A closed in parallel with a closed is closed.	

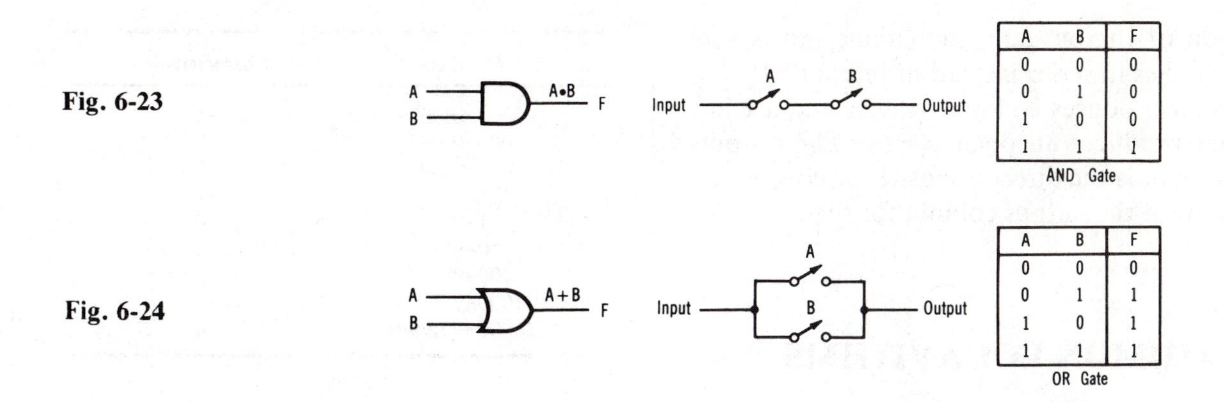

Fig. 6-23

A	B	F
0	0	0
0	1	0
1	0	0
1	1	1

AND Gate

Fig. 6-24

A	B	F
0	0	0
0	1	1
1	0	1
1	1	1

OR Gate

gate shows, then, that an output is obtained in only one case: when both *A* and *B* are 1. If either *A* or *B*, or both, are 0, then the output is 0, as shown by the first three lines of the truth table.

The OR gate (Fig. 6-24) is represented by switches in parallel. The circuit shows that an output is obtained when one or the other, or both, switches are closed. These examples are represented by the last three lines in the truth table. If both switches are open (first line of the table), no output is obtained.

In logic circuits, a statement can be negated or contradicted by any device that inverts the input by 180°. This means that an input that would normally produce an output produces no output, and vice versa. In the gate circuits, this would be represented by a closed switch in place of an open one, or by an open switch in place of a closed one. The negation of an AND gate is a NAND gate (short for NOT AND). A NAND gate is represented in Fig. 6-25. Notice in the circuit that switches are now shown in parallel, instead of in series as they were for the AND gate. Also, do not forget that a 0 in the truth table now produces a closed switch and a 1 produces an open switch.

The output column for the AND gate is the direct opposite or contradiction of the output column for the NAND gate, and vice versa. The input that produced an output with the AND gate now produces no output with the NAND gate, and, conversely, the inputs that produced no output with the AND gate now produce outputs with the NAND gate.

Figure 6-26 shows a NOR gate, a circuit, and a truth table. Since this is a contradic-

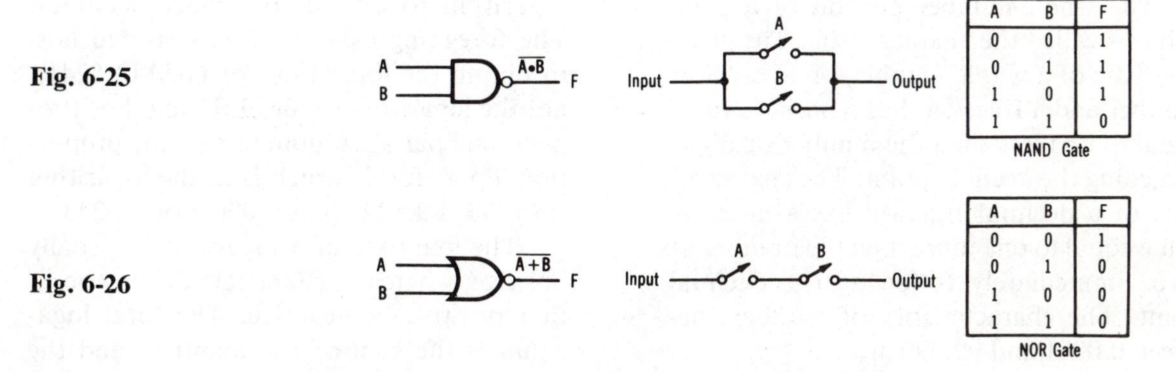

Fig. 6-25

A	B	F
0	0	1
0	1	1
1	0	1
1	1	0

NAND Gate

Fig. 6-26

A	B	F
0	0	1
0	1	0
1	0	0
1	1	0

NOR Gate

tion of the OR gate, the circuit shows the switches in series instead of in parallel. A 0 input produces a closed switch, and a 1 input produces an open switch. The output column is the direct opposite, or contradiction, of the output column for the OR gate.

COMMON LOGARITHMS

The logarithm of a quantity is the power to which a given number (base) must be raised in order to equal that quantity. Thus, any number may be used as the base. The most common system is the base 10. Logarithms with the base 10 are known as common, or Briggs, logarithms; they are written $\log_{10}$, or simply log. When the base is omitted, the base 10 is understood.

A common logarithm of a given number is the number which, when applied to the number 10 as an exponent, will produce the given number. Thus, 2 is the common logarithm of 100, since 10^2 equals 100; 3 is the logarithm of 1000, since 10^3 equals 1000. From this we can see that the logarithm of any number except a whole number power of 10 consists of a whole number and a decimal fraction.

Characteristic of a Logarithm

The whole-number portion of a logarithm is called the characteristic. The characteristic of a whole number, or of a whole number and a fraction, has a positive value equal to one less than the number of digits preceding the decimal point. The characteristic of a decimal fraction has a negative value equal to one more than the number of zeros immediately following the decimal point. The characteristics of numbers between 0.0001 and 99,999 are:

Numbers	Characteristic
0.0001-0.0009	-4
0.001-0.009	-3
0.01-0.09	-2
0.1-0.9	-1
1-9	0
10-99	1
100-999	2
1000-9999	3
10,000-99,999	4

Use of Logarithm Table

The mantissa, or decimal-fracton portion, of a logarithm is obtained from Table 6-9. To find the mantissa for the logarithm of any number, locate the first two figures of the number in the left-hand column (*N*); then, in the column under the third figure of the number, the mantissa for that number will be found.

Example. To find the logarithm of 6673, first locate 66 in the left-hand column (*N*); then follow across to the column numbered 7. The mantissa for 667 (8241) is located at this point. The characteristic for the logarithm of 6673 is 3. Therefore, the logarithm of 6670 is 3.8241.

For most computations, greater accuracy will not be required. If accuracy to four places is desired, the columns labeled "Proportional parts" may be used. These columns list the numbers to be added to the logarithm to obtain four-place accuracy. The foregoing example demonstrated how to obtain the logarithm for 6670 (3.8241), not the logarithm for 6673. Using the "Proportional parts" column to find the proportional part for 3, which is 2, the logarithm for 6673 is 3.8241 plus 0.0002, or 3.8243.

The mantissa of a logarithm is usually positive, whereas a characteristic may be either positive or negative. The total logarithm is the sum of the mantissa and the

TABLE 6-9
Common Logarithms

N	0	1	2	3	4	5	6	7	8	9	Proportional parts								
											1	*2*	*3*	*4*	*5*	*6*	*7*	*8*	*9*
10	0000	0043	0086	0128	0170	0212	0253	0294	0334	0374	4	8	12	17	21	25	29	33	37
11	0414	0453	0492	0531	0569	0607	0645	0682	0719	0755	4	8	11	15	19	23	26	30	34
12	0792	0828	0864	0899	0934	0969	1004	1038	1072	1106	3	7	10	14	17	21	24	28	31
13	1139	1173	1206	1239	1271	1303	1335	1367	1399	1430	3	6	10	13	16	19	23	26	29
14	1461	1492	1523	1553	1584	1614	1644	1673	1703	1732	3	6	9	12	15	18	21	24	27
15	1761	1790	1818	1847	1875	1903	1931	1959	1987	2014	3	6	8	11	14	17	20	22	25
16	2041	2068	2095	2122	2148	2175	2201	2227	2253	2279	3	5	8	11	13	16	18	21	24
17	2304	2330	2355	2380	2405	2430	2455	2480	2504	2529	2	5	7	10	12	15	17	20	22
18	2553	2577	2601	2625	2648	2672	2695	2718	2742	2765	2	5	7	9	12	14	16	19	21
19	2788	2810	2833	2856	2878	2900	2923	2945	2967	2989	2	4	7	9	11	13	16	18	20
20	3010	3032	3054	3075	3096	3118	3139	3160	3181	3201	2	4	6	8	11	13	15	17	19
21	3222	3243	3263	3284	3304	3324	3345	3365	3385	3404	2	4	6	8	10	12	14	16	18
22	3424	3444	3464	3483	3502	3522	3541	3560	3579	3598	2	4	6	8	10	12	14	15	17
23	3617	3636	3655	3674	3692	3711	3729	3747	3766	3784	2	4	6	7	9	11	13	15	17
24	3802	3820	3838	3856	3874	3892	3909	3927	3945	3962	2	4	5	7	9	11	12	14	16
25	3979	3997	4014	4031	4048	4065	4082	4099	4116	4133	2	3	5	7	9	10	12	14	15
26	4150	4166	4183	4200	4216	4232	4249	4265	4281	4298	2	3	5	7	8	10	11	13	15
27	4314	4330	4346	4362	4378	4393	4409	4425	4440	4456	2	3	5	6	8	9	11	13	14
28	4472	4487	4502	4518	4533	4548	4564	4579	4594	4609	2	3	5	6	8	9	11	12	14
29	4624	4639	4654	4669	4683	4698	4713	4728	4742	4757	1	3	4	6	7	9	10	12	13
30	4771	4786	4800	4814	4829	4843	4857	4871	4886	4900	1	3	4	6	7	9	10	11	13
31	4914	4928	4942	4955	4969	4983	4997	5011	5024	5038	1	3	4	6	7	8	10	11	12
32	5051	5065	5079	5092	5105	5119	5132	5145	5159	5172	1	3	4	5	7	8	9	11	12
33	5185	5198	5211	5224	5237	5250	5263	5276	5289	5302	1	3	4	5	6	8	9	10	12
34	5315	5328	5340	5353	5366	5378	5391	5403	5416	5428	1	3	4	5	6	8	9	10	11
35	5441	5453	5465	5478	5490	5502	5514	5527	5539	5551	1	2	4	5	6	7	9	10	11
36	5563	5575	5587	5599	5611	5623	5635	5647	5658	5670	1	2	4	5	6	7	8	10	11
37	5682	5694	5705	5717	5729	5740	5752	5763	5775	5786	1	2	3	5	6	7	8	9	10
38	5798	5809	5821	5832	5843	5855	5866	5877	5888	5899	1	2	3	5	6	7	8	9	10
39	5911	5922	5933	5944	5955	5966	5977	5988	5999	6010	1	2	3	4	5	7	8	9	10
40	6021	6031	6042	6053	6064	6075	6085	6096	6107	6117	1	2	3	4	5	6	8	9	10
41	6128	6138	6149	6160	6170	6180	6191	6201	6212	6222	1	2	3	4	5	6	7	8	9
42	6232	6243	6253	6263	6274	6284	6294	6304	6314	6325	1	2	3	4	5	6	7	8	9
43	6335	6345	6355	6365	6375	6385	6395	6405	6415	6425	1	2	3	4	5	6	7	8	9
44	6435	6444	6454	6464	6474	6484	6493	6503	6513	6522	1	2	3	4	5	6	7	8	9
											1	*2*	*3*	*4*	*5*	*6*	*7*	*8*	*9*
N	0	1	2	3	4	5	6	7	8	9	Proportional parts								

TABLE 6-9 Cont.
Common Logarithms

N	0	1	2	3	4	5	6	7	8	9	Proportional parts								
											1	*2*	*3*	*4*	*5*	*6*	*7*	*8*	*9*
45	6532	6542	6551	6561	6571	6580	6590	6599	6609	6618	1	2	3	4	5	6	7	8	9
46	6628	6637	6646	6656	6665	6675	6684	6693	6702	6712	1	2	3	4	5	6	7	7	8
47	6721	6730	6739	6749	6758	6767	6776	6785	6794	6803	1	2	3	4	5	5	6	7	8
48	6812	6821	6830	6839	6848	6857	6866	6875	6884	6893	1	2	3	4	4	5	6	7	8
49	6902	6911	6920	6928	6937	6946	6955	6964	6972	6981	1	2	3	4	4	5	6	7	8
50	6990	6998	7007	7016	7024	7033	7042	7050	7059	7067	1	2	3	3	4	5	6	7	8
51	7076	7084	7093	7101	7110	7118	7126	7135	7143	7152	1	2	3	3	4	5	6	7	8
52	7160	7168	7177	7185	7193	7202	7210	7218	7226	7235	1	2	2	3	4	5	6	7	7
53	7243	7251	7259	7267	7275	7284	7292	7300	7308	7316	1	2	2	3	4	5	6	6	7
54	7324	7332	7340	7348	7356	7364	7372	7380	7388	7396	1	2	2	3	4	5	6	6	7
55	7404	7412	7419	7427	7435	7443	7451	7459	7466	7474	1	2	2	3	4	5	5	6	7
56	7482	7490	7497	7505	7513	7520	7528	7536	7543	7551	1	2	2	3	4	5	5	6	7
57	7559	7566	7574	7582	7589	7597	7604	7612	7619	7627	1	2	2	3	4	5	5	6	7
58	7634	7642	7649	7657	7664	7672	7679	7686	7694	7701	1	1	2	3	4	4	5	6	7
59	7709	7716	7723	7731	7738	7745	7752	7760	7767	7774	1	1	2	3	4	4	5	6	7
60	7782	7789	7796	7803	7810	7818	7825	7832	7839	7846	1	1	2	3	4	4	5	6	6
61	7853	7860	7868	7875	7882	7889	7896	7903	7910	7917	1	1	2	3	4	4	5	6	6
62	7924	7931	7938	7945	7952	7959	7966	7973	7980	7987	1	1	2	3	3	4	5	6	6
63	7993	8000	8007	8014	8021	8028	8035	8041	8048	8055	1	1	2	3	3	4	5	5	6
64	8062	8069	8075	8082	8089	8096	8102	8109	8116	8122	1	1	2	3	3	4	5	5	6
65	8129	8136	8142	8149	8156	8162	8169	8176	8182	8189	1	1	2	3	3	4	5	5	6
66	8195	8202	8209	8215	8222	8228	8235	8241	8248	8254	1	1	2	3	3	4	5	5	6
67	8261	8267	8274	8280	8287	8293	8299	8306	8312	8319	1	1	2	3	3	4	5	5	6
68	8325	8331	8338	8344	8351	8357	8363	8370	8376	8382	1	1	2	3	3	4	4	5	6
69	8388	8395	8401	8407	8414	8420	8426	8432	8439	8445	1	1	2	2	3	4	4	5	6
70	8451	8457	8463	8470	8476	8482	8488	8494	8500	8506	1	1	2	2	3	4	4	5	6
71	8513	8519	8525	8531	8537	8543	8549	8555	8561	8567	1	1	2	2	3	4	4	5	5
72	8573	8579	8585	8591	8597	8603	8609	8615	8621	8627	1	1	2	2	3	4	4	5	5
73	8633	8639	8645	8651	8657	8663	8669	8675	8681	8686	1	1	2	2	3	4	4	5	5
74	8692	8698	8704	8710	8716	8722	8727	8733	8739	8745	1	1	2	2	3	4	4	5	5
75	8751	8756	8762	8768	8774	8779	8785	8791	8797	8802	1	1	2	2	3	3	4	5	5
76	8808	8814	8820	8825	8831	8837	8842	8848	8854	8859	1	1	2	2	3	3	4	5	5
77	8865	8871	8876	8882	8887	8893	8899	8904	8910	8915	1	1	2	2	3	3	4	4	5
78	8921	8927	8932	8938	8943	8949	8954	8960	8965	8971	1	1	2	2	3	3	4	4	5
79	8976	8982	8987	8993	8998	9004	9009	9015	9020	9025	1	1	2	2	3	3	4	4	5
N	0	1	2	3	4	5	6	7	8	9	*1*	*2*	*3*	*4*	*5*	*6*	*7*	*8*	*9*
											Proportional parts								

TABLE 6-9 Cont.
Common Logarithms

N	0	1	2	3	4	5	6	7	8	9	Proportional parts								
											1	*2*	*3*	*4*	*5*	*6*	*7*	*8*	*9*
80	9031	9036	9042	9047	9053	9058	9063	9069	9074	9079	1	1	2	2	3	3	4	4	5
81	9085	9090	9096	9101	9106	9112	9117	9122	9128	9133	1	1	2	2	3	3	4	4	5
82	9138	9143	9149	9154	9159	9165	9170	9175	9180	9186	1	1	2	2	3	3	4	4	5
83	9191	9196	9201	9206	9212	9217	9222	9227	9232	9238	1	1	2	2	3	3	4	4	5
84	9243	9248	9253	9258	9263	9269	9274	9279	9284	9289	1	1	2	2	3	3	4	4	5
85	9294	9299	9304	9309	9315	9320	9325	9330	9335	9340	1	1	2	2	3	3	4	4	5
86	9345	9350	9355	9360	9365	9370	9375	9380	9385	9390	1	1	2	2	3	3	4	4	5
87	9395	9400	9405	9410	9415	9420	9425	9430	9435	9440	0	1	1	2	2	3	3	4	4
88	9445	9450	9455	9460	9465	9469	9474	9479	9484	9489	0	1	1	2	2	3	3	4	4
89	9494	9499	9504	9509	9513	9518	9523	9528	9533	9538	0	1	1	2	2	3	3	4	4
90	9542	9547	9552	9557	9562	9566	9571	9576	9581	9586	0	1	1	2	2	3	3	4	4
91	9590	9595	9600	9605	9609	9614	9619	9624	9628	9633	0	1	1	2	2	3	3	4	4
92	9638	9643	9647	9652	9657	9661	9666	9671	9675	9680	0	1	1	2	2	3	3	4	4
93	9685	9689	9694	9699	9703	9708	9713	9717	9722	9727	0	1	1	2	2	3	3	4	4
94	9731	9736	9741	9745	9750	9754	9759	9763	9768	9773	0	1	1	2	2	3	3	4	4
95	9777	9782	9786	9791	9795	9800	9805	9809	9814	9818	0	1	1	2	2	3	3	4	4
96	9823	9827	9832	9836	9841	9845	9850	9854	9859	9863	0	1	1	2	2	3	3	4	4
97	9868	9872	9877	9881	9886	9890	9894	9899	9903	9908	0	1	1	2	2	3	3	4	4
98	9912	9917	9921	9926	9930	9934	9939	9943	9948	9952	0	1	1	2	2	3	3	4	4
99	9956	9961	9965	9969	9974	9978	9983	9987	9991	9996	0	1	1	2	2	3	3	3	4
											1	*2*	*3*	*4*	*5*	*6*	*7*	*8*	*9*
N	0	1	2	3	4	5	6	7	8	9	Proportional parts								

characteristic. Thus, the mantissa of 0.0234 is 3692, and the characteristic is −2. The total logarithm is −2 + 0.3692, or −1.6308. A negative logarithm is difficult to use; therefore, it is more convenient to convert the logarithm to a positive number. This is possible by adding 10, or a multiple thereof, to the characteristic when it is negative, and compensating for this by indicating the subtraction of 10 from the entire logarithm. Thus, the logarithm of 0.0234 would be written 8.3692 −10, since −2 + 0.3692 equals 8 + 0.3692 −10. This logarithm may now be used like any other positive logarithm, except that the −10 must be considered in determining the characteristic of the answer.

Antilogarithms

An antilogarithm (abbreviated antilog or $\log^{-1}$) is a number corresponding to a given logarithm. To find an antilog, locate in the logarithm table the mantissa closest to that of the given logarithm. Record the number in the *N* column directly opposite the mantissa located, and annex to this the number on the top line immediately above the mantissa. Next determine where the decimal point is located by counting off the number of places indicated by the characteristic. Starting between the first and second digits, count to the right if the characteristic is positive, and to the left if it is negative. If greater accuracy is desired, the "Proportional parts" columns of the logarithm table can be used, in the same manner already described for finding the mantissa.

Example. Find the antilog of 3.4548. Locate 4548 in Table 6-9. Then read the first two figures of the antilog from the *N* column (28) and the third figure directly above the mantissa (5). Thus, the three figures of the antilog are 285. Locate the decimal point by counting off three places to the right, from the point between the 2 and the 8, to obtain 2850.0—the antilog of 3.4548.

In the foregoing example, if the logarithm had been −2 + 0.4548, the procedure would have been the same except for the location of the decimal point. The decimal point in this example would be located by starting at the point between the 2 and the 8, and counting two places to the left to obtain 0.0285—the antilog of −2 + 0.4548.

Multiplication

Numbers are multiplied by adding their logarithms and finding the antilog of the sum.

Example. To multiply 682 × 497, proceed as follows:

$$\log N = \log 682 + \log 497$$

$$\begin{aligned} \log 682 &= 2.8338 \\ + \log 497 &= \underline{2.6964} \\ \log N &= 5.5302 \end{aligned}$$

$$\text{antilog } 5.5302 = 339{,}000$$

Example. To multiply 0.02 × 0.03 × 0.5, proceed as follows:

$$\log N = \log 0.02 + \log 0.03 + \log 0.5$$

$$\begin{aligned} \log 0.02 &= -2 + 0.3010 = 8.3010 - 10 \\ + \log 0.03 &= -2 + 0.4771 = 8.4771 - 10 \\ + \log 0.5 &= -1 + 0.6990 = \underline{9.6990 - 10} \\ \log N &= 26.4771 - 30 \\ &= -4 + 0.4771 \end{aligned}$$

$$\text{antilog } -4 + 0.4771 = 0.0003$$

Division

Numbers are divided by subtracting the logarithm of the divisor from the logarithm of the dividend and finding the antilog of the difference.

Example. To divide 39,200 by 27.2, proceed as follows:

$$\log N = \log 39{,}200 - \log 27.2$$

$$\begin{aligned} \log 39{,}200 &= 4.5933 \\ -\log 27.2 &= \underline{1.4346} \\ \log N &= 3.1587 \end{aligned}$$

$$\text{antilog } 3.1587 = 1441$$

Example. To divide 0.3 by 0.007, proceed as follows:

$$\log N = \log 0.3 - \log 0.007$$

$$\begin{aligned} \log 0.3 &= -1 + 0.4771 = 9.4771 - 10 \\ - \log 0.007 &= -3 + 0.8451 = \underline{7.8451 - 10} \\ \log N &= 1.6320 - \ \ 0 \end{aligned}$$

$$\text{antilog } 1.6320 = 42.86$$

Raising to Powers

A given number can be raised to any power by multiplying the logarithm of the given number by the power to which the number is to be raised and finding the antilog of the product.

Example. To raise 39.7 to the third power, proceed as follows:

$$\log N = \log 39.7 \times 3$$

$$\begin{aligned} \log 39.7 &= 1.5988 \\ \log N &= 1.5988 \times 3 \\ &= 4.7964 \end{aligned}$$

$$\text{antilog } 4.7964 = 62{,}570$$

Extracting Roots

Any root can be extracted from a given number by dividing the logarithm of the given number by the index of the root and finding the antilog of the quotient.

Example. To extract the cube root of 149, proceed as follows:

$$\log N = \log 149 \div 3$$

$$\begin{aligned} \log 149 &= 2.1732 \\ \log N &= 2.1732 \div 3 \\ &= 0.7244 \end{aligned}$$

$$\text{antilog } 0.7244 = 5.301$$

Natural Logarithms

Natural logarithms are similar to common logarithms, except that a natural logarithm uses the base 2.71828 instead of the base 10. Natural logarithms are important because many desk-top computers process natural logarithms but do not process common logarithms. In turn, the programmer must convert terms with common logarithms into corresponding terms with natural logarithms. Thus, $\log_{10} x = \log_e 10$. Note that $e = 2.71828$.

Example.

dB power gain = 10 * $\log_{10}$ (P2/P1)

Programmer writes:

dB power gain = 10 * $\log_e$ (P2/P1)/$\log_e$ (10)

SQUARES, CUBES, SQUARE ROOTS, CUBE ROOTS, AND RECIPROCALS

Table 6-10 gives, for the natural numbers up to 1000, the squares, cubes, square roots, cube roots, and reciprocals multiplied by 1000.

The "1000/*n*" column contains the product of 1/*n* and 1000. To find the reciprocal of *n* (viz., 1/*n*), divide the entry for *n* in the "1000/*n*" column by 1000 (i.e., move the decimal point three places to the left).

For any number that is not a natural number between 1 and 1000, its square and cube may be quickly, though perhaps approximately, obtained as follows. First, write the given number as the product of a number between 100 and 1000, and a power of 10. Disregard the nonzero digits, if any, to the right of the decimal point of the number between 100 and 1000. (If there are any numbers to the right of the decimal point, the square or cube obtained will be approximate.) Using the equations:

$$(a \times 10^b)^2 = a^2 \times 10^{2b}$$

and

$$(a \times 10^b)^3 = a^3 \times 10^{3b}$$

the square or cube of the given number may be obtained, the value of a^2 or a^3 being read from the table. The term 10^{2b} or 10^{3b} may be calculated mentally.

To find the square root of a number that is not a natural number between 1 and 1000, write the number in the form $a \times 10^b$, where a is between 10 and 1000 and b is even and can be positive or negative. Disregard the nonzero digits, if any, to the right of the decimal point of the number a. (If there are any numbers to the right of the decimal point, the square root obtained will be approximate.) Locate the two- or three-digit number a in the n column, and read its square root. Multiplying this square root by $10^{b/2}$ gives the square root of the given number.

To find the cube root of a number that is not a natural number between 1 and 1000, write the given number in the form $a \times 10^b$, where a is between 1 and 1000 and b is divisible by 3 and can be positive or negative. Disregard the nonzero digits, if any, to the right of the decimal point of the number a. (If there are any, the cube root obtained will be approximate.) Locate the number a in the n column of Table 6-10, and read its cube root. Multiplying this cube root by 10^{b3} gives the cube root of the given number.

Note. Various kinds of numbers are used in electric and electronics formulas. The decimal numbers are 0, 1, 2, 3, 4, 5, 6, 7, 8, and 9; the binary numbers are 0 and 1. The negative numbers are −1, −2, −3, and so on. Although +2 is greater than +1, −2 is less than −1, and the computer processes negative numbers accordingly. The absolute value of −2 is greater than the absolute value of −1, and the computer processes absolute values accordingly. The imaginary numbers are $\sqrt{-1}$, $-\sqrt{-1}$, $2\sqrt{-1}$, $-3\sqrt{-1}$, and so on. The $\sqrt{-1}$ is called the j operator in electricity and electronics; $+j$ denotes inductive reactance, and $-j$ denotes capacitive reactance. Small computers cannot process imaginary numbers, and such formulas must be converted into square, square root, and trigonometric terms.

TABLE 6-10
Squares, Cubes, Square Roots, Cube Roots, and Reciprocals

n	n^2	n^3	$\sqrt{n}$	$\sqrt[3]{n}$	$\frac{1000}{n}$
1	1	1	1.0000	1.0000	1000.000
2	4	8	1.4142	1.2599	500.000
3	9	27	1.7321	1.4422	333.333
4	16	64	2.0000	1.5874	250.000
5	25	125	2.2361	1.7100	200.000
6	36	216	2.4495	1.8171	166.667
7	49	343	2.6458	1.9129	142.857
8	64	512	2.8284	2.0000	125.000
9	81	729	3.0000	2.0801	111.111
10	100	1000	3.1623	2.1544	100.000
11	121	1331	3.3166	2.2240	90.9091
12	144	1728	3.4641	2.2894	83.3333
13	169	2197	3.6056	2.3513	76.9231
14	196	2744	3.7417	2.4101	71.4286
15	225	3375	3.8730	2.4662	66.6667
16	256	4096	4.0000	2.5198	62.5000
17	289	4913	4.1231	2.5713	58.8235
18	324	5832	4.2426	2.6207	55.5556
19	361	6859	4.3589	2.6684	52.6316

TABLE 6-10 Cont.
Squares, Cubes, Square Roots, Cube Roots, and Reciprocals

n	n^2	n^3	$\sqrt{n}$	$\sqrt[3]{n}$	$\frac{1000}{n}$
20	400	8000	4.4721	2.7144	50.0000
21	441	9261	4.5826	2.7589	47.6190
22	484	10648	4.6904	2.8020	45.4545
23	529	12167	4.7958	2.8439	43.4783
24	576	13824	4.8990	2.8845	41.6667
25	625	15625	5.0000	2.9240	40.0000
26	676	17576	5.0990	2.9625	38.4615
27	729	19683	5.1962	3.0000	37.0370
28	784	21952	5.2915	3.0366	35.7143
29	841	24389	5.3852	3.0723	34.4828
30	900	27000	5.4772	3.1072	33.3333
31	961	29791	5.5678	3.1414	32.2581
32	1024	32768	5.6569	3.1748	31.2500
33	1089	35937	5.7446	3.2075	30.3030
34	1156	39304	5.8310	3.2396	29.4118
35	1225	42875	5.9161	3.2711	28.5714
36	1296	46656	6.0000	3.3019	27.7778
37	1369	50653	6.0828	3.3322	27.0270
38	1444	54872	6.1644	3.3620	26.3158
39	1521	59319	6.2450	3.3912	25.6410
40	1600	64000	6.3246	3.4200	25.0000
41	1681	68921	6.4031	3.4482	24.3902
42	1764	74088	6.4807	3.4760	23.8095
43	1849	79507	6.5574	3.5034	23.2558
44	1936	85184	6.6332	3.5303	22.7273
45	2025	91125	6.7082	3.5569	22.2222
46	2116	97336	6.7823	3.5830	21.7391
47	2209	103823	6.8557	3.6088	21.2766
48	2304	110592	6.9282	3.6342	20.8333
49	2401	117649	7.0000	3.6593	20.4082
50	2500	125000	7.0711	3.6840	20.0000
51	2601	132651	7.1414	3.7084	19.6078
52	2704	140608	7.2111	3.7325	19.2308
53	2809	148877	7.2801	3.7563	18.8679
54	2916	157464	7.3485	3.7798	18.5185
55	3025	166375	7.4162	3.8030	18.1818
56	3136	175616	7.4833	3.8259	17.8571
57	3249	185193	7.5498	3.8485	17.5439
58	3364	195112	7.6158	3.8709	17.2414
59	3481	205379	7.6811	3.8930	16.9492
60	3600	216000	7.7460	3.9149	16.6667
61	3721	226981	7.8102	3.9365	16.3934
62	3844	238328	7.8740	3.9579	16.1290
63	3969	250047	7.9373	3.9791	15.8730
64	4096	262144	8.0000	4.0000	15.6250
65	4225	274625	8.0623	4.0207	15.3846

TABLE 6-10 Cont.
Squares, Cubes, Square Roots, Cube Roots, and Reciprocals

n	n^2	n^3	$\sqrt{n}$	$\sqrt[3]{n}$	$\frac{1000}{n}$
66	4356	287496	8.1240	4.0412	15.1515
67	4489	300763	8.1854	4.0615	14.9254
68	4624	314432	8.2462	4.0817	14.7059
69	4761	328509	8.3066	4.1016	14.4928
70	4900	343000	8.3666	4.1213	14.2857
71	5041	357911	8.4261	4.1408	14.0845
72	5184	373248	8.4853	4.1602	13.8889
73	5329	389017	8.5440	4.1793	13.6986
74	5476	405224	8.6023	4.1983	13.5135
75	5625	421875	8.6603	4.2172	13.3333
76	5776	438976	8.7178	4.2358	13.1579
77	5929	456533	8.7750	4.2543	12.9870
78	6084	474552	8.8318	4.2727	12.8205
79	6241	493039	8.8882	4.2908	12.6582
80	6400	512000	8.9443	4.3089	12.5000
81	6561	531441	9.0000	4.3267	12.3457
82	6724	551368	9.0554	4.3445	12.1951
83	6889	571787	9.1104	4.3621	12.0482
84	7056	592704	9.1652	4.3795	11.9048
85	7225	614125	9.2195	4.3968	11.7647
86	7396	636056	9.2736	4.4140	11.6279
87	7569	658503	9.3274	4.4310	11.4943
88	7744	681472	9.3808	4.4480	11.3636
89	7921	704969	9.4340	4.4647	11.2360
90	8100	729000	9.4868	4.4814	11.1111
91	8281	753571	9.5394	4.4979	10.9890
92	8464	778688	9.5917	4.5144	10.8696
93	8649	804357	9.6437	4.5307	10.7527
94	8836	830584	9.6954	4.5468	10.6383
95	9025	857375	9.7468	4.5629	10.5263
96	9216	884736	9.7980	4.5789	10.4167
97	9409	912673	9.8489	4.5947	10.3093
98	9604	941192	9.8995	4.6104	10.2041
99	9801	970299	9.9499	4.6261	10.1010
100	10000	1000000	10.0000	4.6416	10.0000
101	10201	1030301	10.0499	4.6570	9.90099
102	10404	1061208	10.0995	4.6723	9.80392
103	10609	1092727	10.1489	4.6875	9.70874
104	10816	1124864	10.1980	4.7027	9.61538
105	11025	1157625	10.2470	4.7177	9.52381
106	11236	1191016	10.2956	4.7326	9.43396
107	11449	1225043	10.3441	4.7475	9.34579
108	11664	1259712	10.3923	4.7622	9.25926
109	11881	1295029	10.4403	4.7769	9.17431

TABLE 6-10 Cont.
Squares, Cubes, Square Roots, Cube Roots, and Reciprocals

n	n^2	n^3	$\sqrt{n}$	$\sqrt[3]{n}$	$\frac{1000}{n}$
110	12100	1331000	10.4881	4.7914	9.09091
111	12321	1367631	10.5357	4.8059	9.00901
112	12544	1404928	10.5830	4.8203	8.92857
113	12769	1442897	10.6301	4.8346	8.84956
114	12996	1481544	10.6771	4.8488	8.77193
115	13225	1520875	10.7238	4.8629	8.69565
116	13456	1560896	10.7703	4.8770	8.62069
117	13689	1601613	10.8167	4.8910	8.54701
118	13924	1643032	10.8628	4.9049	8.47458
119	14161	1685159	10.9087	4.9187	8.40336
120	14400	1728000	10.9545	4.9324	8.33333
121	14641	1771561	11.0000	4.9461	8.26446
122	14884	1815848	11.0454	4.9597	8.19672
123	15129	1860867	11.0905	4.9732	8.13008
124	15376	1906624	11.1355	4.9866	8.06452
125	15625	1953125	11.1803	5.0000	8.00000
126	15876	2000376	11.2250	5.0133	7.93651
127	16129	2048383	11.2694	5.0265	7.87402
128	16384	2097152	11.3137	5.0397	7.81250
129	16641	2146689	11.3578	5.0528	7.75194
130	16900	2197000	11.4018	5.0658	7.69231
131	17161	2248091	11.4455	5.0788	7.63359
132	17424	2299968	11.4891	5.0916	7.57576
133	17689	2352637	11.5326	5.1045	7.51880
134	17956	2406104	11.5758	5.1172	7.46269
135	18225	2460375	11.6190	5.1299	7.40741
136	18496	2515456	11.6619	5.1426	7.35294
137	18769	2571353	11.7047	5.1551	7.29927
138	19044	2628072	11.7473	5.1676	7.24638
139	19321	2685619	11.7898	5.1801	7.19424
140	19600	2744000	11.8322	5.1925	7.14286
141	19881	2803221	11.8743	5.2048	7.09220
142	20164	2863288	11.9164	5.2171	7.04255
143	20449	2924207	11.9583	5.2293	6.99301
144	20736	2985984	12.0000	5.2415	6.94444
145	21025	3048625	12.0416	5.2536	6.89655
146	21316	3112126	12.0830	5.2656	6.84932
147	21609	3176523	12.1244	5.2776	6.80272
148	21904	3241792	12.1655	5.2896	6.75676
149	22201	3307949	12.2066	5.3015	6.71141
150	22500	3375000	12.2474	5.3133	6.66667
151	22801	3442951	12.2882	5.3251	6.62252
152	23104	3511808	12.3288	5.3368	6.57895
153	23409	3581577	12.3693	5.3485	6.53595
154	23716	3652264	12.4097	5.3601	6.49351
155	24025	3723875	12.4499	5.3717	6.45161

TABLE 6-10 Cont.
Squares, Cubes, Square Roots, Cube Roots, and Reciprocals

n	n^2	n^3	$\sqrt{n}$	$\sqrt[3]{n}$	$\frac{1000}{n}$
156	24336	3796416	12.4900	5.3832	6.41026
157	24649	3869893	12.5300	5.3947	6.36943
158	24964	3944312	12.5698	5.4061	6.32911
159	25281	4019679	12.6095	5.4175	6.28931
160	25600	4096000	12.6491	5.4288	6.25000
161	25921	4173281	12.6886	5.4401	6.21118
162	26244	4251528	12.7279	5.4514	6.17284
163	26569	4330747	12.7671	5.4626	6.13497
164	26896	4410944	12.8062	5.4737	6.09756
165	27225	4492125	12.8452	5.4848	6.06061
166	27556	4574296	12.8841	5.4959	6.02410
167	27889	4657463	12.9228	5.5069	5.98802
168	28224	4741632	12.9615	5.5178	5.95238
169	28561	4826809	13.0000	5.5288	5.91716
170	28900	4913000	13.0384	5.5397	5.88235
171	29241	5000211	13.0767	5.5505	5.84795
172	29584	5088448	13.1149	5.5613	5.81395
173	29929	5177717	13.1529	5.5721	5.78035
174	30276	5268024	13.1909	5.5828	5.74713
175	30625	5359375	13.2288	5.5934	5.71429
176	30976	5451776	13.2665	5.6041	5.68182
177	31329	5545233	13.3041	5.6147	5.64972
178	31684	5639752	13.3417	5.6252	5.61798
179	32041	5735339	13.3791	5.6357	5.58659
180	32400	5832000	13.4164	5.6462	5.55556
181	32761	5929741	13.4536	5.6567	5.52486
182	33124	6028568	13.4907	5.6671	5.49451
183	33489	6128487	13.5277	5.6774	5.46448
184	33856	6229504	13.5647	5.6877	5.43478
185	34225	6331625	13.6015	5.6980	5.40541
186	34596	6434856	13.6382	5.7083	5.37634
187	34969	6539203	13.6748	5.7185	5.34759
188	35344	6644672	13.7113	5.7287	5.31915
189	35721	6751269	13.7477	5.7388	5.29101
190	36100	6859000	13.7840	5.7489	5.26316
191	36481	6967871	13.8203	5.7590	5.23560
192	36864	7077888	13.8564	5.7690	5.20833
193	37249	7189057	13.8924	5.7790	5.18135
194	37636	7301384	13.9284	5.7890	5.15464
195	38025	7414875	13.9642	5.7989	5.12821
196	38416	7529536	14.0000	5.8088	5.10204
197	38809	7645373	14.0357	5.8186	5.07614
198	39204	7762392	14.0712	5.8285	5.05051
199	39601	7880599	14.1067	5.8383	5.02513

TABLE 6-10 Cont.
Squares, Cubes, Square Roots, Cube Roots, and Reciprocals

n	n^2	n^3	$\sqrt{n}$	$\sqrt[3]{n}$	$\frac{1000}{n}$
200	40000	8000000	14.1421	5.8480	5.00000
201	40401	8120601	14.1774	5.8578	4.97512
202	40804	8242408	14.2127	5.8675	4.95050
203	41209	8365427	14.2478	5.8771	4.92611
204	41616	8489664	14.2829	5.8868	4.90196
205	42025	8615125	14.3178	5.8964	4.87805
206	42436	8741816	14.3527	5.9059	4.85437
207	42849	8869743	14.3875	5.9155	4.83092
208	43264	8998912	14.4222	5.9250	4.80769
209	43681	9129329	14.4568	5.9345	4.78469
210	44100	9261000	14.4914	5.9439	4.76190
211	44521	9393931	14.5258	5.9533	4.73934
212	44944	9528128	14.5602	5.9627	4.71698
213	45369	9663597	14.5945	5.9721	4.69484
214	45796	9800344	14.6287	5.9814	4.67290
215	46225	9938375	14.6629	5.9907	4.65116
216	46656	10077696	14.6969	6.0000	4.62963
217	47089	10218313	14.7309	6.0092	4.60829
218	47524	10360232	14.7648	6.0185	4.58716
219	47961	10503459	14.7986	6.0277	4.56621
220	48400	10648000	14.8324	6.0368	4.54545
221	48841	10793861	14.8661	6.0459	4.52489
222	49284	10941048	14.8997	6.0550	4.50450
223	49729	11089567	14.9332	6.0641	4.48431
224	50176	11239424	14.9666	6.0732	4.46429
225	50625	11390625	15.0000	6.0822	4.44444
226	51076	11543176	15.0333	6.0912	4.42478
227	51529	11697083	15.0665	6.1002	4.40529
228	51984	11852352	15.0997	6.1091	4.38596
229	52441	12008989	15.1327	6.1180	4.36681
230	52900	12167000	15.1658	6.1269	4.34783
231	53361	12326391	15.1987	6.1358	4.32900
232	53824	12487168	15.2315	6.1446	4.31034
233	54289	12649337	15.2643	6.1534	4.29185
234	54756	12812904	15.2971	6.1622	4.27350
235	55225	12977875	15.3297	6.1710	4.25532
236	55696	13144256	15.3623	6.1797	4.23729
237	56169	13312053	15.3948	6.1885	4.21941
238	56644	13481272	15.4272	6.1972	4.20168
239	57121	13651919	15.4596	6.2058	4.18410
240	57600	13824000	15.4919	6.2145	4.16667
241	58081	13997521	15.5242	6.2231	4.14938
242	58564	14172488	15.5563	6.2317	4.13223
243	59049	14348907	15.5885	6.2403	4.11523
244	59536	14526784	15.6205	6.2488	4.09836
245	60025	14706125	15.6525	6.2573	4.08163

TABLE 6-10 Cont.
Squares, Cubes, Square Roots, Cube Roots, and Reciprocals

n	n^2	n^3	$\sqrt{n}$	$\sqrt[3]{n}$	$\frac{1000}{n}$
246	60516	14886936	15.6844	6.2658	4.06504
247	61009	15069223	15.7162	6.2743	4.04858
248	61504	15252992	15.7480	6.2828	4.03226
249	62001	15438249	15.7797	6.2912	4.01606
250	62500	15625000	15.8114	6.2996	4.00000
251	63001	15813251	15.8430	6.3080	3.98406
252	63504	16003008	15.8745	6.3164	3.96825
253	64009	16194277	15.9060	6.3247	3.95257
254	64516	16387064	15.9374	6.3330	3.93701
255	65025	16581375	15.9687	6.3413	3.92157
256	65536	16777216	16.0000	6.3496	3.90625
257	66049	16974593	16.0312	6.3579	3.89105
258	66564	17173512	16.0624	6.3661	3.87597
259	67081	17373979	16.0935	6.3743	3.86100
260	67600	17576000	16.1245	6.3825	3.84615
261	68121	17779581	16.1555	6.3907	3.83142
262	68644	17984728	16.1864	6.3988	3.81679
263	69169	18191447	16.2173	6.4070	3.80228
264	69696	18399744	16.2481	6.4151	3.78788
265	70225	18609625	16.2788	6.4232	3.77358
266	70756	18821096	16.3095	6.4312	3.75940
267	71289	19034163	16.3401	6.4393	3.74532
268	71824	19248832	16.3707	6.4473	3.73134
269	72361	19465109	16.4012	6.4553	3.71747
270	72900	19683000	16.4317	6.4633	3.70370
271	73441	19902511	16.4621	6.4713	3.69004
272	73984	20123648	16.4924	6.4792	3.67647
273	74529	20346417	16.5227	6.4872	3.66300
274	75076	20570824	16.5529	6.4951	3.64964
275	75625	20796875	16.5831	6.5030	3.63636
276	76176	21024576	16.6132	6.5108	3.62319
277	76729	21253933	16.6433	6.5187	3.61011
278	77284	21484952	16.6733	6.5265	3.59712
279	77841	21717639	16.7033	6.5343	3.58423
280	78400	21952000	16.7332	6.5421	3.57143
281	78961	22188041	16.7631	6.5499	3.55872
282	79524	22425768	16.7929	6.5577	3.54610
283	80089	22665187	16.8226	6.5654	3.53357
284	80656	22906304	16.8523	6.5731	3.52113
285	81225	23149125	16.8819	6.5808	3.50877
286	81796	23393656	16.9115	6.5885	3.49650
287	82369	23639903	16.9411	6.5962	3.48432
288	82944	23887872	16.9706	6.6039	3.47222
289	83521	24137569	17.0000	6.6115	3.46021

TABLE 6-10 Cont.
Squares, Cubes, Square Roots, and Reciprocals

n	n^2	n^3	$\sqrt{n}$	$\sqrt[3]{n}$	$\frac{1000}{n}$
290	84100	24389000	17.0294	6.6191	3.44828
291	84681	24642171	17.0587	6.6267	3.43643
292	85264	24897088	17.0880	6.6343	3.42466
293	85849	25153757	17.1172	6.6419	3.41297
294	86436	25412184	17.1464	6.6494	3.40136
295	87025	25672375	17.1756	6.6569	3.38983
296	87616	25934336	17.2047	6.6644	3.37838
297	88209	26198073	17.2337	6.6719	3.36700
298	88804	26463592	17.2627	6.6794	3.35570
299	89401	26730899	17.2916	6.6869	3.34448
300	90000	27000000	17.3205	6.6943	3.33333
301	90601	27270901	17.3494	6.7018	3.32226
302	91204	27543608	17.3781	6.7092	3.31126
303	91809	27818127	17.4069	6.7166	3.30033
304	92416	28094464	17.4356	6.7240	3.28947
305	93025	28372625	17.4642	6.7313	3.27869
306	93636	28652616	17.4929	6.7387	3.26797
307	94249	28934443	17.5214	6.7460	3.25733
308	94864	29218112	17.5499	6.7533	3.24675
309	95481	29503629	17.5784	6.7606	3.23625
310	96100	29791000	17.6068	6.7679	3.22581
311	96721	30080231	17.6352	6.7752	3.21543
312	97344	30371328	17.6635	6.7824	3.20513
313	97969	30664297	17.6918	6.7897	3.19489
314	98596	30959144	17.7200	6.7969	3.18471
315	99225	31255875	17.7482	6.8041	3.17460
316	99856	31554496	17.7764	6.8113	3.16456
317	100489	31855013	17.8045	6.8185	3.15457
318	101124	32157432	17.8326	6.8256	3.14465
319	101761	32461759	17.8606	6.8328	3.13480
320	102400	32768000	17.8885	6.8399	3.12500
321	103041	33076161	17.9165	6.8470	3.11527
322	103684	33386284	17.9444	6.8541	3.10559
323	104329	33698267	17.9722	6.8612	3.09598
324	104976	34012224	18.0000	6.8683	3.08642
325	105625	34328125	18.0278	6.8753	3.07692
326	106276	34645976	18.0555	6.8824	3.06749
327	106929	34965783	18.0831	6.8894	3.05810
328	107584	35287552	18.1108	6.8964	3.04878
329	108241	35611289	18.1384	6.9034	3.03951
330	108900	35937000	18.1659	6.9104	3.03030
331	109561	36264691	18.1934	6.9174	3.02115
332	110224	36594368	18.2209	6.9244	3.01205
333	110889	36926037	18.2483	6.9313	3.00300
334	111556	37259704	18.2757	6.9382	2.99401
335	112225	37595375	18.3030	6.9451	2.98507

TABLE 6-10 Cont.
Squares, Cubes, Square Roots, Cube Roots, and Reciprocals

n	n^2	n^3	$\sqrt{n}$	$\sqrt[3]{n}$	$\frac{1000}{n}$
336	112896	37933056	18.3303	6.9521	2.97619
337	113569	38272753	18.3576	6.9589	2.96736
338	114244	38614472	18.3848	6.9658	2.95858
339	114921	38958219	18.4120	6.9727	2.94985
340	115600	39304000	18.4391	6.9795	2.94118
341	116281	39651821	18.4662	6.9864	2.93255
342	116964	40001688	18.4932	6.9932	2.92398
343	117649	40353607	18.5203	7.0000	2.91545
344	118336	40707584	18.5472	7.0068	2.90698
345	119025	41063625	18.5742	7.0136	2.89855
346	119716	41421736	18.6011	7.0203	2.89017
347	120409	41781923	18.6279	7.0271	2.88184
348	121104	42144192	18.6548	7.0338	2.87356
349	121801	42508549	18.6815	7.0406	2.86533
350	122500	42875000	18.7083	7.0473	2.85714
351	123201	43243551	18.7350	7.0540	2.84900
352	123904	43614208	18.7617	7.0607	2.84091
353	124609	43986977	18.7883	7.0674	2.83286
354	125316	44361864	18.8149	7.0740	2.82486
355	126025	44738875	18.8414	7.0807	2.81690
356	126736	45118016	18.8680	7.0873	2.80899
357	127449	45499293	18.8944	7.0940	2.80112
358	128164	45882712	18.9209	7.1006	2.79330
359	128881	46268279	18.9473	7.1072	2.78552
360	129600	46656000	18.9737	7.1138	2.77778
361	130321	47045881	19.0000	7.1204	2.77008
362	131044	47437928	19.0263	7.1269	2.76243
363	131769	47832147	19.0526	7.1335	2.75482
364	132496	48228544	19.0788	7.1400	2.74725
365	133225	48627125	19.1050	7.1466	2.73973
366	133956	49027896	19.1311	7.1531	2.73224
367	134689	49430863	19.1572	7.1596	2.72480
368	135424	49836032	19.1833	7.1661	2.71739
369	136161	50243409	19.2094	7.1726	2.71003
370	136900	50653000	19.2354	7.1791	2.70270
371	137641	51064811	19.2614	7.1855	2.69542
372	138384	51478848	19.2873	7.1920	2.68817
373	139129	51895117	19.3132	7.1984	2.68097
374	139876	52313624	19.3391	7.2048	2.67380
375	140625	52734375	19.3649	7.2112	2.66667
376	141376	53157376	19.3907	7.2177	2.65957
377	142129	53582633	19.4165	7.2240	2.65252
378	142884	54010152	19.4422	7.2304	2.64550
379	143641	54439939	19.4679	7.2368	2.63852

TABLE 6-10 Cont.
Squares, Cubes, Square Roots, Cube Roots, and Reciprocals

n	n^2	n^3	$\sqrt{n}$	$\sqrt[3]{n}$	$\frac{1000}{n}$
380	144400	54872000	19.4936	7.2432	2.63158
381	145161	55306341	19.5192	7.2495	2.62467
382	145924	55742968	19.5448	7.2558	2.61780
383	146689	56181887	19.5704	7.2622	2.61097
384	147456	56623104	19.5959	7.2685	2.60417
385	148225	57066625	19.6214	7.2748	2.59740
386	148996	57512456	19.6469	7.2811	2.59067
387	149769	57960603	19.6723	7.2874	2.58398
388	150544	58411072	19.6977	7.2936	2.57732
389	151321	58863869	19.7231	7.2999	2.57069
390	152100	59319000	19.7484	7.3061	2.56410
391	152881	59776471	19.7737	7.3124	2.55755
392	153664	60236288	19.7990	7.3186	2.55102
393	154449	60698457	19.8242	7.3248	2.54453
394	155236	61162984	19.8494	7.3310	2.53807
395	156025	61629875	19.8746	7.3372	2.53165
396	156816	62099136	19.8997	7.3434	2.52525
397	157609	62570773	19.9249	7.3496	2.51889
398	158404	63044792	19.9499	7.3558	2.51256
399	159201	63521199	19.9750	7.3619	2.50627
400	160000	64000000	20.0000	7.3681	2.50000
401	160801	64481201	20.0250	7.3742	2.49377
402	161604	64964808	20.0499	7.3803	2.48756
403	162409	65450827	20.0749	7.3864	2.48139
404	163216	65939264	20.0998	7.3925	2.47525
405	164025	66430125	20.1246	7.3986	2.46914
406	164836	66923416	20.1494	7.4047	2.46305
407	165649	67419143	20.1742	7.4108	2.45700
408	166464	67917312	20.1990	7.4169	2.45098
409	167281	68417929	20.2237	7.4229	2.44499
410	168100	68921000	20.2485	7.4290	2.43902
411	168921	69426531	20.2731	7.4350	2.43309
412	169744	69934528	20.2978	7.4410	2.42718
413	170569	70444997	20.3224	7.4470	2.42131
414	171396	70957944	20.3470	7.4530	2.41546
415	172225	71473375	20.3715	7.4590	2.40964
416	173056	71991296	20.3961	7.4650	2.40385
417	173889	72511713	20.4206	7.4710	2.39808
418	174724	73034632	20.4450	7.4770	2.39234
419	175561	73560059	20.4695	7.4829	2.38664
420	176400	74088000	20.4939	7.4889	2.38095
421	177241	74618461	20.5183	7.4948	2.37530
422	178084	45151448	20.5426	7.5007	2.36967
423	178929	75686967	20.5670	7.5067	2.36407
424	179776	76225024	20.5913	7.5126	2.35849
425	180625	76765625	20.6155	7.5185	2.35294

TABLE 6-10 Cont.
Squares, Cubes, Square Roots, Cube Roots, and Reciprocals

n	n^2	n^3	$\sqrt{n}$	$\sqrt[3]{n}$	$\frac{1000}{n}$
426	181476	77308776	20.6398	7.5244	2.34742
427	182329	77854483	20.6640	7.5302	2.34192
428	183184	78402752	20.6882	7.5361	2.33645
429	184041	78953589	20.7123	7.5420	2.33100
430	184900	79507000	20.7364	7.5478	2.32558
431	185761	80062991	20.7605	7.5537	2.32019
432	186624	80621568	20.7846	7.5595	2.31482
433	187489	81182737	20.8087	7.5654	2.30947
434	188356	81746504	20.8237	7.5712	2.30415
435	189225	82312875	20.8567	7.5770	2.29885
436	190096	82881856	20.8806	7.5828	2.29358
437	190969	83453453	20.9045	7.5886	2.28833
438	191844	84027672	20.9284	7.5944	2.28311
439	192721	84604519	20.9523	7.6001	2.27790
440	193600	85184000	20.9762	7.6059	2.27273
441	194481	85766121	21.0000	7.6117	2.26757
442	195364	86350888	21.0238	7.6174	2.26244
443	196249	86938307	21.0476	7.6232	2.25734
444	197136	87528384	21.0713	7.6289	2.25225
445	198025	88121125	21.0950	7.6346	2.24719
446	198916	88716536	21.1187	7.6403	2.24215
447	199809	89314623	21.1424	7.6460	2.23714
448	200704	89915392	21.1660	7.6517	2.23214
449	201601	90518849	21.1896	7.6574	2.22717
450	202500	91125000	21.2132	7.6631	2.22222
451	203401	91733851	21.2368	7.6688	2.21730
452	204304	92345408	21.2603	7.6744	2.21239
453	205209	92959677	21.2838	7.6801	2.20751
454	206116	93576664	21.3073	7.6857	2.20264
455	207025	94196375	21.3307	7.6914	2.19780
456	207936	94818816	21.3542	7.6970	2.19298
457	208849	95443993	21.3776	7.7026	2.18818
458	209764	96071912	21.4009	7.7082	2.18341
459	210681	96702579	21.4243	7.7138	2.17865
460	211600	97336000	21.4476	7.7194	2.17391
461	212521	97972181	21.4709	7.7250	2.16920
462	213444	98611128	21.4942	7.7306	2.16450
463	214369	99252847	21.5174	7.7362	2.15983
464	215296	99897344	21.5407	7.7418	2.15517
465	216225	100544625	21.5639	7.7473	2.15054
466	217156	101194696	21.5870	7.7529	2.14592
467	218089	101847563	21.6102	7.7584	2.14133
468	219024	102503232	21.6333	7.7639	2.13675
469	219961	103161709	21.6564	7.7695	2.13220

TABLE 6-10 Cont.
Squares, Cubes, Square Roots, Cube Roots, and Reciprocals

n	n^2	n^3	$\sqrt{n}$	$\sqrt[3]{n}$	$\frac{1000}{n}$
470	220900	103823000	21.6795	7.7750	2.12766
471	221841	104487111	21.7025	7.7805	2.12314
472	222784	105154048	21.7256	7.7860	2.11864
473	223729	105823817	21.7486	7.7915	2.11417
474	224676	106496424	21.7715	7.7970	2.10971
475	225625	107171875	21.7945	7.8025	2.10526
476	226576	107850176	21.8174	7.8079	2.10084
477	227529	108531333	21.8403	7.8134	2.09644
478	228484	109215352	21.8632	7.8188	2.09205
479	229441	109902239	21.8861	7.8243	2.08768
480	230400	110592000	21.9089	7.8297	2.08333
481	231361	111284641	21.9317	7.8352	2.07900
482	232324	111980168	21.9545	7.8406	2.07469
483	233289	112678587	21.9773	7.8460	2.07039
484	234256	113379904	22.0000	7.8514	2.06612
485	235225	114084125	22.0227	7.8568	2.06186
486	236196	114791256	22.0454	7.8622	2.05761
487	237169	115501303	22.0681	7.8676	2.05339
488	238144	116214272	22.0907	7.8730	2.04918
489	239121	116930169	22.1133	7.8784	2.04499
490	240100	117649000	22.1359	7.8837	2.04082
491	241081	118370771	22.1585	7.8891	2.03666
492	242064	119095488	22.1811	7.8944	2.03252
493	243049	119823157	22.2036	7.8998	2.02840
494	244036	120553784	22.2261	7.9051	2.02429
495	245025	121287375	22.2486	7.9105	2.02020
496	246016	122023936	22.2711	7.9158	2.01613
497	247009	122763473	22.2935	7.9211	2.01207
498	248004	123505992	22.3159	7.9264	2.00803
499	249001	124251499	22.3383	7.9317	2.00401
500	250000	125000000	22.3607	7.9370	2.00000
501	251001	125751501	22.3830	7.9423	1.99601
502	252004	126506008	22.4054	7.9476	1.99203
503	253009	127263527	22.4277	7.9528	1.98807
504	254016	128024064	22.4499	7.9581	1.98413
505	255025	128787625	22.4722	7.9634	1.98020
506	256036	129554216	22.4944	7.9686	1.97629
507	257049	130323843	22.5167	7.9739	1.97239
508	258064	131096512	22.5389	7.9791	1.96850
509	259081	131872229	22.5610	7.9843	1.96464
510	260100	132651000	22.5832	7.9896	1.96078
511	261121	133432831	22.6053	7.9948	1.95695
512	262144	134217728	22.6274	8.0000	1.95312
513	263169	135005697	22.6495	8.0052	1.94932
514	264196	135796744	22.6716	8.0104	1.94553
515	265225	136590875	22.6936	8.0156	1.94175

TABLE 6-10 Cont.
Squares, Cubes, Square Roots, and Reciprocals

n	n^2	n^3	$\sqrt{n}$	$\sqrt[3]{n}$	$\frac{1000}{n}$
516	266256	137388096	22.7156	8.0208	1.93798
517	267289	138188413	22.7376	8.0260	1.93424
518	268324	138991832	22.7596	8.0311	1.93050
519	269361	139798359	22.7816	8.0363	1.92678
520	270400	140608000	22.8035	8.0415	1.92308
521	271441	141420761	22.8254	8.0466	1.91939
522	272484	142236648	22.8473	8.0517	1.91571
523	273529	143055667	22.8692	8.0569	1.91205
524	274576	143877824	22.8910	8.0620	1.90840
525	275625	144703125	22.9129	8.0671	1.90476
526	276676	145531576	22.9347	8.0723	1.90114
527	277729	146363183	22.9565	8.0774	1.89753
528	278784	147197952	22.9783	8.0825	1.89394
529	279841	148035889	23.0000	8.0876	1.89036
530	280900	148877000	23.0217	8.0927	1.88679
531	281961	149721291	23.0434	8.0978	1.88324
532	283024	150568768	23.0651	8.1028	1.87970
533	284089	151419437	23.0868	8.1079	1.87617
534	285156	152273304	23.1084	8.1130	1.87266
535	286225	153130375	23.1301	8.1180	1.86916
536	287296	153990656	23.1517	8.1231	1.86567
537	288369	154854153	23.1733	8.1281	1.86220
538	289444	155720872	23.1948	8.1332	1.85874
539	290521	156590819	23.2164	8.1382	1.85529
540	291600	157464000	23.2379	8.1433	1.85185
541	292681	158340421	23.2594	8.1483	1.84843
542	293764	159220088	23.2809	8.1533	1.84502
543	294849	160103007	23.3024	8.1583	1.84162
544	295936	160989184	23.3238	8.1633	1.83824
545	297025	161878625	23.3452	8.1683	1.83486
546	298116	162771336	23.3666	8.1733	1.83150
547	299209	163667323	23.3880	8.1783	1.82815
548	300304	164566592	23.4094	8.1833	1.82482
549	301401	165469149	23.4307	8.1882	1.82149
550	302500	166375000	23.4521	8.1932	1.81818
551	303601	167284151	23.4734	8.1982	1.81488
552	304704	168196608	23.4947	8.2031	1.81159
553	305809	169112377	23.5160	8.2081	1.80832
554	306916	170031464	23.5372	8.2130	1.80505
555	308025	170953875	23.5584	8.2180	1.80180
556	309136	171879616	23.5797	8.2229	1.79856
557	310249	172808693	23.6008	8.2278	1.79533
558	311364	173741112	23.6220	8.2327	1.79211
559	312481	174676879	23.6432	8.2377	1.78891

TABLE 6-10 Cont.
Squares, Cubes, Square Roots, and Reciprocals

n	n^2	n^3	$\sqrt{n}$	$\sqrt[3]{n}$	$\frac{1000}{n}$
560	313600	175616000	23.6643	8.2426	1.78571
561	314721	176558481	23.6854	8.2475	1.78253
562	315844	177504328	23.7065	8.2524	1.77936
563	316969	178453547	23.7276	8.2573	1.77620
564	318096	179406144	23.7487	8.2621	1.77305
565	319225	180362125	23.7697	8.2670	1.76991
566	320356	181321496	23.7908	8.2719	1.76678
567	321489	182284263	23.8118	8.2768	1.76367
568	322624	183250432	23.8328	8.2816	1.76056
569	323761	184220009	23.8537	8.2865	1.75747
570	324900	185193000	23.8747	8.2913	1.75439
571	326041	186169411	23.8956	8.2962	1.75131
572	327184	187149248	23.9165	8.3010	1.74825
573	328329	188132517	23.9374	8.3059	1.74520
574	329476	189119224	23.9583	8.3107	1.74216
575	330625	190109375	23.9792	8.3155	1.73913
576	331776	191102976	24.0000	8.3203	1.73611
577	332929	192100033	24.0208	8.3251	1.73310
578	334084	193100552	24.0416	8.3300	1.73010
579	335241	194104539	24.0624	8.3348	1.72712
580	336400	195112000	24.0832	8.3396	1.72414
581	337561	196122941	24.1039	8.3443	1.72117
582	338724	197137368	24.1247	8.3491	1.71821
583	339889	198155287	24.1454	8.3539	1.71527
584	341056	199176704	24.1661	8.3587	1.71233
585	342225	200201625	24.1868	8.3634	1.70940
586	343396	201230056	24.2074	8.3682	1.70649
587	344569	202262003	24.2281	8.3730	1.70358
588	345744	203297472	24.2487	8.3777	1.70068
589	346921	204336469	24.2693	8.3825	1.69779
590	348100	205379000	24.2899	8.3872	1.69492
591	349281	206425071	24.3105	8.3919	1.69205
592	350464	207474688	24.3311	8.3967	1.68919
593	351649	208527857	24.3516	8.4014	1.68634
594	352836	209584584	24.3721	8.4061	1.68350
595	354025	210644875	24.3926	8.4108	1.68067
596	355216	211708736	24.4131	8.4155	1.67785
597	356409	212776173	24.4336	8.4202	1.67504
598	357604	213847192	24.4540	8.4249	1.67224
599	358801	214921799	24.4745	8.4296	1.66945
600	360000	216000000	24.4949	8.4343	1.66667
601	361201	217081801	24.5153	8.4390	1.66389
602	362404	218167208	24.5357	8.4437	1.66113
603	363609	219256227	24.5561	8.4484	1.65837
604	364816	220348864	24.5764	8.4530	1.65563
605	366025	221445125	24.5967	8.4577	1.65289

TABLE 6-10 Cont.
Squares, Cubes, Square Roots, Cube Roots, and Reciprocals

n	n^2	n^3	$\sqrt{n}$	$\sqrt[3]{n}$	$\frac{1000}{n}$
606	367236	222545016	24.6171	8.4623	1.65017
607	368449	223648543	24.6374	8.4670	1.64745
608	369664	224755712	24.6577	8.4716	1.64474
609	370881	225866529	24.6779	8.4763	1.64204
610	372100	226981000	24.6982	8.4809	1.63934
611	373321	228099131	24.7184	8.4856	1.63666
612	374544	229220928	24.7386	8.4902	1.63399
613	375769	230346397	24.7588	8.4948	1.63132
614	376996	231475544	24.7790	8.4994	1.62866
615	378225	232608375	24.7992	8.5040	1.62602
616	379456	233744896	24.8193	8.5086	1.62338
617	380689	234885113	24.8395	8.5132	1.62075
618	381924	236029032	24.8506	8.5178	1.61812
619	383161	237176659	24.8797	8.5224	1.61551
620	384400	238328000	24.8998	8.5270	1.61290
621	385641	239483061	24.9199	8.5316	1.61031
622	386884	240641848	24.9399	8.5362	1.60772
623	388129	241804367	24.9600	8.5408	1.60514
624	389376	242970624	24.9800	8.5453	1.60256
625	390625	244140625	25.0000	8.5499	1.60000
626	391876	245314376	25.0200	8.5544	1.59744
627	393129	246491883	25.0400	8.5590	1.59490
628	394384	247673152	25.0599	8.5635	1.59236
629	395641	248858189	25.0799	8.5681	1.58983
630	396900	250047000	25.0998	8.5726	1.58730
631	398161	251239591	25.1197	8.5772	1.58479
632	399424	252435968	25.1396	8.5817	1.58228
633	400689	253636137	25.1595	8.5862	1.57978
634	401956	254840104	25.1794	8.5907	1.57729
635	403225	256047875	25.1992	8.5952	1.57480
636	404496	257259456	25.2190	8.5997	1.57233
637	405769	258474853	25.2389	8.6043	1.56986
638	407044	259694072	25.2587	8.6088	1.56740
639	408321	260917119	25.2784	8.6132	1.56495
640	409600	262144000	25.2982	8.6177	1.56250
641	410881	263374721	25.3180	8.6222	1.56006
642	412164	264609288	25.3377	8.6267	1.55763
643	413449	265847707	25.3574	8.6312	1.55521
644	414736	267089984	25.3772	8.6357	1.55280
645	416025	268336125	25.3969	8.6401	1.55039
646	417316	269586136	25.4165	8.6446	1.54799
647	418609	270840023	25.4362	8.6490	1.54560
648	419904	272097792	25.4558	8.6535	1.54321
649	421201	273359449	25.4755	8.6579	1.54083

TABLE 6-10 Cont.
Squares, Cubes, Square Roots, Cube Roots, and Reciprocals

n	n^2	n^3	$\sqrt{n}$	$\sqrt[3]{n}$	$\frac{1000}{n}$
650	422500	274625000	25.4951	8.6624	1.53846
651	423801	275894451	25.5147	8.6668	1.53610
652	425104	277167808	25.5343	8.6713	1.53374
653	426409	278445077	25.5539	8.6757	1.53139
654	427716	279726264	25.5734	8.6801	1.52905
655	429025	281011375	25.5930	8.6845	1.52672
656	430336	282300416	25.6125	8.6890	1.52439
657	431649	283593393	25.6320	8.6934	1.52207
658	432964	284890312	25.6515	8.6978	1.51976
659	434281	286191179	25.6710	8.7022	1.51745
660	435600	287496000	25.6905	8.7066	1.51515
661	436921	288804781	25.7099	8.7110	1.51286
662	438244	290117528	25.7294	8.7154	1.51057
663	439569	291434247	25.7488	8.7198	1.50830
664	440896	292754944	25.7682	8.7241	1.50602
665	442225	294079625	25.7876	8.7285	1.50376
666	443556	295408296	25.8070	8.7329	1.50150
667	444889	296740963	25.8263	8.7373	1.49925
668	446224	298077632	25.8457	8.7416	1.49701
669	447561	299418309	25.8650	8.7460	1.49477
670	448900	300763000	25.8844	8.7503	1.49254
671	450241	302111711	25.9037	8.7547	1.49031
672	451584	303464448	25.9230	8.7590	1.48810
673	452929	304821217	25.9422	8.7634	1.48588
674	454276	306182024	25.9615	8.7677	1.48368
675	455625	307546875	25.9808	8.7721	1.48148
676	456976	308915776	26.0000	8.7764	1.47929
677	458329	310288733	26.0192	8.7807	1.47711
678	459684	311665752	26.0384	8.7850	1.47493
679	461041	313046839	26.0576	8.7893	1.47275
680	462400	314432000	26.0768	8.7937	1.47059
681	463761	315821241	26.0960	8.7980	1.46843
682	465124	317214568	26.1151	8.8023	1.46628
683	466489	318611987	26.1343	8.8066	1.46413
684	467856	320013504	26.1534	8.8109	1.46199
685	469225	321419125	26.1725	8.8152	1.45985
686	470596	322828856	26.1916	8.8194	1.45773
687	471969	324242703	26.2107	8.8237	1.45560
688	473344	325660672	26.2298	8.8280	1.45349
689	474721	327082769	26.2488	8.8323	1.45138
690	476100	328509000	26.2679	8.8366	1.44928
691	477481	329939371	26.2869	8.8408	1.44718
692	478864	331373888	26.3059	8.8451	1.44509
693	480249	332812557	26.3249	8.8493	1.44300
694	481636	334255384	26.3439	8.8536	1.44092
695	483025	335702375	26.3629	8.8578	1.43885

TABLE 6-10 Cont.
Squares, Cubes, Square Roots, Cube roots, and Reciprocals

n	n^2	n^3	$\sqrt{n}$	$\sqrt[3]{n}$	$\frac{1000}{n}$
696	484416	337153536	26.3818	8.8621	1.43678
697	485809	338608873	26.4008	8.8663	1.43472
698	487204	340068392	26.4197	8.8706	1.43267
699	488601	341532099	26.4386	8.8748	1.43062
700	490000	343000000	26.4575	8.8790	1.42857
701	491401	344472101	26.4764	8.8833	1.42653
702	492804	345948408	26.4953	8.8875	1.42450
703	494209	347428927	26.5141	8.8917	1.42248
704	495616	348913664	26.5330	8.8959	1.42046
705	497025	350402625	26.5518	8.9001	1.41844
706	498436	351895816	26.5707	8.9043	1.41643
707	499849	353393243	26.5895	8.9085	1.41443
708	501264	354894912	26.6083	8.9127	1.41243
709	502681	356400829	26.6271	8.9169	1.41044
710	504100	357911000	26.6458	8.9211	1.40845
711	505521	359425431	26.6646	8.9253	1.40647
712	506944	360944128	26.6833	8.9295	1.40449
713	508369	362467097	26.7021	8.9337	1.40253
714	509796	363994344	26.7208	8.9378	1.40056
715	511225	365525875	26.7395	8.9420	1.39860
716	512656	367061696	26.7582	8.9462	1.39665
717	514089	368601813	26.7769	8.9503	1.39470
718	515524	370146232	26.7955	8.9545	1.39276
719	516961	371694959	26.8142	8.9587	1.39082
720	518400	373248000	26.8328	8.9628	1.38889
721	519841	374805361	26.8514	8.9670	1.38696
722	521284	376367048	26.8701	8.9711	1.38504
723	522729	377933067	26.8887	8.9752	1.38313
724	524176	379503424	26.9072	8.9794	1.38122
725	525625	381078125	26.9258	8.9835	1.37931
726	527076	382657176	26.9444	8.9876	1.37741
727	528529	384240583	26.9629	8.9918	1.37552
728	529984	385828352	26.9815	8.9959	1.37363
729	531441	387420489	27.0000	9.0000	1.37174
730	532900	389017000	27.0185	9.0041	1.36986
731	534361	390617891	27.0370	9.0082	1.36799
732	535824	392223168	27.0555	9.0123	1.36612
733	537289	393832837	27.0740	9.0164	1.36426
734	538756	395446904	27.0924	9.0205	1.36240
735	540225	397065375	27.1109	9.0246	1.36054
736	541696	398688256	27.1293	9.0287	1.35870
737	543169	400315553	27.1477	9.0328	1.35685
738	544644	401947272	27.1662	9.0369	1.35501
739	546121	403583419	27.1846	9.0410	1.35318

TABLE 6-10 Cont.
Squares, Cubes, Square Roots, Cube Roots, and Reciprocals

n	n^2	n^3	$\sqrt{n}$	$\sqrt[3]{n}$	$\frac{1000}{n}$
740	547600	405224000	27.2029	9.0450	1.35135
741	549081	406869021	27.2213	9.0491	1.34953
742	550564	408518488	27.2397	9.0532	1.34771
743	552049	410172407	27.2580	9.0572	1.34590
744	553536	411830784	27.2764	9.0613	1.34409
745	555025	413493625	27.2947	9.0654	1.34228
746	556516	415160936	27.3130	9.0694	1.34048
747	558009	416832723	27.3313	9.0735	1.33869
748	559504	418508992	27.3496	9.0775	1.33690
749	561001	420189749	27.3679	9.0816	1.33511
750	562500	421875000	27.3861	9.0856	1.33333
751	564001	423564751	27.4044	9.0896	1.33156
752	565504	425259008	27.4226	9.0937	1.32979
753	567009	426957777	27.4408	9.0977	1.32802
754	568516	428661064	27.4591	9.1017	1.32626
755	570025	430368875	27.4773	9.1057	1.32450
756	571536	432081216	27.4955	9.1098	1.32275
757	573049	433798093	27.5136	9.1138	1.32100
758	574564	435519512	27.5318	9.1178	1.31926
759	576081	437245479	27.5500	9.1218	1.31752
760	577600	438976000	27.5681	9.1258	1.31579
761	579121	440711081	27.5862	9.1298	1.31406
762	580644	442450728	27.6043	9.1338	1.31234
763	582169	444194947	27.6225	9.1378	1.31062
764	583696	445943744	27.6405	9.1418	1.30890
765	585225	447697125	27.6586	9.1458	1.30719
766	586756	449455096	27.6767	9.1498	1.30548
767	588289	451217663	27.6948	9.1537	1.30378
768	589824	452984832	27.7128	9.1577	1.30208
769	591361	454756609	27.7308	9.1617	1.30039
770	592900	456533000	27.7489	9.1657	1.29870
771	594441	458314011	27.7669	9.1696	1.29702
772	595984	460099648	27.7849	9.1736	1.29534
773	597529	461889917	27.8029	9.1775	1.29366
774	599076	463684824	27.8209	9.1815	1.29199
775	600625	465484375	27.8388	9.1855	1.29032
776	602176	467288576	27.8568	9.1894	1.28866
777	603729	469097433	27.8747	9.1933	1.28700
778	605284	470910952	27.8927	9.1973	1.28535
779	606841	472729139	27.9106	9.2012	1.28370
780	608400	474552000	27.9285	9.2052	1.28205
781	609961	476379541	27.9464	9.2091	1.28041
782	611524	478211768	27.9643	9.2130	1.27877
783	613089	480048687	27.9821	9.2170	1.27714
784	614656	481890304	28.0000	9.2209	1.27551
785	616225	483736625	28.0179	9.2248	1.27389

TABLE 6-10 Cont.
Squares, Cubes, Square Roots, Cube Roots, and Reciprocals

n	n^2	n^3	$\sqrt{n}$	$\sqrt[3]{n}$	$\frac{1000}{n}$
786	617796	485587656	28.0357	9.2287	1.27226
787	619369	487443403	28.0535	9.2326	1.27065
788	620944	489303872	28.0713	9.2365	1.26904
789	622521	491169069	28.0891	9.2404	1.26743
790	624100	493039000	28.1069	9.2443	1.26582
791	625681	494913671	28.1247	9.2482	1.26422
792	627264	496793088	28.1425	9.2521	1.26263
793	628849	498677257	28.1603	9.2560	1.26103
794	630436	500566184	28.1780	9.2599	1.25945
795	632025	502459875	28.1957	9.2638	1.25786
796	633616	504358336	28.2135	9.2677	1.25628
797	635209	506261573	28.2312	9.2716	1.25471
798	636804	508169592	28.2489	9.2754	1.25313
799	638401	510082399	28.2666	9.2793	1.25156
800	640000	512000000	28.2843	9.2832	1.25000
801	641601	513922401	28.3019	9.2870	1.24844
802	643204	515849608	28.3196	9.2909	1.24688
803	644809	517781627	28.3373	9.2948	1.24533
804	646416	519718464	28.3549	9.2986	1.24378
805	648025	521660125	28.3725	9.3025	1.24224
806	649636	523606616	28.3901	9.3063	1.24069
807	651249	525557943	28.4077	9.3102	1.23916
808	652864	527514112	28.4253	9.3140	1.23762
809	654481	529475129	28.4429	9.3179	1.23609
810	656100	531441000	28.4605	9.3217	1.23457
811	657721	533411731	28.4781	9.3255	1.23305
812	659344	535387328	28.4956	9.3294	1.23153
813	660969	537367797	28.5132	9.3332	1.23001
814	662596	539353144	28.5307	9.3370	1.22850
815	664225	541343375	28.5482	9.3408	1.22699
816	665856	543338496	28.5657	9.3447	1.22549
817	667489	545338513	28.5832	9.3485	1.22399
818	669124	547343432	28.6007	9.3523	1.22249
819	670761	549353259	28.6182	9.3561	1.22100
820	672400	551368000	28.6356	9.3599	1.21951
821	674041	553387661	28.6531	9.3637	1.21803
822	675684	555412248	28.6705	9.3675	1.21655
823	677329	557441767	28.6880	9.3713	1.21507
824	678976	559476224	28.7054	9.3751	1.21359
825	680625	561515625	28.7228	9.3789	1.21212
826	682276	563559976	28.7402	9.3827	1.21065
827	683929	565609283	28.7576	9.3865	1.20919
828	685584	567663552	28.7750	9.3902	1.20773
829	687241	569722789	28.7924	9.3940	1.20627

TABLE 6-10 Cont.
Squares, Cubes, Square Roots, Cube Roots, and Reciprocals

n	n^2	n^3	$\sqrt{n}$	$\sqrt[3]{n}$	$\frac{1000}{n}$
830	688900	571787000	28.8097	9.3978	1.20482
831	690561	573856191	28.8271	9.4016	1.20337
832	692224	575930368	28.8444	9.4053	1.20192
833	693889	578009537	28.8617	9.4091	1.20048
834	695556	580093704	28.8791	9.4129	1.19904
835	697225	582182875	28.8964	9.4166	1.19760
836	698896	584277056	28.9137	9.4204	1.19617
837	700569	586376253	28.9310	9.4241	1.19474
838	702244	588480472	28.9482	9.4279	1.19332
839	703921	590589719	28.9655	9.4316	1.19189
840	705600	592704000	28.9828	9.4354	1.19048
841	707281	594823321	29.0000	9.4391	1.18906
842	708964	596947688	29.0172	9.4429	1.18765
843	710649	599077107	29.0345	9.4466	1.18624
844	712336	601211584	29.0517	9.4503	1.18483
845	714025	603351125	29.0689	9.4541	1.18343
846	715716	605495736	29.0861	9.4578	1.18203
847	717409	607645423	29.1033	9.4615	1.18064
848	719104	609800192	29.1204	9.4652	1.17925
849	720801	611960049	29.1376	9.4690	1.17786
850	722500	614125000	29.1548	9.4727	1.17647
851	724201	616295051	29.1719	9.4764	1.17509
852	725904	618470208	29.1890	9.4801	1.17371
853	727609	620650477	29.2062	9.4838	1.17233
854	729316	622835864	29.2233	9.4875	1.17096
855	731025	625026375	29.2404	9.4912	1.16959
856	732736	627222016	29.2575	9.4949	1.16822
857	734449	629422793	29.2746	9.4986	1.16686
858	736164	631628712	29.2916	9.5023	1.16550
859	737881	633839779	29.3087	9.5060	1.16414
860	739600	636056000	29.3258	9.5097	1.16279
861	741321	638277381	29.3428	9.5134	1.16144
862	743044	640503928	29.3598	9.5171	1.16009
863	744769	642735647	29.3769	9.5207	1.15875
864	746496	644972544	29.3939	9.5244	1.15741
865	748225	647214625	29.4109	9.5281	1.15607
866	749956	649461896	29.4279	9.5317	1.15473
867	751689	651714363	29.4449	9.5354	1.15340
868	753424	653972032	29.4618	9.5391	1.15207
869	755161	656234909	29.4788	9.5427	1.15075
870	756900	658503000	29.4958	9.5464	1.14943
871	758641	660776311	29.5127	9.5501	1.14811
872	760384	663054848	29.5296	9.5537	1.14679
873	762129	665338617	29.5466	9.5574	1.14548
874	763876	667627624	29.5635	9.5610	1.14416
875	765625	669921875	29.5804	9.5647	1.14286

TABLE 6-10 Cont.
Squares, Cubes, Square Roots, Cube Roots, and Reciprocals

n	n^2	n^3	$\sqrt{n}$	$\sqrt[3]{n}$	$\frac{1000}{n}$
876	767376	672221376	29.5973	9.5683	1.14155
877	769129	674526133	29.6142	9.5719	1.14025
878	770884	676836152	29.6311	9.5756	1.13895
879	772641	679151439	29.6479	9.5792	1.13766
880	774400	681472000	29.6648	9.5828	1.13636
881	776161	683797841	29.6816	9.5865	1.13507
882	777924	686128968	29.6985	9.5901	1.13379
883	779689	688465387	29.7153	9.5937	1.13250
884	781456	690807104	29.7321	9.5973	1.13122
885	783225	693154125	29.7489	9.6010	1.12994
886	784996	695506456	29.7658	9.6046	1.12867
887	786769	697864103	29.7825	9.6082	1.12740
888	788544	700227072	29.7993	9.6118	1.12613
889	790321	702595369	29.8161	9.6154	1.12486
890	792100	704969000	29.8329	9.6190	1.12360
891	793881	707347971	29.8496	9.6226	1.12233
892	795664	709732288	29.8664	9.6262	1.12108
893	797449	712121957	29.8831	9.6298	1.11982
894	799236	714516984	29.8998	9.6334	1.11857
895	801025	716917375	29.9166	9.6370	1.11732
896	802815	719323136	29.9333	9.6406	1.11607
897	804609	721734273	29.9500	9.6442	1.11483
898	806404	724150792	29.9666	9.6477	1.11359
899	808201	726572699	29.9833	9.6513	1.11235
900	810000	729000000	30.0000	9.6549	1.11111
901	811801	731432701	30.0167	9.6585	1.10988
902	813604	733870808	30.0333	9.6620	1.10865
903	815409	736314327	30.0500	9.6656	1.10742
904	817216	738763264	30.0666	9.6692	1.10619
905	819025	741217625	30.0832	9.6727	1.10497
906	820836	743677416	30.0998	9.6763	1.10375
907	822649	746142643	30.1164	9.6799	1.10254
908	824464	748613312	30.1330	9.6834	1.10132
909	826281	751089429	30.1496	9.6870	1.10011
910	828100	753571000	30.1662	9.6905	1.09890
911	829921	756058031	30.1828	9.6941	1.09769
912	831744	758550528	30.1993	9.6976	1.09649
913	833569	761048497	30.2159	9.7012	1.09529
914	835396	763551944	30.2324	9.7047	1.09409
915	837225	766060875	30.2490	9.7082	1.09290
916	839056	768575296	30.2655	9.7118	1.09170
917	840889	771095213	30.2820	9.7153	1.09051
918	842724	773620632	30.2985	9.7188	1.08932
919	844561	776151559	30.3150	9.7224	1.08814

TABLE 6-10 Cont.
Squares, Cubes, Square Roots, Cube Roots, and Reciprocals

n	n^2	n^3	$\sqrt{n}$	$\sqrt[3]{n}$	$\frac{1000}{n}$
920	846400	778688000	30.3315	9.7259	1.08696
921	848241	781229961	30.3480	9.7294	1.08578
922	850084	783777448	30.3645	9.7329	1.08460
923	851929	786330467	30.3809	9.7364	1.08342
924	853776	788889024	30.3974	9.7400	1.08225
925	855625	791453125	30.4138	9.7435	1.08108
926	857476	794022776	30.4302	9.7470	1.07991
927	859329	796597983	30.4467	9.7505	1.07875
928	861184	799178752	30.4631	9.7540	1.07759
929	863041	801765089	30.4795	9.7575	1.07643
930	864900	804357000	30.4959	9.7610	1.07527
931	866761	806954491	30.5123	9.7645	1.07411
932	868624	809557568	30.5287	9.7680	1.07296
933	870489	812166237	30.5450	9.7715	1.07181
934	872356	814780504	30.5614	9.7750	1.07066
935	874225	817400375	30.5778	9.7785	1.06952
936	876096	820025856	30.5941	9.7819	1.06838
937	877969	822656953	30.6105	9.7854	1.06724
938	879844	825293672	30.6268	9.7889	1.06610
939	881721	827936019	30.6431	9.7924	1.06496
940	883600	830584000	30.6594	9.7959	1.06383
941	885481	833237621	30.6757	9.7993	1.06270
942	887364	835896888	30.6920	9.8028	1.06157
943	889249	838561807	30.7083	9.8063	1.06045
944	891136	841232384	30.7246	9.8097	1.05932
945	893025	843908625	30.7409	9.8132	1.05820
946	894916	846590536	30.7571	9.8167	1.05708
947	896809	849278123	30.7734	9.8201	1.05597
948	898704	851971392	30.7896	9.8236	1.05485
949	900601	854670349	30.8058	9.8270	1.05374
950	902500	857375000	30.8221	9.8305	1.05263
951	904401	860085351	30.8383	9.8339	1.05152
952	906304	862801408	30.8545	9.8374	1.05042
953	908209	865523177	30.8707	9.8408	1.04932
954	910116	868250664	30.8869	9.8443	1.04822
955	912025	870983875	30.9031	9.8477	1.04712
956	913936	873722816	30.9192	9.8511	1.04603
957	915849	876467493	30.9354	9.8546	1.04493
958	917764	879217912	30.9516	9.8580	1.04384
959	919681	881974079	30.9677	9.8614	1.04275
960	921600	884736000	30.9839	9.8648	1.04167
961	923521	887503681	31.0000	9.8683	1.04058
962	925444	890277128	31.0161	9.8717	1.03950
963	927369	893056347	31.0322	9.8751	1.03842
964	929296	895841344	31.0483	9.8785	1.03734
965	931225	898632125	31.0644	9.8819	1.03627

TABLE 6-10 Cont.
Squares, Cubes, Square Roots, Cube Roots, and Reciprocals

n	n^2	n^3	$\sqrt{n}$	$\sqrt[3]{n}$	$\frac{1000}{n}$
966	933156	901428696	31.0805	9.8854	1.03520
967	935089	904231063	31.0966	9.8888	1.03413
968	937024	907039232	31.1127	9.8922	1.03306
969	938961	909853209	31.1288	9.8956	1.03199
970	940900	912673000	31.1448	9.8990	1.03093
971	942841	915498611	31.1609	9.9024	1.02987
972	944784	918330048	31.1769	9.9058	1.02881
973	946729	921167317	31.1929	9.9092	1.02775
974	948676	924010424	31.2090	9.9126	1.02669
975	950625	926859375	31.2250	9.9160	1.02564
976	952576	929714176	31.2410	9.9194	1.02459
977	954529	932574833	31.2570	9.9227	1.02354
978	956484	935441352	31.2730	9.9261	1.02249
979	958441	938313739	31.2890	9.9295	1.02145
980	960400	941192000	31.3050	9.9329	1.02041
981	962361	944076141	31.3209	9.9363	1.01937
982	964324	946966168	31.3369	9.9396	1.01833
983	966289	949862087	31.3528	9.9430	1.01729
984	968256	952763904	31.3688	9.9464	1.01626
985	970225	955671625	31.3847	9.9497	1.01523
986	972196	958585256	31.4006	9.9531	1.01420
987	974169	961504803	31.4166	9.9565	1.01317
988	976144	964430272	31.4325	9.9598	1.01215
989	978121	967361669	31.4484	9.9632	1.01112
990	980100	970299000	31.4643	9.9666	1.01010
991	982081	973242271	31.4802	9.9699	1.00908
992	984064	976191488	31.4960	9.9733	1.00806
993	986049	979146657	31.5119	9.9766	1.00705
994	988036	982107784	31.5278	9.9800	1.00604
995	990025	985074875	31.5436	9.9833	1.00503
996	992016	988047936	31.5595	9.9866	1.00402
997	994009	991026973	31.5753	9.9900	1.00301
998	996004	994011992	31.5911	9.9933	1.00200
999	998001	997002999	31.6070	9.9967	1.00100
1000	1000000	1000000000	31.6228	10.0000	1.00000

Chapter 7

MISCELLANEOUS

TEMPERATURE CONVERSION

The nomograph in Fig. 7-1 can be used to convert from degrees Fahrenheit to degrees Celsius (or vice versa) for any temperature between absolute zero and 540°F (281°C). The term Celsius was officially adopted, in place of centigrade, by international agreement in 1948. Actually, Celsius and centigrade scales differ slightly—the Celsius scale is based on 0° at the triple point of water (0.01°C), and centigrade has 0° at the freezing point of water. For all practical purposes, though, the two terms are interchangeable.

Two absolute temperature scales are also in use. The Fahrenheit absolute scale is called the Rankine—0°R = -459.67°F. The Celsius absolute scale is the Kelvin—0 K = - 273.16°C. Note the degree sign (°) is not used with Kelvin, the SI unit of temperature.

The following formulas can be used to convert from any temperature to the other:

$$°F = (°C \times 9/5) + 32$$

$$°F = °R - 459.67$$

$$°F = 9/5\,(K - 273.16) + 32$$

$$°C = 5/9\,(°F - 32)$$

$$°C = K - 273.16$$

$$°C = 5/9\,(°R - 491.67)$$

$$°R = °F + 459.67$$

$$°R = (°C \times 9/5) + 491.67$$

$$°R = 9/5\,(K - 273.16) + 491.67$$

$$K = °C + 273.16$$

$$K = 5/9\,(°F - 32) + 273.16$$

$$K = 5/9\,(°R - 491.67) + 273.16$$

TELEPRINTER CODES

Letter and figure assignments for teleprinter codes are given in Table 7-1.

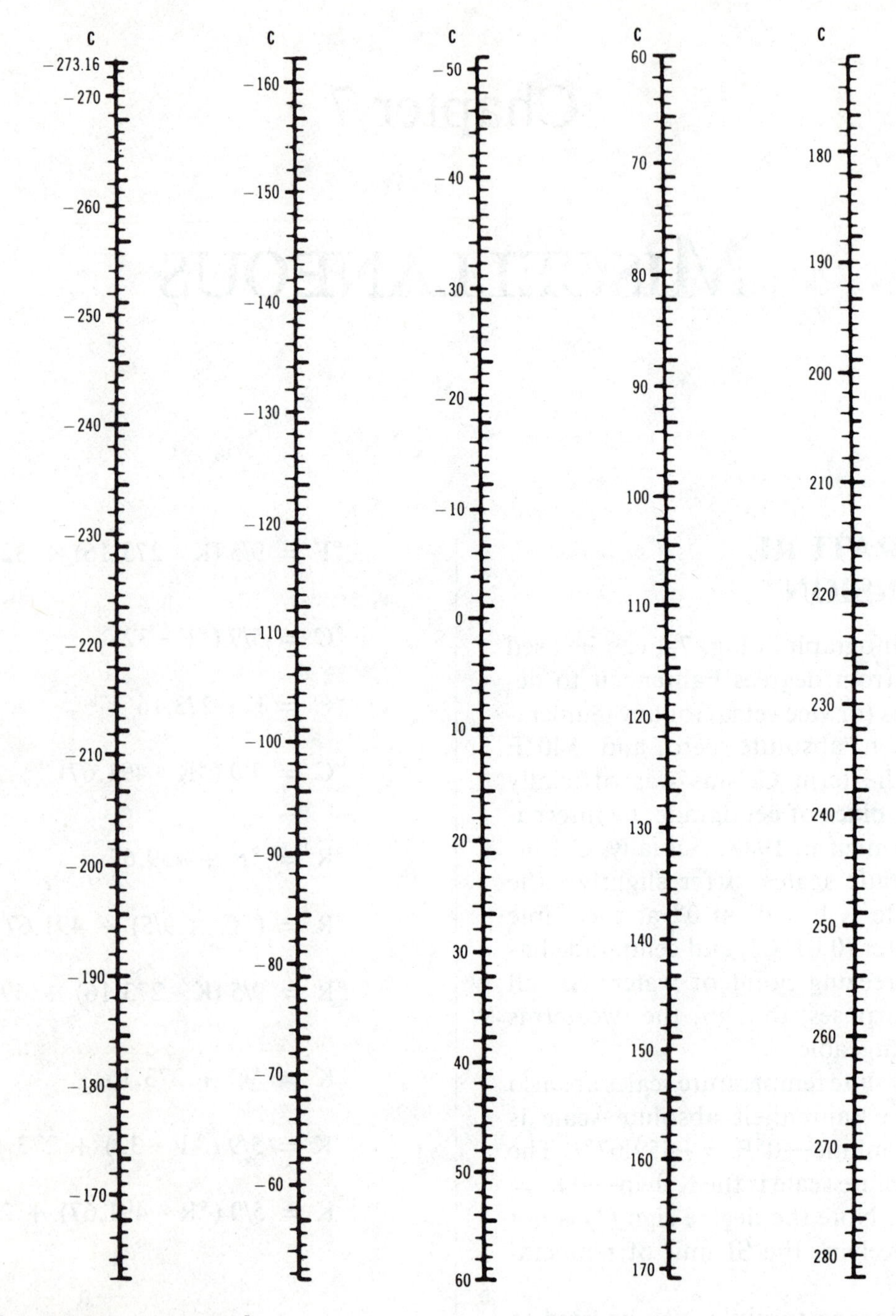

Fig. 7-1. Temperature nomograph.

TABLE 7-1
Moore ARQ Code (Compared with Five-Unit Teleprinter Code)

Code assignments		Moore ARQ code	Five-unit TTY code
Letters case	*Figures case*	*Bit numbers 7 6 5 4 3 2 1*	*Bit numbers 5 4 3 2 1*
blank	blank	1 1 1 0 0 0 0	0 0 0 0 0
E	3	0 0 0 1 1 1 0	0 0 0 0 1
line feed	line feed	0 0 0 1 1 0 1	0 0 0 1 0
A	—	0 1 0 1 1 0 0	0 0 0 1 1
space	space	0 0 0 1 0 1 1	0 0 1 0 0
S	apostrophe	0 1 0 1 0 1 0	0 0 1 0 1
I	8	0 0 0 0 1 1 1	0 0 1 1 0
U	7	0 1 0 0 1 1 0	0 0 1 1 1
carriage return	carriage return	1 1 0 0 0 0 1	0 1 0 0 0
D	⊕	0 0 1 1 1 0 0	0 1 0 0 1
R	4	0 0 1 0 0 1 1	0 1 0 1 0
J	bell	1 1 0 0 0 1 0	0 1 0 1 1
N	comma	0 0 1 0 1 0 1	0 1 1 0 0
F	□	1 1 0 0 1 0 0	0 1 1 0 1
C	:	0 0 1 1 0 0 1	0 1 1 1 0
K	(	1 1 0 1 0 0 0	0 1 1 1 1
T	5	1 0 1 0 0 0 1	1 0 0 0 0
Z	+	1 0 0 0 1 1 0	1 0 0 0 1
L	)	0 1 0 0 0 1 1	1 0 0 1 0
W	2	1 0 1 0 0 1 0	1 0 0 1 1
H	□	0 1 0 0 1 0 1	1 0 1 0 0
Y	6	1 0 1 0 1 0 0	1 0 1 0 1
P	0	0 1 0 1 0 0 1	1 0 1 1 0
Q	1	1 0 1 1 0 0 0	1 0 1 1 1
O	9	0 1 1 0 0 0 1	1 1 0 0 0
B	?	1 0 0 1 1 0 0	1 1 0 0 1
G	□	1 0 0 0 0 1 1	1 1 0 1 0
figures	figures	0 1 1 0 0 1 0	1 1 0 1 1
M	.	1 0 0 0 1 0 1	1 1 1 0 0
X	/	0 1 1 0 1 0 0	1 1 1 0 1
V	=	1 0 0 1 0 0 1	1 1 1 1 0
letters	letters	0 1 1 1 0 0 0	1 1 1 1 1
signal 1	signal 1	0 0 1 0 1 1 0	
idle α	idle α	1 0 0 1 0 1 0	
idle β	idle β	0 0 1 1 0 1 0	

Note: Transmission Order: Bit 1→Bit 7.

ASCII CODE

The American Standard Code for Information Interchange (ASCII) Code is used extensively in computer data transmission. The ASCII Code, which is produced by most computer keyboards, is shown in Table 7-2.

KANSAS CITY STANDARD

The Kansas City standard is a widely used digital tape format, consisting of 1's represented by eight cycles of 2400 Hz, and 0's represented by four cycles of 1200 Hz (Fig. 7-2). It is a frequency-shift keying (FSK) mode of operation using tone bursts. A variation of the Kansas City standard employs the same frequencies for 1 and 0, but with different durations. Each bit starts with a 3700-Hz frequency and ends with a 2400-Hz frequency, and each bit has the same duration (7.452 ms). However, a 1 has its frequency-transition point one-third from the start of the burst, whereas a 0 has its frequency-transition point two-thirds from the start of the burst.

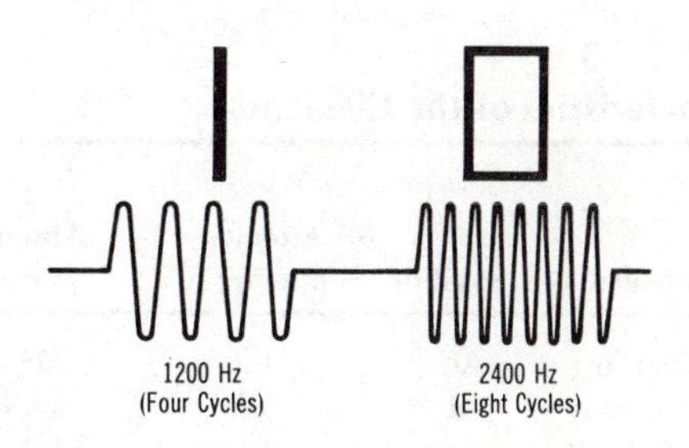

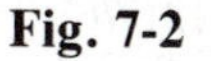
Fig. 7-2

TABLE 7-2
The ASCII Code

Bit numbers b_7					0	0	0	0	1	1	1	1
b_6					0	0	1	1	0	0	1	1
b_5					0	1	0	1	0	1	0	1
b_4 ↓	b_3 ↓	b_2 ↓	b_1 ↓	Column → Row ↓	0	1	2	3	4	5	6	7
0	0	0	0	0	NUL	DLE	SP	0	@	P		p
0	0	0	1	1	SOH	DC1	!	1	A	Q	a	q
0	0	1	0	2	STX	DC2	‖	2	B	R	b	r
0	0	1	1	3	ETX	DC3	#	3	C	S	c	s
0	1	0	0	4	EOT	DC4	$	4	D	T	d	t
0	1	0	1	5	ENQ	NAK	%	5	E	U	e	u
0	1	1	0	6	ACK	SYN	&	6	F	V	f	v
0	1	1	1	7	BEL	ETB	'	7	G	W	g	w
1	0	0	0	8	BS	CAN	(	8	H	X	h	x
1	0	0	1	9	HT	EM	)	9	I	Y	i	y
1	0	1	0	A	LF	SUB	*	:	J	Z	j	z
1	0	1	1	b	VT	ESC	+	;	K	[	k	{
1	1	0	0	C	FF	FS	,	<	L	\	l	:
1	1	0	1	d	CR	GS	—	=	M	]	m	}
1	1	1	0	E	SO	RS	.	>	N	∧	n	~
1	1	1	1	F	SI	US	/	?	O	—	o	DEL

CHARACTERISTICS OF THE ELEMENTS

A list of all the known elements (105) is given in Table 7-3. The symbol, atomic number, and atomic weight are included for each element. Where known, the melting and boiling points of each element are also given.

TABLE 7-3
Characteristics of the Elements

Element	Symbol	Atomic number	Atomic weight	Melting point (°C)	Boiling point (°C)	Density (20 °C) (g/cm³)
actinium	Ac	89	227*	1050	3220	
aluminum	Al	13	26.97	660.1	2467	2.70
americium	Am	95	243*	1000	7600	
antimony	Sb	51	121.76	630.5	1380	6.62
argon	A	18	39.944	−189.2	−185.7	1.78[†]
arsenic	As	33	74.91	(820)[‡]	615	5.73
astatine	At	85	210*	—	—	
barium	Ba	56	137.36	725	1140	3.50
berkelium	Bk	97	247*	—	—	
beryllium	Be	4	9.013	1350	2970	1.82
bismuth	Bi	83	209.00	271.3	1560	9.80
boron	B	5	10.82	2300	2550	2.30

TABLE 7-3 Cont.
Characteristics of the Elements

Element	Symbol	Atomic number	Atomic weight	Melting point (°C)	Boiling point (°C)	Density (20 °C) (g/cm³)
bromine	Br	35	79.916	-7.2	58.8	3.12
cadmium	Cd	48	112.41	320.9	766	8.65
calcium	Ca	20	40.08	850	1487	1.54
californium	Cf	98	251*	—	—	—
carbon	C	6	12.01	>3500§	4827	2.22
cerium	Ce	58	140.13	795	3468	6.90
cesium	Cs	55	132.91	28	670	1.87
chlorine	Cl	17	35.457	-101.6	-34.7	3.21†
chromium	Cr	24	52.01	1890	2482	7.14
cobalt	Co	27	58.94	1492	3000	8.90
copper	Cu	29	63.54	1083	2595	8.96
curium	Cm	96	247	—	—	—
dysprosium	Dy	66	162.46	1400	2600	—
einsteinium	E	99	254*	—	—	—
erbium	Er	68	167.2	1497	2900	—
europium	Eu	63	152.0	826	1439	—
fermium	Fm	100	255*	—	—	—
fluorine	F	9	19.00	-223	-188.14	1.69(15°)†
francium	Fr	87	233*	—	—	—
gadolinium	Gd	64	156.9	1312	3000	—
gallium	Ga	31	69.72	29.7	2403	5.91
germanium	Ge	32	72.60	958.5	(2700)‡	5.36
gold	Au	79	197.0	1063	2966	19.30
hafnium	Hf	72	178.6	2150	5400	11.40
hahmium	Ha	105	262*	—	—	—
helium	He	2	4.003	<-271.4§	-268.94	0.164†
holmium	Ho	67	164.94	1461	2600	—
hydrogen	H	1	1.0080	-259.14	-252.8	0.08375†
indium	In	49	114.76	155	2000	7.31
iodine	I	53	126.91	113.5	184.35	4.93
iridium	Ir	77	192.2	2443	4500	22.4
iron	Fe	26	55.85	1533	3000	7.87
krypton	Kr	36	83.8	-156.6	-151.8	3.448†
lanthanum	La	57	138.92	920	3469	6.15
lawrencium	Lw	103	257*	—	—	—
lead	Pb	82	207.21	327.4	1744	11.34
lithium	Li	3	6.940	186	1317	0.53
lutetium	Lu	71	174.99	1652	3327	—
magnesium	Mg	12	24.32	651	1100	1.74
manganese	Mn	25	54.94	1260	2097	7.44
mendelevium	Mv	101	256*	—	—	—
mercury	Hg	80	200.61	-38.87	356.9	13.55
molybdenum	Mo	42	95.95	2620	5660	10.20
neodymium	Nd	60	144.27	1024	3027	7.05
neon	Ne	10	20.183	-248.67	-245.9	0.8387†
neptunium	Np	93	237*	639	—	—
nickel	Ni	28	58.69	1453	2900	8.90
niobium	Nb	41	92.91	2500	4927	8.57
nitrogen	N	7	14.008	-209.86	-195.81	1.1649†
nobelium	No	102	253	—	—	

TABLE 7-3 Cont.
Characteristics of the Elements

Element	Symbol	Atomic number	Atomic weight	Melting point (°C)	Boiling point (°C)	Density (20 °C) (g/cm³)
osmium	Os	76	190.2	2700	(>5300)‡§	22.48
oxygen	O	8	16.000	-218.4	-183	1.3318†
palladium	Pd	46	106.7	1552	2927	12.00
phosphorus	P	15	30.975	44.1	280	1.82
platinum	Pt	78	195.23	1769	3800	21.45
plutonium	Pu	94	244	639	3250	—
polonium	Po	84	210	250	960	—
potassium	K	19	39.100	62.3	760	0.86
praseodymium	Pr	59	140.92	940	3127	6.63
promethium	Pm	61	145*	1035	2730	—
protactinium	Pa	91	231*	1225	—	—
radium	Ra	88	226.05	700	1140	5.00
radon	Rn	86	222	-76	-62	4.40†
rhenium	Re	75	186.31	(3000)‡	5627	20.00
rhodium	Rh	45	102.91	1960	3960	12.44
rubidium	Rb	37	85.48	38.5	700	1.53
ruthenium	Ru	44	101.1	2500	4111	12.2
rutherfordium or kurchatonium	Rf or Ku	104	260*	—	—	—
samarium	Sm	62	150.43	>1300§	1900	7.70
scandium	Sc	21	44.96	1539	2727	2.50
selenium	Se	34	78.96	220	688	4.81
silicon	Si	14	28.09	1420	2355	2.40
silver	Ag	47	107.880	960.8	2212	10.49
sodium	Na	11	22.997	97.5	880	0.97
strontium	Sr	38	87.63	880	1384	2.60
sulfur	S	16	32.066	112.8	444.6	2.07
tantalum	Ta	73	180.95	3005	5425	16.60
technetium	Tc	43	97*	2140	—	—
tellurium	Te	52	127.61	452	989.8	6.24
terbium	Tb	65	158.93	1356	2550	—
thallium	Tl	81	204.39	303.5	1457	11.85
thorium	Th	90	232.12	1845	>3000§	11.50
thulium	Tm	69	168.94	1545	1727	—
tin	Sn	50	118.70	231.9	2260	7.30
titanium	Ti	22	47.90	1820	(>3000)‡§	4.54
tungsten	W	74	183.92	3380	5900	19.30
uranium	U	92	238.07	1133	3818	18.70
vanadium	V	23	50.95	1735	(3000)‡	5.68
xenon	Xe	54	131.3	-111.9	-109.1	5.495†
ytterbium	Yb	70	173.04	875	1450	—
yttrium	Y	39	88.92	1490	(2500)‡	5.51
zinc	Zn	30	65.38	419.47	907	7.14
zirconium	Zr	40	91.22	1852	3578	6.40

*Mass number of the longest-lived of the known available forms of the element, usually synthetic.
†Grams per liter.
‡Values in parentheses indicate an approximate value.
§< indicates that the value may be lower; > indicates that the value may be higher.

MEASURES AND WEIGHTS

Linear Measure

1 inch	= 1000 mils
1 hand	= 4 inches
1 foot	= 12 inches
1 yard	= 3 feet
1 fathom	= 6 feet
1 rod	= 5½ yards
1 furlong	= 40 rods
1 statute mile	= 8 furlongs
1 statute mile	= 5280 feet
1 nautical mile	= 6076.1 feet
1 nautical mile	= 1.1508 statute miles
1 league	= 3 miles

Square Measure

1 square foot	= 144 square inches
1 square yard	= 9 square feet
1 square rod	= 30¼ square yards
1 section (of land)	= 1 square mile
1 township	= 6 miles square (36 square miles)
1 acre	= 160 square rods
1 acre	= 43,560 square feet
1 square mile	= 640 acres

Volume Measure

1 cubic foot	= 1728 cubic inches
1 cubic yard	= 27 cubic feet
1 U.S. gallon	= 231 cubic inches

Liquid Measure

1 pint	= 4 gills
1 quart	= 2 pints
1 gallon	= 4 quarts
1 barrel (petroleum)	= 42 gallons
1 barrel	= 31½ gallons
1 hogshead	= 2 barrels (63 gallons)
1 tun	= 252 gallons

Dry Measure

1 quart	= 2 pints = 67.2006 cubic inches
1 peck	= 8 quarts = 537.605 cubic inches
1 bushel	= 4 pecks = 2150.419 cubic inches
1 barrel	= 3.281 bushels = 7056 cubic inches

Avoirdupois Weight

(for other than drugs, gold, silver, etc.)

1 dram (dr)	= 27.3437 grains*
1 ounce (oz)	= 16 drams
1 pound (lb)	= 16 ounces
1 quarter	= 25 pounds
1 hundredweight (cwt)	= 4 quarters
1 ton (tn)	= 20 hundredweights
1 short ton	= 2000 pounds
1 long ton	= 2240 pounds

Troy Weight

(for gold, silver, etc.)

1 pennyweight (dwt)	= 24 grains*
1 ounce troy (oz t)	= 20 pennyweights
1 pound troy (lb t)	= 12 ounces troy = 240 pennyweights = 5760 grains

Apothecaries' Weight

(for drugs)

1 dram apoth (dr ap)	= 3 scruples
1 ounce apoth (oz ap)	= 8 drams apoth
1 pound apoth (lb ap)	= 12 ounces apoth = 96 drams apoth = 288 scruples = 5760 drams
1 scruple (s ap)	= 20 grains*

*1 grain = 0.0648 gram

METRIC SYSTEM

Linear Measure

10 millimeters	= 1 centimeter
10 centimeters	= 1 decimeter
10 decimeters	= 1 meter
1000 meters	= 1 kilometer

Area Measure

100 square millimeters	= 1 square centimeter
100 square centimeters	= 1 square decimeter
100 square decimeters	= 1 square meter

Volume Measure

1000 cubic millimeters	= 1 cubic centimeter
1000 cubic centimeters	= 1 cubic decimeter
1000 cubic decimeters	= 1 cubic meter

Liquid Measure

10 milliliters	= 1 centiliter
10 centiliters	= 1 deciliter
10 deciliters	= 1 liter

Weight Measure

10 milligrams	= 1 centigram
10 centigrams	= 1 decigram
10 decigrams	= 1 gram
10 grams	= 1 dekagram
10 dekagrams	= 1 hectogram
10 hectograms	= 1 kilogram
1000 kilograms	= 1 metric ton

WINDS

Designation	Miles per hour
calm	less than 1
light air	1–3
light breeze	4–7
gentle breeze	8–12
moderate breeze	13–18
fresh breeze	19–24
strong breeze	25–31
moderate gale	32–38
fresh gale	39–46
strong gale	47–54
whole gale	55–63
storm	64–72
hurricane	above 72

WEIGHT OF WATER

1 cubic inch	= 0.0360 pound
12 cubic inches	= 0.433 pound
1 cubic foot	= 62.4 pounds
1 cubic foot	= 7.48052 U.S. gallons
1.8 cubic feet	= 112.0 pounds
35.96 cubic feet	= 2240.0 pounds
1 imperial gallon	= 10.0 pounds
11.2 imperial gallons	= 112.0 pounds
224 imperial gallons	= 2240.0 pounds
1 U.S. gallon	= 8.33 pounds
13.45 U.S. gallons	= 112.0 pounds
269.0 U.S. gallons	= 2240.0 pounds

HYDRAULIC EQUATIONS

pounds per square inch	= 0.434 × head of water in feet
head in feet	= 2.31 × pounds per square inch

Approximate loss of head due to friction in clean iron pipes is:

$$\frac{0.02 \times L \times V^2}{64.4D} \text{ ft}$$

where

L is the length of pipe, in feet,

V is the velocity of flow, in feet per second,

D is the diameter, in feet.

In calculating the total head to be pumped against, it is common to consider this value as being equal to the sum of the friction head and the actual head:

$$\text{Horsepower of waterfall} = \frac{62 \times A \times V \times H}{33{,}000}$$

where
A is the cross section of water, in square feet,
V is the velocity of flow, in feet per minute,
H is the head of fall, in feet.

FALLING OBJECT

The speed acquired by a falling object is determined by the formula:

$$V = 32t$$

where
V is the velocity, in feet per second,
t is the time, in seconds.

The distance traveled by a falling object is determined by the formula:

$$d = 16t^2$$

where
d is the distance traveled, in feet,
t is the time, in seconds.

SPEED OF SOUND

The speed of sound through air at 0 °C is usually considered to be 1087.42 ft/s, and at normal temperature, 1130 ft/s. The speed of sound through any given temperature of air is determined by the formula:

$$V = \frac{1087\sqrt{(273 + t)}}{16.52}$$

where
V is the speed, in feet per second,
t is the temperature, in degrees Celsius.

PROPERTIES OF FREE SPACE

$$\begin{aligned} \text{velocity of light} &= c \\ &= \frac{1}{(\mu_r \varepsilon_r)^{1/2}} \\ &= 2.998 \times 10^8 \text{ m/s} \\ &= 186{,}280 \text{ mi/s} \\ &= 984 \times 10^6 \text{ ft/s} \end{aligned}$$

$$\begin{aligned} \text{permeability} &= \mu_v = 4\pi \times 10^{-7} \\ &= 1.257 \times 10^{-6} \text{ H/m} \end{aligned}$$

$$\begin{aligned} \text{permittivity} &= \varepsilon_v = 8.85 \times 10^{-12} \\ &\approx (36\pi \times 10^9)^{-1} \text{ F/m} \end{aligned}$$

$$\begin{aligned} \text{characteristic impedance} &= Z_0 = \left(\frac{\mu_v}{\varepsilon_r}\right)^{1/2} \\ &= 376.7 \\ &\approx 120\pi\ \Omega \end{aligned}$$

COST OF OPERATION

The cost of operation of an electrical device is determined by the formula:

$$C = \frac{Wtc}{1000}$$

where
C is the cost of operation,
W is the wattage of the device, in watts,
t is the time, in hours,
c is the cost per kilowatt-hour of electricity.

CONVERSION OF MATTER INTO ENERGY

The conversion of matter into energy (Einstein's theorem) is expressed by:

$$E = mc^2$$

where

E is the energy, in ergs,
m is the mass of the matter, in grams,
c is the speed of light, in centimeters per second ($c^2 = 9 \times 10^{20}$).

ATOMIC SECOND

The atomic second was permanently adopted as the International Unit of Time by the 13th General Conference on Weights and Measures, in Paris on October 13, 1967. The atomic second is defined as the duration of 9,192,631,770 periods of the radiation corresponding to the transition between two specific hyperfine levels of the fundamental state of the cesium-133 atom. It was chosen to be identical with the ephemeris second.

INTERNATIONAL AND ABSOLUTE UNITS

The following list shows the international unit values compared to the absolute values. The values used will vary from one country to another.

1 international volt = 1.00033 absolute volt

1 international ohm = 1.000495 absolute ohm

1 international coulomb = 0.999835 absolute coulomb

1 international henry = 1.00049 absolute henry

1 international farad = 0.999505 absolute farad

1 international joule = 1.000165 absolute joule

1 international watt = 1.000165 absolute watt

DEGREES, MINUTES, AND SECONDS OF A CIRCLE

A complete circle consists of 360 equal divisions called degrees. Each degree is made up of 60 equal parts called minutes, and each minute is made up of 60 seconds. Thus, a circle consists of 360 degrees or 21,600 minutes, or 1,296,000 seconds. Table 7-4 converts minutes and seconds to decimal parts of a degree.

GRAD

A grad is equal to 0.01 of a right angle. Computers designed for engineering applications may provide a choice of degrees, radians, or grads. There are 2 π (6.2832 . . .) radians in 360°.

TABLE 7-4
Minutes and Seconds in Decimal Parts of a Degree

Minutes	Degrees	Minutes	Degrees	Seconds	Degrees	Seconds	Degrees
1	0.01667	31	0.51667	1	0.00028	31	0.00861
2	0.03333	32	0.53333	2	0.00056	32	0.00889
3	0.05	33	0.55	3	0.00083	33	0.00917
4	0.06667	34	0.56667	4	0.00111	34	0.00944
5	0.08333	35	0.58333	5	0.00139	35	0.00972
6	0.10	36	0.60	6	0.00167	36	0.01
7	0.11667	37	0.61667	7	0.00194	37	0.01028
8	0.13333	38	0.63333	8	0.00222	38	0.01056
9	0.15	39	0.65	9	0.0025	39	0.01083
10	0.16667	40	0.66667	10	0.00278	40	0.01111
11	0.18333	41	0.68333	11	0.00306	41	0.01139
12	0.20	42	0.70	12	0.00333	42	0.01167
13	0.21667	43	0.71667	13	0.00361	43	0.01194
14	0.23333	44	0.73333	14	0.00389	44	0.01222
15	0.25	45	0.75	15	0.00417	45	0.0125
16	0.26667	46	0.76667	16	0.00444	46	0.01278
17	0.28333	47	0.78333	17	0.00472	47	0.01306
18	0.30	48	0.80	18	0.005	48	0.01333
19	0.31667	49	0.81667	19	0.00528	49	0.01361
20	0.33333	50	0.83333	20	0.00556	50	0.01389
21	0.35	51	0.85	21	0.00583	51	0.01417
22	0.36667	52	0.86667	22	0.00611	52	0.01444
23	0.38333	53	0.88333	23	0.00639	53	0.01472
24	0.40	54	0.90	24	0.00667	54	0.015
25	0.41667	55	0.91667	25	0.00694	55	0.01528
26	0.43333	56	0.93333	26	0.00722	56	0.01556
27	0.45	57	0.95	27	0.0075	57	0.01583
28	0.46667	58	0.96667	28	0.00778	58	0.01611
29	0.48333	59	0.98333	29	0.00806	59	0.01639
30	0.50	60	1.00	30	0.00833	60	0.01667

Appendix A

Calculations Using Commodore 64® Computer

Robert L. Kruse

The following pages contain programs which may be run on the Commodore 64® computer to perform the following calculations:

1. Conversion of impedance to admittance and admittance to impedance.
2. Conversion of vectors from rectangular to polar form and polar form to rectangular form.
3. Impedance and phase angle of resultant for two vectors (phasors) in parallel.
4. Input impedance and phase angle of RLC parallel resonant circuit.
5. Unsymmetrical two-section lag circuit (with or without resistive load).
6. Unsymmetrical two-section lead circuit (with or without resistive load).

For each application two programs are included. The first is for use without a printer and the second for use with a printer.

PROGRAM 1—Conversion of impedance to admittance and admittance to impedance

Without Printer

```
5 REM PRG 1 W/O PRINTER
10 PRINT "*CONVERSION OF IMPEDANCE TO ADMITTANCE*"
20 PRINT "      *OR ADMITTANCE TO IMPEDANCE*"
30 PRINT:PRINT:R=0:X=0:G=0:B=0:Y=0:GG=0:BB=0:PH=0:TH=0:Z=0
40 PRINT "IF IMPEDANCE DATA,RUN 90:IF ADMITTANCE DATA,RUN 240":END
90 PRINT:INPUT "R (OHMS)=";R
110 INPUT "X (OHMS)=";X
120 G=R/(R↑2+X↑2):B=X/(R↑2+X↑2):PRINT
130 Y=(G↑2+B↑2)↑.5:PH=ATN(B/G):PH=-PH
140 PRINT "G CONDUCTANCE (SIEMENS)=";G
160 PRINT "B SUSCEPTANCE (SIEMENS)=";B
180 PRINT "(RECTANGULAR FORM)":PRINT
190 PRINT "Y ADMITTANCE (SIEMENS)=";Y
195 PZ=.01*INT((PH*360/6.2832)*100)
210 PRINT "Y PHASE ANGLE (DEGREES)=";PZ
230 PRINT "(POLAR FORM)":PRINT:END
240 INPUT "G (SIEMENS)=";GG
260 INPUT "B (SIEMENS)=";BB:PRINT
280 R=GG/(GG↑2+BB↑2):X=BB/(GG↑2+BB↑2)
290 Z=(R↑2+X↑2)↑.5:TH=ATN(X/R)
295 TP=.01*INT(R*100)
300 PRINT "R RESISTANCE (OHMS)=";TP
310 LP=.01*INT(X*100)
320 PRINT "X REACTANCE (OHMS)=";LP
340 PRINT "(RECTANGULAR FORM)":PRINT
345 PL=.01*INT(Z*100)
350 PRINT "Z IMPEDANCE (OHMS)=";PL
355 LA=.01*INT((TH*360/6.2832)*100)
370 PRINT "Z PHASE ANGLE (DEGREES)=";LA
390 PRINT "(POLAR FORM)":PRINT
400 PRINT "****************************************":PRINT
410 END
```

With Printer

```
5 REM PRG 1 WITH PRINTER
10 PRINT "*CONVERSION OF IMPEDANCE TO ADMITTANCE*"
15 OPEN1,4:PRINT#1,"*CONVERSION OF IMPEDANCE TO ADMITTANCE*":CLOSE1,4
20 PRINT "      *OR ADMITTANCE TO IMPEDANCE*"
25 OPEN1,4:PRINT#1,"      *OR ADMITTANCE TO IMPEDANCE*"
27 PRINT#1,:PRINT#1,:CLOSE1,4
30 PRINT:PRINT:R=0:X=0:G=0:B=0:Y=0:GG=0:BB=0:PH=0:TH=0:Z=0
40 PRINT "IF IMPEDANCE DATA,RUN 90:IF ADMITTANCE DATA,RUN 240"
50 OPEN1,4:PRINT#1,"IF IMPEDANCE DATA,RUN 90:IF ADMITTANCE DATA,RUN 240"
60 PRINT#1,:CLOSE1,4:END
90 PRINT:OPEN1,4:PRINT#1,:CLOSE1,4:INPUT "R (OHMS)=";R
100 OPEN1,4:PRINT#1,"R (OHMS)=";R:CLOSE1,4
110 INPUT "X (OHMS)=";X
115 OPEN1,4:PRINT#1,"X (OHMS)=";X:PRINT#1,:CLOSE1,4
120 G=R/(R↑2+X↑2):B=X/(R↑2+X↑2):PRINT
130 Y=(G↑2+B↑2)↑.5:PH=ATN(B/G):PH=-PH
140 PRINT "G CONDUCTANCE (SIEMENS)=";G
150 OPEN1,4:PRINT#1,"G CONDUCTANCE (SIEMENS)=";G:CLOSE1,4
160 PRINT "B SUSCEPTANCE (SIEMENS)=";B
170 OPEN1,4:PRINT#1,"B SUSCEPTANCE (SIEMENS)=";B:CLOSE1,4
```

```
180 PRINT "(RECTANGULAR FORM)":PRINT
185 OPEN1,4:PRINT#1,"(RECTANGULAR FORM)":PRINT#1,:CLOSE1,4
190 PRINT "Y ADMITTANCE (SIEMENS)=";Y
193 OPEN1,4:PRINT#1,"Y ADMITTANCE (SIEMENS)=";Y:CLOSE1,4
195 PZ=.01*INT((PH*360/6.2832)*100)
210 PRINT "Y PHASE ANGLE (DEGREES)=";PZ
220 OPEN1,4:PRINT#1,"Y PHASE ANGLE (DEGREES)=";PZ:CLOSE1,4
230 PRINT "(POLAR FORM)"
235 OPEN1,4:PRINT#1,"(POLAR FORM)":END:CLOSE1,4
240 PRINT:OPEN1,4:PRINT#1,:CLOSE1,4:INPUT "G (SIEMENS)=";GG
250 OPEN1,4:PRINT#1,"G (SIEMENS)=";GG:CLOSE1,4
260 INPUT "B (SIEMENS)=";BB
270 OPEN1,4:PRINT#1,"B (SIEMENS)=";BB:PRINT#1,:CLOSE1,4
280 R=GG/(GG↑2+BB↑2):X=BB/(GG↑2+BB↑2)
290 Z=(R↑2+X↑2)↑.5:TH=ATN(X/R):PRINT
295 TP=.01*INT(R*100)
300 PRINT "R RESISTANCE (OHMS)=";TP
305 OPEN1,4:PRINT#1,"R RESISTANCE (OHMS)=";TP:CLOSE1,4
310 LP=.01*INT(X*100)
320 PRINT "X REACTANCE (OHMS)=";LP
330 OPEN1,4:PRINT#1,"X REACTANCE (OHMS)=";LP:CLOSE1,4
340 PRINT "(RECTANGULAR FORM)":PRINT
342 OPEN1,4:PRINT#1,"(RECTANGULAR FORM)":PRINT#1,:CLOSE1,4
345 PL=.01*INT(Z*100)
350 PRINT "Z IMPEDANCE (OHMS)=";PL
353 OPEN1,4:PRINT#1,"Z IMPEDANCE (OHMS)=";PL:CLOSE1,4
355 LA=.01*INT((TH*360/6.2832)*100)
370 PRINT "Z PHASE ANGLE (DEGREES)=";LA
380 OPEN1,4:PRINT#1,"Z PHASE ANGLE (DEGREES)=";LA:CLOSE1,4
390 PRINT "(POLAR FORM)"
395 OPEN 1,4:PRINT#1,"(POLAR FORM)":CLOSE1,4
400 PRINT "*************************************":PRINT
405 OPEN1,4:PRINT#1,"*********************************************":PRINT#1,:CLOSE1,4
410 END
```

Sample Run

```
*CONVERSION OF IMPEDANCE TO ADMITTANCE*
      *OR ADMITTANCE TO IMPEDANCE*


IF IMPEDANCE DATA,RUN 90:IF ADMITTANCE DATA,RUN 240


R (OHMS)= 75
X (OHMS)= 100

G CONDUCTANCE (SIEMENS)= 4.8E-03
B SUSCEPTANCE (SIEMENS)= 6.4E-03
(RECTANGULAR FORM)

Y ADMITTANCE (SIEMENS)= 8E-03
Y PHASE ANGLE (DEGREES)=-53.13
(POLAR FORM)

G (SIEMENS)= 4.8E-03
B (SIEMENS)= 6.4E-03
```

```
R RESISTANCE (OHMS)= 74.99
X REACTANCE (OHMS)= 99.99
(RECTANGULAR FORM)

Z IMPEDANCE (OHMS)= 124.99
Z PHASE ANGLE (DEGREES)= 53.12
(POLAR FORM)
****************************************
```

PROGRAM 2—Conversion of vectors from rectangular to polar form and from polar to rectangular form

Without Printer

```
5 REM PRG 2 W/O PRINTER
10 PRINT "*CONVERSION OF VECTORS FROM RECTANGULAR TO POLAR FORM,"
15 PRINT "AND FROM POLAR TO RECTANGULAR FORM*":PRINT
30 PRINT "IF POLAR TO RECTANGULAR, RUN 70:IF RECTANGULAR TO POLAR, RUN 170":END
70 PRINT:INPUT "Z (RESISTIVE COMPONENT, OHMS)=";A
90 INPUT "Z (REACTIVE COMPONENT, OHMS)=";B
110 PQ=(A↑2+B↑2)↑.5:QP=ATN(B/A)
115 LM=.01*INT(PQ*100)
120 PRINT "Z (MAGNITUDE, OHMS)=";LM
130 ML=.01*INT((QP*360/6.2832)*100)
140 PRINT "Z (ANGLE, DEGREES)=";ML
150 PRINT "****************************************"
160 END
170 PRINT:INPUT "Z (MAGNITUDE, OHMS)=";C
190 INPUT "Z (ANGLE, DEGREES)=";D
210 JK=C*COS(D*6.2832/360):KJ=C*SIN(D*6.2832/360)
215 QQ=.01*INT(JK*100)
220 PRINT "Z (RESISTIVE COMPONENT, OHMS)=";QQ
230 QZ=.01*INT(KJ*100)
240 PRINT "Z (REACTIVE COMPONENT, OHMS)=";QZ
260 PRINT "****************************************":PRINT
270 END
```

With Printer

```
5 REM PRG 2 WITH PRINTER
10 PRINT "*CONVERSION OF VECTORS FROM RECTANGULAR TO POLAR FORM,"
12 OPEN1,4
13 PRINT#1,"*CONVERSION OF VECTORS FROM RECTANGULAR TO POLAR FORM,"
14 CLOSE1,4
15 PRINT "AND FROM POLAR TO RECTANGULAR FORM*":PRINT
20 OPEN1,4
25 PRINT#1,"AND FROM POLAR TO RECTANGULAR FORM*":PRINT#1,:CLOSE1,4
30 PRINT "IF POLAR TO RECTANGULAR, RUN 70:IF RECTANGULAR TO POLAR, RUN 170"
40 OPEN1,4
50 PRINT#1,"IF POLAR TO RECTANGULAR, RUN 70:IF RECTANGULAR TO POLAR, RUN 170"
60 PRINT#1,:CLOSE1,4:END
70 PRINT:INPUT "Z (RESISTIVE COMPONENT, OHMS)=";A
80 OPEN1,4:PRINT#1,"Z (RESISTIVE COMPONENT, OHMS)=";A:CLOSE1,4
90 INPUT "Z (REACTIVE COMPONENT, OHMS)=";B
100 OPEN1,4:PRINT#1,"Z (REACTIVE COMPONENT, OHMS)=";B:CLOSE1,4
110 PQ=(A↑2+B↑2)↑.5:QP=ATN(B/A)
```

```
115 LM=.01*INT(PQ*100)
120 PRINT "Z (MAGNITUDE, OHMS)=";LM
125 OPEN1,4:PRINT#1,"Z (MAGNITUDE, OHMS)=";LM:CLOSE1,4
130 ML=.01*INT((QP*360/6.2832)*100)
140 PRINT "Z (ANGLE, DEGREES)=";ML
145 OPEN1,4:PRINT#1,"Z (ANGLE, DEGREES)=";ML:CLOSE1,4
150 PRINT "****************************************"
155 OPEN1,4:PRINT#1,"****************************************"
160 PRINT#1,:CLOSE1,4:END
170 PRINT:INPUT "Z (MAGNITUDE, OHMS)=";C
180 OPEN1,4:PRINT#1,"Z (MAGNITUDE, OHMS)=";C:CLOSE1,4
190 INPUT "Z (ANGLE, DEGREES)=";D
200 OPEN1,4:PRINT#1,"Z (ANGLE, DEGREES)=";D:CLOSE1,4
210 JK=C*COS(D*6.2832/360):KJ=C*SIN(D*6.2832/360)
215 QQ=.01*INT(JK*100)
220 PRINT "Z (RESISTIVE COMPONENT, OHMS)=";QQ
225 OPEN1,4:PRINT#1,"Z (RESISTIVE COMPONENT, OHMS)=";QQ:CLOSE1,4
230 QZ=.01*INT(KJ*100)
240 PRINT "Z (REACTIVE COMPONENT, OHMS)=";QZ
250 OPEN1,4:PRINT#1,"Z (REACTIVE COMPONENT, OHMS)=";QZ:CLOSE1,4
260 PRINT "****************************************":PRINT
265 OPEN1,4:PRINT#1,"****************************************":CLOSE1,4
270 END
```

Sample Run

```
*CONVERSION OF VECTORS FROM RECTANGULAR TO POLAR FORM,
AND FROM POLAR TO RECTANGULAR FORM*

IF POLAR TO RECTANGULAR, RUN 70:IF RECTANGULAR TO POLAR, RUN 170

Z (RESISTIVE COMPONENT, OHMS)= 64.95
Z (REACTIVE COMPONENT, OHMS)= 37.5
Z (MAGNITUDE, OHMS)= 74.99
Z (ANGLE, DEGREES)= 30
****************************************

Z (MAGNITUDE, OHMS)= 75
Z (ANGLE, DEGREES)= 30
Z (RESISTIVE COMPONENT, OHMS)= 64.95
Z (REACTIVE COMPONENT, OHMS)= 37.5
****************************************
```

PROGRAM 3—Impedance and phase angle of resultant for two vectors (phasors) in parallel

Without Printer

```
5 REM PRG 3 W/O PRINTER
10 PRINT "IMPEDANCE AND PHASE ANGLE OF RESULTANT FOR TWO VECTORS (PHASORS)"
12 PRINT
30 PRINT "IF POLAR DATA, RUN 170: IF RECTANGULAR DATA, RUN 80":END
80 PRINT:PRINT
90 INPUT "Z1 RESISTANCE (OHMS)=";A
110 INPUT "Z1 REACTANCE (OHMS)=";B
130 INPUT "Z2 RESISTANCE (OHMS)=";C
150 INPUT "Z2 REACTANCE (OHMS)=";D:PRINT:GOTO 265
```

```
170 INPUT "Z1 MAGNITUDE (OHMS)=";E
190 INPUT "Z1 PHASE ANGLE (DEGREES)=";F
210 INPUT "Z2 MAGNITUDE (OHMS)=";G
230 INPUT "Z2 PHASE ANGLE (DEGREES)=";H:PRINT
250 A=E*COS(F*6.2832/360):B=E*SIN(F*6.2832/360):PRINT:PRINT
260 C=G*COS(H*6.2832/360):D=G*SIN(H*6.2832/360)
265 WA=.01*INT(A*100)
270 PRINT "R1 (OHMS)=";WA
275 WB=.01*INT(B*100)
280 PRINT "X1 (OHMS)=";WB
285 WC=.01*INT(C*100)
290 PRINT "R2 (OHMS)=";WC
295 WD=.01*INT(D*100)
300 PRINT "X2 (OHMS)=";WD
310 PRINT "(RECTANGULAR COMPONENTS)":PRINT
320 E=(A↑2+B↑2)↑.5:NJ=B/A:PH=ATN(NJ)
325 WE=.01*INT(E*100)
330 PRINT "Z1 (OHMS)=";WE
335 WF=.01*INT((PH*360/6.2832)*100)
340 PRINT "PHASE ANGLE (DEGREES)=";WF
360 PRINT "(POLAR FORM)":PRINT
370 G=(C↑2+D↑2)↑.5:JJ=D/C:HP=ATN(JJ)
375 WG=.01*INT(G*100)
380 PRINT "Z2 (OHMS)=";WG
385 WH=.01*INT((HP*360/6.2832)*100)
390 PRINT "PHASE ANGLE (DEGREES)=";WH
410 PRINT "(POLAR FORM)":PRINT
420 Q=A+C:QQ=B+D
425 WI=.01*INT(Q*100)
430 PRINT "Z1+Z2 (OHMS RESISTANCE)=";WI
435 WJ=.01*INT(QQ*100)
440 PRINT "Z1+Z2 (OHMS REACTANCE)=";WJ
460 PRINT "(RECTANGULAR FORM)":PRINT
470 NN=E*G:MM=PH+HP
475 WK=.01*INT(NN*100)
480 PRINT "Z1*Z2 (OHMS)=";WK
485 WL=.01*INT((MM*360/6.2832)*100)
490 PRINT "PHASE ANGLE (DEGREES)=";WL
510 PRINT "(POLAR FORM)":PRINT
520 Y=(Q↑2+QQ↑2)↑.5:XY=QQ/Q:TH=ATN(XY)
525 WM=.01*INT(Y*100)
530 PRINT "Z1+Z2 (OHMS)=";WM
535 WN=.01*INT((TH*360/6.2832)*100)
540 PRINT "PHASE ANGLE (DEGREES)=";WN
560 PRINT "(POLAR FORM)":PRINT
570 RZ=NN*COS(MM):XZ=NN*SIN(MM)
575 WO=.01*INT(RZ*100)
580 PRINT "Z1*Z2 (OHMS RESISTANCE)=";WO
585 WP=.01*INT(XZ*100)
590 PRINT "Z1*Z2 (OHMS REACTANCE)=";WP
610 PRINT "(RECTANGULAR FORM)":PRINT
620 AQ=NN/Y:QA=MM-TH
625 WQ=.01*INT(AQ*100)
630 PRINT "Z1 AND Z2 IN PARALLEL (OHMS)=";WQ
635 WR=.01*INT((QA*360/6.2832)*100)
640 PRINT "PHASE ANGLE (DEGREES)=";WR
660 PRINT "(POLAR FORM)":PRINT
670 BS=AQ*COS(QA):SB=AQ*SIN(QA)
675 WS=.01*INT(BS*100)
680 PRINT "Z1 AND Z2 IN PARALLEL (OHMS RESISTANCE)=";WS
685 WT=.01*INT(SB*100)
```

```
700 PRINT "Z1 AND Z2 IN PARALLEL (OHMS REACTANCE)=";WT
720 PRINT "(RECTANGULAR FORM)":PRINT
730 PRINT "**************************************":PRINT
750 PRINT "WHEN ANGLES GREATER THAN 90 DEGREES ARE BEING PROCESSED,"
760 PRINT "THERE IS A POSSIBLE 180 DEGREE AMBIGUITY IN THE FINAL ANSWER."
770 PRINT "TO CHECK FOR 180 DEGREE AMBIGUITY, MAKE A ROUGH SKETCH OF"
780 PRINT "THE VECTOR DIAGRAM.":PRINT
790 PRINT "SLIGHT INACCURACIES IN COMPUTED VALUES MAY OCCUR DUE TO"
800 PRINT "SINGLE-PRECISION AND ROUNDING-OFF PROGRAMMING AND PROCESSING."
810 END
```

With Printer

```
5 REM PRG 3 WITH PRINTER
10 PRINT "IMPEDANCE AND PHASE ANGLE OF RESULTANT FOR TWO VECTORS (PHASORS)"
15 OPEN1,4:PRINT#1,"IMPEDANCE AND PHASE ANGLE OF RESULTANT FOR TWO VECTORS"
20 PRINT#1,"                              (PHASORS)":PRINT#1,:CLOSE1,4
30 PRINT "IF POLAR DATA, RUN 170: IF RECTANGULAR DATA, RUN 80"
40 OPEN1,4:PRINT#1,"IF POLAR DATA, RUN 170: IF RECTANGULAR DATA, RUN 80"
50 PRINT#1,:PRINT#1,:CLOSE1,4:END
80 PRINT:PRINT
90 INPUT "Z1 RESISTANCE (OHMS)=";A
100 OPEN1,4:PRINT#1,"Z1 RESISTANCE (OHMS)=";A:CLOSE1,4
110 INPUT "Z1 REACTANCE (OHMS)=";B
120 OPEN1,4:PRINT#1,"Z1 REACTANCE (OHMS)=";B:CLOSE1,4
130 INPUT "Z2 RESISTANCE (OHMS)=";C
140 OPEN1,4:PRINT#1,"Z2 RESISTANCE (OHMS)=";C:CLOSE1,4
150 INPUT "Z2 REACTANCE (OHMS)=";D:PRINT
160 OPEN1,4:PRINT#1,"Z2 REACTANCE (OHMS)=";D:PRINT#1,:CLOSE1,4:GOTO 265
170 INPUT "Z1 MAGNITUDE (OHMS)=";E
180 OPEN1,4:PRINT#1,"Z1 MAGNITUDE (OHMS)=";E:CLOSE1,4
190 INPUT "Z1 PHASE ANGLE (DEGREES)=";F
200 OPEN1,4:PRINT#1,"Z1 PHASE ANGLE (DEGREES)=";F:CLOSE1,4
210 INPUT "Z2 MAGNITUDE (OHMS)=";G
220 OPEN1,4:PRINT#1,"Z2 MAGNITUDE (OHMS)=";G:CLOSE1,4
230 INPUT "Z2 PHASE ANGLE (DEGREES)=";H:PRINT
240 OPEN1,4:PRINT#1,"Z2 PHASE ANGLE (DEGREES)=";H:PRINT#1,:CLOSE1,4
250 A=E*COS(F*6.2832/360):B=E*SIN(F*6.2832/360):PRINT:PRINT
260 C=G*COS(H*6.2832/360):D=G*SIN(H*6.2832/360)
265 WA=.01*INT(A*100)
270 PRINT "R1 (OHMS)=";WA
273 OPEN1,4:PRINT#1,"R1 (OHMS)=";WA:CLOSE1,4
275 WB=.01*INT(B*100)
280 PRINT "X1 (OHMS)=";WB
283 OPEN1,4:PRINT#1,"X1 (OHMS)=";WB:CLOSE1,4
285 WC=.01*INT(C*100)
290 PRINT "R2 (OHMS)=";WC
293 OPEN1,4:PRINT#1,"R2 (OHMS)=";WC:CLOSE1,4
295 WD=.01*INT(D*100)
300 PRINT "X2 (OHMS)=";WD
305 OPEN1,4:PRINT#1,"X2 (OHMS)=";WD:CLOSE1,4
310 PRINT "(RECTANGULAR COMPONENTS)":PRINT
315 OPEN1,4:PRINT#1,"(RECTANGULAR COMPONENTS)":PRINT#1,:CLOSE1,4
320 E=(A↑2+B↑2)↑.5:NJ=B/A:PH=ATN(NJ)
325 WE=.01*INT(E*100)
330 PRINT "Z1 (OHMS)=";WE
333 OPEN1,4:PRINT#1,"Z1 (OHMS)=";WE:CLOSE1,4
335 WF=.01*INT((PH*360/6.2832)*100)
340 PRINT "PHASE ANGLE (DEGREES)=";WF
350 OPEN1,4:PRINT#1,"PHASE ANGLE (DEGREES)=";WF:CLOSE1,4
```

```
360 PRINT "(POLAR FORM)":PRINT
365 OPEN1,4:PRINT#1,"(POLAR FORM)":PRINT#1,:CLOSE1,4
370 G=(C↑2+D↑2)↑.5:JJ=D/C:HP=ATN(JJ)
375 WG=.01*INT(G*100)
380 PRINT "Z2 (OHMS)=";WG
383 OPEN1,4:PRINT#1,"Z2 (OHMS)=";WG:CLOSE1,4
385 WH=.01*INT((HP*360/6.2832)*100)
390 PRINT "PHASE ANGLE (DEGREES)=";WH
400 OPEN1,4:PRINT#1,"PHASE ANGLE (DEGREES)=";WH:CLOSE1,4
410 PRINT "(POLAR FORM)":PRINT
415 OPEN1,4:PRINT#1,"(POLAR FORM)":PRINT#1,:CLOSE1,4
420 Q=A+C:QQ=B+D
425 WI=.01*INT(Q*100)
430 PRINT "Z1+Z2 (OHMS RESISTANCE)=";WI
433 OPEN1,4:PRINT#1,"Z1+Z2 (OHMS RESISTANCE)=";WI:CLOSE1,4
435 WJ=.01*INT(QQ*100)
440 PRINT "Z1+Z2 (OHMS REACTANCE)=";WJ
450 OPEN1,4:PRINT#1,"Z1+Z2 (OHMS REACTANCE)=";WJ:CLOSE1,4
460 PRINT "(RECTANGULAR FORM)":PRINT
465 OPEN1,4:PRINT#1,"(RECTANGULAR FORM)":PRINT#1,:CLOSE1,4
470 NN=E*G:MM=PH+HP
475 WK=.01*INT(NN*100)
480 PRINT "Z1*Z2 (OHMS)=";WK
483 OPEN1,4:PRINT#1,"Z1*Z2 (OHMS)=";WK:CLOSE1,4
485 WL=.01*INT((MM*360/6.2832)*100)
490 PRINT "PHASE ANGLE (DEGREES)=";WL
500 OPEN1,4:PRINT#1,"PHASE ANGLE (DEGREES)=";WL:CLOSE1,4
510 PRINT "(POLAR FORM)":PRINT
515 OPEN1,4:PRINT#1,"(POLAR FORM)":PRINT#1,:CLOSE1,4
520 Y=(Q↑2+QQ↑2)↑.5:XY=QQ/Q:TH=ATN(XY)
525 WM=.01*INT(Y*100)
530 PRINT "Z1+Z2 (OHMS)=";WM
533 OPEN1,4:PRINT#1,"Z1+Z2 (OHMS)=";WM:CLOSE1,4
535 WN=.01*INT((TH*360/6.2832)*100)
540 PRINT "PHASE ANGLE (DEGREES)=";WN
550 OPEN1,4:PRINT#1,"PHASE ANGLE (DEGREES)=";WN:CLOSE1,4
560 PRINT "(POLAR FORM)":PRINT
565 OPEN1,4:PRINT#1,"(POLAR FORM)":PRINT#1,:CLOSE1,4
570 RZ=NN*COS(MM):XZ=NN*SIN(MM)
575 WO=.01*INT(RZ*100)
580 PRINT "Z1*Z2 (OHMS RESISTANCE)=";WO
583 OPEN1,4:PRINT#1,"Z1*Z2 (OHMS RESISTANCE)=";WO:CLOSE1,4
585 WP=.01*INT(XZ*100)
590 PRINT "Z1*Z2 (OHMS REACTANCE)=";WP
600 OPEN1,4:PRINT#1,"Z1*Z2 (OHMS REACTANCE)=";WP:CLOSE1,4
610 PRINT "(RECTANGULAR FORM)":PRINT
615 OPEN1,4:PRINT#1,"(RECTANGULAR FORM)":PRINT#1,:CLOSE1,4
620 AQ=NN/Y:QA=MM-TH
625 WQ=.01*INT(AQ*100)
630 PRINT "Z1 AND Z2 IN PARALLEL (OHMS)=";WQ
633 OPEN1,4:PRINT#1,"Z1 AND Z2 IN PARALLEL (OHMS)=";WQ:CLOSE1,4
635 WR=.01*INT((QA*360/6.2832)*100)
640 PRINT "PHASE ANGLE (DEGREES)=";WR
650 OPEN1,4:PRINT#1,"PHASE ANGLE (DEGREES)=";WR:CLOSE1,4
660 PRINT "(POLAR FORM)":PRINT
665 OPEN1,4:PRINT#1,"(POLAR FORM)":PRINT#1,:CLOSE1,4
670 BS=AQ*COS(QA):SB=AQ*SIN(QA)
675 WS=.01*INT(BS*100)
680 PRINT "Z1 AND Z2 IN PARALLEL (OHMS RESISTANCE)=";WS
683 OPEN1,4:PRINT#1,"Z1 AND Z2 IN PARALLEL (OHMS RESISTANCE)=";WS:CLOSE1,4
685 WT=.01*INT(SB*100)
```

```
700 PRINT "Z1 AND Z2 IN PARALLEL (OHMS REACTANCE)=";WT
710 OPEN1,4:PRINT#1,"Z1 AND Z2 IN PARALLEL (OHMS REACTANCE)=";WT:CLOSE1,4
720 PRINT "(RECTANGULAR FORM)":PRINT
725 OPEN1,4:PRINT#1,"(RECTANGULAR FORM)":PRINT#1,:PRINT#1,:CLOSE1,4
730 PRINT:PRINT
750 PRINT "WHEN ANGLES GREATER THAN 90 DEGREES ARE BEING PROCESSED,"
755 OPEN1,4:PRINT#1,"WHEN ANGLES GREATER THAN 90 DEGREES ARE BEING PROCESSED,"
757 CLOSE1,4
760 PRINT "THERE IS A POSSIBLE 180 DEGREE AMBIGUITY IN THE FINAL ANSWER."
763 OPEN1,4
765 PRINT#1,"THERE IS A POSSIBLE 180 DEGREE AMBIGUITY IN THE FINAL ANSWER."
767 CLOSE1,4
770 PRINT "TO CHECK FOR 180 DEGREE AMBIGUITY, MAKE A ROUGH SKETCH OF"
773 OPEN1,4
775 PRINT#1,"TO CHECK FOR 180 DEGREE AMBIGUITY, MAKE A ROUGH SKETCH OF"
777 CLOSE1,4
780 PRINT "THE VECTOR DIAGRAM.":PRINT
785 OPEN1,4:PRINT#1,"THE VECTOR DIAGRAM.":PRINT#1,:CLOSE1,4
790 PRINT "SLIGHT INACCURACIES IN COMPUTED VALUES MAY OCCUR DUE TO"
793 OPEN1,4
795 PRINT#1,"SLIGHT INACCURACIES IN COMPUTED VALUES MAY OCCUR DUE TO"
797 CLOSE1,4
800 PRINT "SINGLE-PRECISION AND ROUNDING-OFF PROGRAMMING AND PROCESSING."
803 OPEN1,4
805 PRINT#1,"SINGLE-PRECISION AND ROUNDING-OFF PROGRAMMING AND PROCESSING."
807 CLOSE1,4
810 PRINT "****************************************"
815 OPEN1,4:PRINT#1,"****************************************":CLOSE1,4
820 END
```

Sample Run

```
IMPEDANCE AND PHASE ANGLE OF RESULTANT FOR TWO VECTORS
                     (PHASORS)

IF POLAR DATA, RUN 170: IF RECTANGULAR DATA, RUN 80


Z1 MAGNITUDE (OHMS)= 50
Z1 PHASE ANGLE (DEGREES)= 60
Z2 MAGNITUDE (OHMS)= 75
Z2 PHASE ANGLE (DEGREES)= 20

R1 (OHMS)= 24.99
X1 (OHMS)= 43.3
R2 (OHMS)= 70.47
X2 (OHMS)= 25.65
(RECTANGULAR COMPONENTS)

Z1 (OHMS)= 50
PHASE ANGLE (DEGREES)= 59.99
(POLAR FORM)

Z2 (OHMS)= 75
PHASE ANGLE (DEGREES)= 20
(POLAR FORM)
```

```
Z1+Z2 (OHMS RESISTANCE)= 95.47
Z1+Z2 (OHMS REACTANCE)= 68.95
(RECTANGULAR FORM)

Z1*Z2 (OHMS)= 3750
PHASE ANGLE (DEGREES)= 80
(POLAR FORM)

Z1+Z2 (OHMS)= 117.77
PHASE ANGLE (DEGREES)= 35.83
(POLAR FORM)

Z1*Z2 (OHMS RESISTANCE)= 651.16
Z1*Z2 (OHMS REACTANCE)= 3693.03
(RECTANGULAR FORM)

Z1 AND Z2 IN PARALLEL (OHMS)= 31.84
PHASE ANGLE (DEGREES)= 44.16
(POLAR FORM)

Z1 AND Z2 IN PARALLEL (OHMS RESISTANCE)= 22.84
Z1 AND Z2 IN PARALLEL (OHMS REACTANCE)= 22.18
(RECTANGULAR FORM)


WHEN ANGLES GREATER THAN 90 DEGREES ARE BEING PROCESSED,
THERE IS A POSSIBLE 180 DEGREE AMBIGUITY IN THE FINAL ANSWER.
TO CHECK FOR 180 DEGREE AMBIGUITY, MAKE A ROUGH SKETCH OF
THE VECTOR DIAGRAM.

SLIGHT INACCURACIES IN COMPUTED VALUES MAY OCCUR DUE TO
SINGLE-PRECISION AND ROUNDING-OFF PROGRAMMING AND PROCESSING.
****************************************
Z1 RESISTANCE (OHMS)= 25
Z1 REACTANCE (OHMS)= 43.3
Z2 RESISTANCE (OHMS)= 70.48
Z2 REACTANCE (OHMS)= 25.65

R1 (OHMS)= 25
X1 (OHMS)= 43.29
R2 (OHMS)= 70.47
X2 (OHMS)= 25.64
(RECTANGULAR COMPONENTS)

Z1 (OHMS)= 49.99
PHASE ANGLE (DEGREES)= 59.99
(POLAR FORM)

Z2 (OHMS)= 75
PHASE ANGLE (DEGREES)= 19.99
(POLAR FORM)

Z1+Z2 (OHMS RESISTANCE)= 95.47
Z1+Z2 (OHMS REACTANCE)= 68.94
(RECTANGULAR FORM)

Z1*Z2 (OHMS)= 3750.03
PHASE ANGLE (DEGREES)= 79.99
(POLAR FORM)
```

```
Z1+Z2 (OHMS)= 117.77
PHASE ANGLE (DEGREES)= 35.83
(POLAR FORM)

Z1*Z2 (OHMS RESISTANCE)= 651.35
Z1*Z2 (OHMS REACTANCE)= 3693.03
(RECTANGULAR FORM)

Z1 AND Z2 IN PARALLEL (OHMS)= 31.84
PHASE ANGLE (DEGREES)= 44.16
(POLAR FORM)

Z1 AND Z2 IN PARALLEL (OHMS RESISTANCE)= 22.84
Z1 AND Z2 IN PARALLEL (OHMS REACTANCE)= 22.18
(RECTANGULAR FORM)


WHEN ANGLES GREATER THAN 90 DEGREES ARE BEING PROCESSED,
THERE IS A POSSIBLE 180 DEGREE AMBIGUITY IN THE FINAL ANSWER.
TO CHECK FOR 180 DEGREE AMBIGUITY, MAKE A ROUGH SKETCH OF
THE VECTOR DIAGRAM.

SLIGHT INACCURACIES IN COMPUTED VALUES MAY OCCUR DUE TO
SINGLE-PRECISION AND ROUNDING-OFF PROGRAMMING AND PROCESSING.
*******************************************
```

PROGRAM 4—Input impedance and phase angle of RLC parallel resonant circuit

Without Printer

```
5 REM PRG 4 W/O PRINTER
10 PRINT "INPUT IMPEDANCE AND PHASE ANGLE OF RLC PARALLEL RESONANT CIRCUIT"
20 PRINT
30 INPUT "L (MH)=";L
40 INPUT "C (MFD)=";C
50 INPUT "RL (OHMS)=";RL
60 INPUT "RC (OHMS)=";RC
70 INPUT "F (HZ)=";F
80 PRINT
90 XL=6.2832*F*L*.001:XC=1/(6.2832*F*C*10↑-6)
100 ZL=(RL↑2+XL↑2)↑.5:ZC=(RC↑2+XC↑2)↑.5:LZ=ATN(XL/RL):CZ=-ATN(XC/RC)
110 RT=RL+RC:XT=XL-XC:DE=(RT↑2+XT↑2)↑.5:ED=ATN(XT/RT)
120 BS=ZL*ZC:SB=LZ+CZ:HS=BS/DE:SH=SB-ED
125 QQ=INT(HS)
130 PRINT "ZIN (OHMS)=";QQ
135 QP=INT(SH*360/6.2832)
140 PRINT "PHASE ANGLE (DEGREES)=";QP
150 END
```

With Printer

```
5 REM PRG 4 WITH PRINTER
10 PRINT "INPUT IMPEDANCE AND PHASE ANGLE OF RLC PARALLEL RESONANT CIRCUIT"
12 OPEN1,4
15 PRINT#1,"INPUT IMPEDANCE AND PHASE ANGLE OF RLC PARALLEL RESONANT CIRCUIT"
17 PRINT#1,:CLOSE1,4
20 PRINT
```

```
30 INPUT "L (MH)=";L
35 OPEN1,4:PRINT#1,"L (MH)=";L:CLOSE1,4
40 INPUT "C (MFD)=";C
45 OPEN1,4:PRINT#1,"C (MFD)=";C:CLOSE1,4
50 INPUT "RL (OHMS)=";RL
55 OPEN1,4:PRINT#1,"RL (OHMS)=";RL:CLOSE1,4
60 INPUT "RC (OHMS)=";RC
65 OPEN1,4:PRINT#1,"RC (OHMS)=";RC:CLOSE1,4
70 INPUT "F (HZ)=";F
75 OPEN1,4:PRINT#1,"F (HZ)=";F:PRINT#1,:CLOSE1,4
80 PRINT
90 XL=6.2832*F*L*.001:XC=1/(6.2832*F*C*10↑-6)
100 ZL=(RL↑2+XL↑2)↑.5:ZC=(RC↑2+XC↑2)↑.5:LZ=ATN(XL/RL):CZ=-ATN(XC/RC)
110 RT=RL+RC:XT=XL-XC:DE=(RT↑2+XT↑2)↑.5:ED=ATN(XT/RT)
120 BS=ZL*ZC:SB=LZ+CZ:HS=BS/DE:SH=SB-ED
125 QQ=INT(HS)
130 PRINT "ZIN (OHMS)=";QQ
132 OPEN1,4:PRINT#1,"ZIN (OHMS)=";QQ:CLOSE1,4
135 QP=INT(SH*360/6.2832)
140 PRINT "PHASE ANGLE (DEGREES)=";QP
142 OPEN1,4:PRINT#1,"PHASE ANGLE (DEGREES)=";QP:CLOSE1,4
150 END
```

Sample Run

```
INPUT IMPEDANCE AND PHASE ANGLE OF RLC PARALLEL RESONANT CIRCUIT

L (MH)= 160
C (MFD)= .15
RL (OHMS)= 3
RC (OHMS)= 1
F (HZ)= 1000

ZIN (OHMS)= 19094
PHASE ANGLE (DEGREES)= 85
```

PROGRAM 5—Unsymmetrical two-section lag circuit, with and without resistive load

Without Printer

```
5 REM PRG 5 W/O PRINTER
10 PRINT "UNSYMMETRICAL 2-SECTION LAG CIRCUIT"
20 PRINT "   (WITH/WITHOUT RESISTIVE LOAD)"
30 PRINT "***********************************":PRINT
40 PRINT"COMPUTES UNLOADED OUTPUT IMPEDANCE AND PHASE ANGLE"
50 PRINT "(THEVENIN IMPEDANCE AND PHASE ANGLE)":PRINT
60 PRINT"COMPUTES UNLOADED EOUT/EIN AND PHASE ANGLE;"
70 PRINT "LOADED EOUT/EIN AND PHASE ANGLE":PRINT
80 INPUT "R1 (OHMS)=";RO
100 INPUT "R2 (OHMS)=";RT
120 INPUT "C1 (MFD)=";CO
140 INPUT "C2 (MFD)=";CT
160 INPUT "RL (OHMS)=";RL
180 INPUT "F (HZ)=";F:PRINT
200 XO=1/(6.2832*F*CO*10↑-6)
202 XT=1/(6.2832*F*CT*10↑-6)
```

```
205 AB=RO*XO:BA=-6.2832/4
210 AC=(RO↑2+XO↑2)↑.5:CA=-ATN(XO/RO)
215 AD=AB/AC:DA=BA-CA:AE=AD*COS(DA)
220 AF=AD*SIN(DA):AG=AE+RT:AH=(AG↑2+AF↑2)↑.5
225 HA=-ATN(AF/AG):AJ=AH*XT
230 JA=HA-6.2832/4:AK=AF-XT:KA=FA-6.2832/4
235 AL=(AG↑2+AK↑2)↑.5:LA=-ATN(AK/AG)
240 AM=AJ/AL:MA=JA-LA:AN=AM:NA=MA:PF=F*NA
245 YZ=(RO↑2+XO↑2)↑.5:ZY=-ATN(XO/RO)
247 WA=INT(AN)
250 PRINT "ZOUT (OHMS)=";WA
255 XY=XO+XT:WX=(RT↑2+XY↑2)↑.5:XW=-ATN(XY/RT)
260 VW=WX*YZ:WV=XW+ZY:UV=VW/XO:VU=WV+6.2832/4
265 WB=INT(-NA*360/6.2832)
270 PRINT "PHASE ANGLE,DEGREES (RL OPEN)=";WB
290 TU=UV*COS(VU):UT=UV*SIN(VU):ZW=UT+XO
295 RS=(TU↑2+ZW↑2)↑.5:SR=-ATN(ZW/TU)
300 PU=XT/RS:UP=-SR-6.2832/4:GU=PU*RL
305 HU=AN*COS(NA):JU=AN*SIN(NA):UG=UP
307 WC=.01*INT(PU*100)
310 PRINT"EOUT/EIN (RL OPEN)=";WC
315 KU=HU-RL:LU=(KU↑2+JU↑2)↑.5
320 UL=-ATN(JU/KU):NU=GU/LU:UN=UG-UL
325 WD=.01*INT((-UP*360/6.2832)*100)
330 PRINT "PHASE ANGLE (DEGREES)=";WD
340 WE=INT(AN)
350 PRINT "THEVENIN IMPEDANCE (OHMS)=";WE
360 WF=INT(-NA*360/6.2832)
370 PRINT "PHASE ANGLE (DEGREES)=";WF
380 WG=.01*INT(NU*100)
390 PRINT "EOUT/EIN (LOADED)=";WG
400 WH=.01*INT((-UN*360/6.2832)*100)
410 PRINT "PHASE ANGLE (DEGREES)=";WH:PRINT
430 PRINT "****************************************":PRINT
440 END
```

With Printer

```
5 REM PRG 5 WITH PRINTER
10 PRINT "UNSYMMETRICAL 2-SECTION LAG CIRCUIT"
15 OPEN1,4:PRINT#1,"UNSYMMETRICAL 2-SECTION LAG CIRCUIT":CLOSE1,4
20 PRINT "   (WITH/WITHOUT RESISTIVE LOAD)"
25 OPEN1,4:PRINT#1,"   (WITH/WITHOUT RESISTIVE LOAD)":CLOSE1,4
30 PRINT "*************************************":PRINT
35 OPEN1,4:PRINT#1,"*************************************":PRINT#1,:CLOSE1,4
40 PRINT"COMPUTES UNLOADED OUTPUT IMPEDANCE AND PHASE ANGLE"
45 OPEN1,4:PRINT#1,"COMPUTES UNLOADED OUTPUT IMPEDANCE AND PHASE ANGLE"
46 CLOSE1,4
50 PRINT "(THEVENIN IMPEDANCE AND PHASE ANGLE)":PRINT
55 OPEN1,4:PRINT#1,"(THEVENIN IMPEDANCE AND PHASE ANGLE)":PRINT#1,:CLOSE1,4
60 PRINT"COMPUTES UNLOADED EOUT/EIN AND PHASE ANGLE;"
65 OPEN1,4:PRINT#1,"COMPUTES UNLOADED EOUT/EIN AND PHASE ANGLE;":CLOSE1,4
70 PRINT "LOADED EOUT/EIN AND PHASE ANGLE":PRINT
75 OPEN1,4:PRINT#1,"LOADED EOUT/EIN AND PHASE ANGLE":PRINT#1,:CLOSE1,4
80 INPUT "R1 (OHMS)=";RO
90 OPEN1,4:PRINT#1,"R1 (OHMS)=";RO:CLOSE1,4
100 INPUT "R2 (OHMS)=";RT
110 OPEN1,4:PRINT#1,"R2 (OHMS)=";RT:CLOSE1,4
120 INPUT "C1 (MFD)=";CO
130 OPEN1,4:PRINT#1,"C1 (MFD)=";CO:CLOSE1,4
```

```
140 INPUT "C2 (MFD)=";CT
150 OPEN1,4:PRINT#1,"C2 (MFD)=";CT:CLOSE1,4
160 INPUT "RL (OHMS)=";RL
170 OPEN1,4:PRINT#1,"RL (OHMS)=";RL:CLOSE1,4
180 INPUT "F (HZ)=";F:PRINT
190 OPEN1,4:PRINT#1,"F (HZ)=";F:PRINT#1,:CLOSE1,4
200 XO=1/(6.2832*F*CO*10↑-6)
202 XT=1/(6.2832*F*CT*10↑-6)
205 AB=RO*XO:BA=-6.2832/4
210 AC=(RO↑2+XO↑2)↑.5:CA=-ATN(XO/RO)
215 AD=AB/AC:DA=BA-CA:AE=AD*COS(DA)
220 AF=AD*SIN(DA):AG=AE+RT:AH=(AG↑2+AF↑2)↑.5
225 HA=-ATN(AF/AG):AJ=AH*XT
230 JA=HA-6.2832/4:AK=AF-XT:KA=FA-6.2832/4
235 AL=(AG↑2+AK↑2)↑.5:LA=-ATN(AK/AG)
240 AM=AJ/AL:MA=JA-LA:AN=AM:NA=MA:PF=F*NA
245 YZ=(RO↑2+XO↑2)↑.5:ZY=-ATN(XO/RO)
247 WA=INT(AN)
250 PRINT "ZOUT (OHMS)=";WA
253 OPEN1,4:PRINT#1,"ZOUT (OHMS)=";WA:CLOSE1,4
255 XY=XO+XT:WX=(RT↑2+XY↑2)↑.5:XW=-ATN(XY/RT)
260 VW=WX*YZ:WV=XW+ZY:UV=VW/XO:VU=WV+6.2832/4
265 WB=INT(-NA*360/6.2832)
270 PRINT "PHASE ANGLE, DEGREES (RL OPEN)=";WB
280 OPEN1,4:PRINT#1,"PHASE ANGLE, DEGREES (RL OPEN)=";WB:CLOSE1,4
290 TU=UV*COS(VU):UT=UV*SIN(VU):ZW=UT+XO
295 RS=(TU↑2+ZW↑2)↑.5:SR=-ATN(ZW/TU)
300 PU=XT/RS:UP=-SR-6.2832/4:GU=PU*RL
305 HU=AN*COS(NA):JU=AN*SIN(NA):UG=UP
307 WC=.01*INT(PU*100)
310 PRINT"EOUT/EIN (RL OPEN)=";WC
312 OPEN1,4:PRINT#1,"EOUT/EIN (RL OPEN)=";WC:CLOSE1,4
315 KU=HU-RL:LU=(KU↑2+JU↑2)↑.5
320 UL=-ATN(JU/KU):NU=GU/LU:UN=UG-UL
325 WD=.01*INT((-UP*360/6.2832)*100)
330 PRINT "PHASE ANGLE (DEGREES)=";WD
335 OPEN1,4:PRINT#1,"PHASE ANGLE (DEGREES)=";WD:CLOSE1,4
340 WE=INT(AN)
350 PRINT "THEVENIN IMPEDANCE (OHMS)=";WE
355 OPEN1,4:PRINT#1,"THEVENIN IMPEDANCE (OHMS)=";WE:CLOSE1,4
360 WF=INT(-NA*360/6.2832)
370 PRINT "PHASE ANGLE (DEGREES)=";WF
375 OPEN1,4:PRINT#1,"PHASE ANGLE (DEGREES)=";WF:CLOSE1,4
380 WG=.01*INT(NU*100)
390 PRINT "EOUT/EIN (LOADED)=";WG
395 OPEN1,4:PRINT#1,"EOUT/EIN (LOADED)=";WG:CLOSE1,4
400 WH=.01*INT((-UN*360/6.2832)*100)
410 PRINT "PHASE ANGLE (DEGREES)=";WH:PRINT
420 OPEN1,4:PRINT#1,"PHASE ANGLE (DEGREES)=";WH:PRINT#1,:CLOSE1,4
430 PRINT "***************************************":PRINT
435 OPEN1,4:PRINT#1,"***************************************":PRINT#1,
440 CLOSE1,4:END
```

Sample Run

```
UNSYMMETRICAL 2-SECTION LAG CIRCUIT
   (WITH/WITHOUT RESISTIVE LOAD)
************************************

COMPUTES UNLOADED OUTPUT IMPEDANCE AND PHASE ANGLE
(THEVENIN IMPEDANCE AND PHASE ANGLE)

COMPUTES UNLOADED EOUT/EIN AND PHASE ANGLE;
LOADED EOUT/EIN AND PHASE ANGLE

R1 (OHMS)= 5000
R2 (OHMS)= 50000
C1 (MFD)= .015
C2 (MFD)= 1.5E-03
RL (OHMS)= 15000
F (HZ)= 1500

ZOUT (OHMS)= 41735
PHASE ANGLE, DEGREES (RL OPEN)= 141
EOUT/EIN (RL OPEN)= .63
PHASE ANGLE (DEGREES)= 108.62
THEVENIN IMPEDANCE (OHMS)= 41735
PHASE ANGLE (DEGREES)= 141
EOUT/EIN (LOADED)= .17
PHASE ANGLE (DEGREES)= 79.92

*****************************************
```

PROGRAM 6—Unsymmetrical two-section lead circuit, with and without resistive load

Without Printer

```
5 REM PRG 6 W/O PRINTER
10 PRINT "UNSYMMETRICAL 2-SECTION LEAD CIRCUIT"
20 PRINT"  (WITH OR WITHOUT RESISTIVE LOAD)"
30 PRINT "************************************":PRINT
50 PRINT "COMPUTES UNLOADED EOUT/EIN AND  PHASE ANGLE;"
60 PRINT "LOADED EOUT/EIN AND PHASE ANGLE"
70 PRINT:PRINT
80 INPUT "R1 (OHMS)=";RO
100 INPUT "R2 (OHMS)=";RT
120 INPUT "C1 (MFD)=";CO
140 INPUT "C2 (MFD)=";CT
160 INPUT "RL (OHMS)=";RL
180 INPUT "F (HZ)=";F:PRINT:N=0
200 XO=1/(6.2832*F*CO*10↑-6)
203 XT=1/(6.2832*F*CT*10↑-6)
205 RS=RO+RT:AB=(RS↑2+XT↑2)↑.5
210 BA=ATN(XT/RS):AC=(RO↑2+XO↑2)↑.5
215 CA=ATN(XO/RO):AD=AB*AC:DA=BA+CA:AE=AD/RO
220 EA=DA:AF=AE*COS(EA):AG=AE*SIN(EA)
225 AH=AF-RO:AJ=(AH↑2+AG↑2)↑.5:JA=ATN(AG/AH)
230 AK=RT/AJ:KA=-JA:KA=KA*360/6.2832:KA=KA-180
240 IF ABS(KA)>180 THEN KA=KA+180
250 IF N=1 THEN 295
255 WA=.01*INT(AK*100)
```

```
260 PRINT "UNLOADED EOUT/EIN=";WA
265 WB=.01*INT(KA*100)
270 PRINT "PHASE ANGLE (DEGREES)=";WB:N=N+1
290 RT=RT*RL/(RT+RL):GOTO 200
295 WC=.01*INT(AK*100)
300 PRINT "LOADED EOUT/EIN=";WC
305 WD=.01*INT(KA*100)
310 PRINT "PHASE ANGLE (DEGREES)=";WD
330 PRINT "****************************************":PRINT
340 END
```

With Printer

```
5 REM PRG 6 WITH PRINTER
10 PRINT "UNSYMMETRICAL 2-SECTION LEAD CIRCUIT"
15 OPEN1,4:PRINT#1,"UNSYMMETRICAL 2-SECTION LEAD CIRCUIT":CLOSE1,4
20 PRINT"  (WITH OR WITHOUT RESISTIVE LOAD)"
25 OPEN1,4:PRINT#1," (WITH OR WITHOUT RESISTIVE LOAD)":CLOSE1,4
30 PRINT "****************************************":PRINT
40 OPEN1,4:PRINT#1,"****************************************":PRINT#1,:CLOSE1,4
50 PRINT "COMPUTES UNLOADED EOUT/EIN AND  PHASE ANGLE;"
55 OPEN1,4:PRINT#1,"COMPUTES UNLOADED EOUT/EIN AND PHASE ANGLE;":CLOSE1,4
60 PRINT "LOADED EOUT/EIN AND PHASE ANGLE."
65 OPEN1,4:PRINT#1,"LOADED EOUT/EIN AND PHASE ANGLE.":PRINT#1,:PRINT#1,
68 CLOSE1,4
70 PRINT:PRINT
80 INPUT "R1 (OHMS)=";RO
90 OPEN1,4:PRINT#1,"R1 (OHMS)=";RO:CLOSE1,4
100 INPUT "R2 (OHMS)=";RT
110 OPEN1,4:PRINT#1,"R2 (OHMS)=";RT:CLOSE1,4
120 INPUT "C1 (MFD)=";CO
130 OPEN1,4:PRINT#1,"C1 (MFD)=";CO:CLOSE1,4
140 INPUT "C2 (MFD)=";CT
150 OPEN1,4:PRINT#1,"C2 (MFD)=";CT:CLOSE1,4
160 INPUT "RL (OHMS)=";RL
170 OPEN1,4:PRINT#1,"RL (OHMS)=";RL:CLOSE1,4
180 INPUT "F (HZ)=";F:PRINT:N=0
190 OPEN1,4:PRINT#1,"F (HZ)=";F:PRINT#1,:CLOSE1,4
200 XO=1/(6.2832*F*CO*10↑-6)
203 XT=1/(6.2832*F*CT*10↑-6)
205 RS=RO+RT:AB=(RS↑2+XT↑2)↑.5
210 BA=ATN(XT/RS):AC=(RO↑2+XO↑2)↑.5
215 CA=ATN(XO/RO):AD=AB*AC:DA=BA+CA:AE=AD/RO
220 EA=DA:AF=AE*COS(EA):AG=AE*SIN(EA)
225 AH=AF-RO:AJ=(AH↑2+AG↑2)↑.5:JA=ATN(AG/AH)
230 AK=RT/AJ:KA=-JA:KA=KA*360/6.2832:KA=KA-180
240 IF ABS(KA)>180 THEN KA=KA+180
250 IF N=1 THEN 295
255 WA=.01*INT(AK*100)
260 PRINT "UNLOADED EOUT/EIN=";WA
263 OPEN1,4:PRINT#1,"UNLOADED EOUT/EIN=";WA:CLOSE1,4
265 WB=.01*INT(KA*100)
270 PRINT "PHASE ANGLE (DEGREES)=";WB:N=N+1
280 OPEN1,4:PRINT#1,"PHASE ANGLE (DEGREES)=";WB:CLOSE1,4
290 RT=RT*RL/(RT+RL):GOTO 200
295 WC=.01*INT(AK*100)
300 PRINT "LOADED EOUT/EIN=";WC
303 OPEN1,4:PRINT#1,"LOADED EOUT/EIN=";WC:CLOSE1,4
305 WD=.01*INT(KA*100)
310 PRINT "PHASE ANGLE (DEGREES)=";WD
```

```
320 OPEN1,4:PRINT#1,"PHASE ANGLE (DEGREES)=";WD:CLOSE1,4
330 PRINT "*****************************************":PRINT
335 OPEN1,4:PRINT#1,"*****************************************":PRINT#1,
337 CLOSE1,4
340 END
```

Sample Run

```
UNSYMMETRICAL 2-SECTION LEAD CIRCUIT
 (WITH OR WITHOUT RESISTIVE LOAD)
**************************************

COMPUTES UNLOADED EOUT/EIN AND PHASE ANGLE;
LOADED EOUT/EIN AND PHASE ANGLE.


R1 (OHMS)= 5000
R2 (OHMS)= 50000
C1 (MFD)= .015
C2 (MFD)= 1.5E-03
RL (OHMS)= 15000
F (HZ)= 1500

UNLOADED EOUT/EIN= .31
PHASE ANGLE (DEGREES)=-108.63
LOADED EOUT/EIN= .08
PHASE ANGLE (DEGREES)=-133.25
*****************************************
```

Appendix B

PROGRAM CONVERSIONS

Robert L. Kruse

IBM® PC AND PC JR.™

A conversion of the Commodore 64 Program No. 5 (Appendix A) for the IBM® PC or PC Jr.™ may be written as shown in Fig. B-1. Observe the following points:

1. The programs provided in Appendix A illustrate the distinctions that are involved in running routines with a printer, and without a printer. Note that when a printer is used, a duplicate line will be required to the input and print command lines, with its proper coding. This duplicate line permits the program to be displayed on both the video monitor and on the printer. Printer coding for the Commodore is more complex than for the IBM, as seen in the examples. Because the Commodore program opens with a file and device number (open1,4), print the file number (print#1,) and then close the file (close1,4), the free memory is diminished and the amount of data that may be processed is limited in a long program. By way of comparison, th IBM PC merely requires addition of a duplicate line to the input and print lines, starting with an LPRINT code as shown in Fig. B-1.
2. Typically, values will be processed to the seventh decimal place. To control the number of decimal places that will be printed out (rounding-off process), a subroutine using string functions is employed to accommodate the IBM PC. This is shown in line 8 of Fig. B-1. Here the entry A$ = "######" indicates that when a PRINT USING A$ or LPRINT USING A$ statement follows, a whole number is to be printed. Changing to "######.#" indicates one decimal place; "######.##" indicates two decimal places, etc. This is illustrated at lines 13, 14, and 15 of Fig. B-1. The LPRINT USING A$ and PRINT USING A$ produce the whole numbers 19095 and 86 in the results. Without the rounding off of the results, the numbers would have been 19094.82 and 85.77663. Thus in the conversion of any of the programs for use on the IBM PC, insert a string function to indicate the number of decimal places desired.

```
1 LPRINT "INPUT IMPEDANCE AND PHASE ANGLE OF RLC PARALLEL RESONANT CIRCUIT"
2 PRINT "INPUT IMPEDANCE AND PHASE ANGLE OF RLC PARALLEL RESONANT CIRCUIT"
3 LPRINT"":PRINT"":INPUT "L (mH)=";L
4 LPRINT "L (mH)=";L:INPUT "C (Mfd)=";C
5 LPRINT "C (Mfd)=";C:INPUT "RL (Ohms)=";RL
6 LPRINT "RL (Ohms)=";RL:INPUT "RC (Ohms)=";RC
7 LPRINT "RC (Ohms)=";RC:INPUT "f (Hz)=";F
8 LPRINT "f (Hz)=";F:LPRINT"":PRINT"":A$="######"
9 XL=6.2832*F*L*.001:XC=1/(6.2832*F*C*10^-6)
10 ZL=(RL^2+XL^2)^.5:ZC=(RC^2+XC^2)^.5:LZ=ATN(XL/RL):CZ=-ATN(XC/RC)
11 RT=RL+RC:XT=XL-XC:DE=(RT^2+XT^2)^.5:ED=ATN(XT/RT)
12 BS=ZL*ZC:SB=LZ+CZ:HS=BS/DE:SH=SB-ED
13 LPRINT "Zin (Ohms)=";USING A$;HS:PRINT "Zin (Ohms)=";USING A$;HS
14 LPRINT "Phase Angle (Degrees)=";USING A$;SH*360/6.2832
15 PRINT "Phase Angle (Degrees)=";USING A$;SH*360/6.2832
16 END
```

```
INPUT IMPEDANCE AND PHASE ANGLE OF RLC PARALLEL RESONANT CIRCUIT

L (mH)= 160
C (Mfd)= .15
RL (Ohms)= 3
RC (Ohms)= 1
f (Hz)= 1000

Zin (Ohms)= 19095
Phase Angle (Degrees)=     86
```

Fig. B-1

APPLE® IIe AND II+

Apple® II+ conversions for the Commodore 64® programs (Appendix A) may be written as shown in Fig. B-2. Observe the following points:

1. Comparatively, Apple II+ printer coding is somewhat similar to that of the Commodore 64 in that the input and print command lines are duplicated with an opening command (PR#1) and a closing command (PR#0).
2. Rounding-off printout (number of displayed decimal places) is controlled by means of a subroutine employing the INT function when coding Apple IIe and II+ programs. Note that the Commodore 64 also uses the INT function for this purpose. The proper entries to obtain the desired number of decimal places are:

Q=INT(R)	[Whole Number]
Q=.1*INT(R*10)	[One Place]
Q=.01*INT(R*100)	[Two Places]
Q=.001*INT(R*1000)	[Three Places]
Etc.	

Thus the statement:

60 PRINT ″R1 (OHMS) = ″;R

should be written

55 Q = .01*INT(R*100)
60 PRINT ″R1 (OHMS) = ″;Q

to obtain a result to two decimal places.

```
10  GOSUB 145
20  PRINT "": PRINT "":
30  INPUT "L (mH)=";L
40  INPUT "C (Mfd)=";C
50  INPUT "RL (Ohms)=";RL
60  INPUT "RC (Ohms)=";RC
70  INPUT "f (Hz)=;F
80  PRINT "":PRINT "": HOME : GOSUB 147
81  PR# 1: PRINT "": PRINT "": PRINT "L (mH)=";L
82  PRINT "C (Mfd)=;C
83  PRINT "RL (Ohms)=";RL
84  PRINT "RC (Ohms)=";RC
85  PRINT "f (Hz)=";F
86  PRINT "": PRINT "": PR# 0
90  XL = 6.2832 * F * L * .001: XC = 1 / (6.2832 * F * C 10 ^ -6)
100 ZL = (RL ^ 2 + XL ^ 2) ^ .5 : ZC = (RC ^ 2 + XC ^ 2) ^.5 : LZ = ATN (XL /
    RL) : CZ = - ATN (XC / RC)
110 RT = RL + RC : XT = XL - XC : DE = (RT ^ 2 + XT ^ 2) ^ .5 : ED = ATN (XT /
    RT)
120 BS = ZL * ZC : SB = LZ + CZ : HS = BS / DE : SH = SB - ED
125 QQ = INT (HS)
130 PR# 1 : PRINT "Zin (Ohms)=";QQ
135 QP = INT (SH * 360 / 6.2832)
140 PRINT "Phase Angle (Degrees)=";QP
144 END
145 PRINT "INPUT IMPEDANCE AND PHASE ANGLE OF RLC PARALLEL RESONANT CIRCUIT" :
    PRINT ""
146 RETURN
147 PR# 1: PRINT "INPUT IMPEDANCE AND PHASE ANGLE OF RLC PARALLEL RESONANT CIRC
    UIT" :  PRINT "" : PR# 0
148 RETURN
```

```
]INPUT IMPEDANCE AND PHASE ANGLE OF RLC PARALLEL RESONANT CIRCUIT


L (mH)=160
C (Mfd)=.15
RL (Ohms)=3
RC (Ohms)=1
f (Hz)=1000


Zin (Ohms)=19094
Phase Angle (Degrees)=85
```

Fig. B-2

TYPICAL CONVERSION "BUGS"

Error messages resulting from incorrect coding are frequently vague, and the programmer must carefully proofread the routine. Inasmuch as programmers tend to repeat "pet" coding errors, someone else should also proofread the routine. Some common "bugs" are:

1. Numeral 0 typed in instead of capital O.
2. Letters in a two-letter variable reversed (e.g., PQ for QP).
3. Semicolon typed in instead of a colon (or vice versa).
4. Complete program line omitted.
5. "Bug" hidden in the program memory caused by "illegal" word-processing operation. (Retype the complete line if this trouble is suspected.)
6. Plus sign erroneously used for a required minus sign, or plus sign inserted in a coded data line that requires a blank space to imply a plus sign.
7. Improper units employed in assignment of INPUT variables. Numerical values can, for example, be specified within permissible ranges by using compatible units in coding of programs (e.g., the programmer has a choice of farad, microfarad, or picofarad units).
8. Reserved word used illegally for variables. For example, if the programmer attempts to use OR, AND, COM, or INT as a variable, the program will not run.
9. Factors used incorrectly (e.g., 10^6 for 10^{-6} or 6.2832/360 for 360/6.2832). Note also that logarithms of negative numbers will not be processed.

When a RUN stops at some point during the processing interval and an error message is displayed (or when a RUN stops with no error message), the programmer can operate the computer in its calculator (direct) mode to display successively the value of variables that have been processed up to the "bug" point. Accordingly, errors often become obvious. For example, the programmer may find a zero value for a variable or an extremely large value for a variable indicating (division by zero). Or, the programmer may note that the computed value for the variable is greater than one, although its correct value must be less than one (or vice versa). Patience and reasoning will help the programmer identify the coding error.

Programs sometimes appear to have coding "bugs" when the difficulty is actually an erroneous INPUT. Consider, as an illustration, the programmer who accidentally INPUTs 1500 instead of 15000. Because of this small error, a "bug" will appear to be in the program. It is good practice to re-RUN such a program, to ensure that the trouble is actually in the coding and not in an erroneous INPUT.

LINE-BY-LINE CHECKOUT

Although a program may RUN without any error messages, an incorrect answer is sometimes printed out. This difficulty requires a careful line-by-line checkout. Incorrect variables are often responsible—this involves "slips" such as R for RE, or VU for UV. A more subtle error in variable specification is encountered when a heuristic program is written with "recycled" equations.

In this situation, the INPUT variables may be A, B, C, and D. Then, the values of these INPUT variables may be redefined in following equations, and redefined again in following loops. Accordingly, the INPUT values must be kept separate from the redefined values; this is accomplished by coding AA = A*f(X), instead of A = A*f(X).

When an initial line-by-line checkout does not identify the "bug," remember that a PRINT command can be inserted into the program following each equation or logical operation. In turn, the programmer can review the processing action in a printout and find the error in the program. This is a particularly helpful procedure when equations are "recycled" in a survey or heuristic routine.

Sometimes, the programmer is unable to identify the "bug(s)" in a long and involved routine. In this situation, it is advisable to ask someone else to retype the program. This procedure allows a fresh viewpoint, as well as eliminates the programmer's favorite and frequently repeated typing errors.

INDEX

Note: Pages listed in **bold** type indicate coverage in charts or tables

-A-

-D-

-E-

-F-

-G-

-H-

-I-

-N-

-O-

-P-

-Q-

-R-

-S-

ASSEMBLY LANGUAGE/PROGRAMMER'S REFERENCE GUIDES

☐ **Apple® IIc Programmer's Reference Guide** *David L. Heiserman*
This comprehensive user's guide will help you use all the programming capabilities of the Apple IIc. Following a brief introduction, the author describes the four principal programming languages and operating systems for the Apple IIc: Applesoft BASIC, the monitor, Pro-DOS®, and 65C02 machine-language coding. Key topics such as text screen, keyboard input, and low- and high-resolution graphics are covered in separate chapters. A complete memory map is included, with procedures for managing all 128K of memory. Valuable for beginners as well as seasoned programmers.
ISBN: 0-672-22422-4, $24.95

☐ **Apple® IIe Programmer's Reference Guide** *David L. Heiserman*
Explore new programming ideas and take advantage of powerful programming procedures on the Apple IIe. This book provides needed facts, applications, and other technical information at your fingertips. Also contains many short application and demonstration programs in BASIC and assembly language.
ISBN: 0-672-22299-X, $21.95

☐ **Commodore 64® Programmer's Reference Guide** *Commodore Computer*
Includes a complete dictionary of all Commodore BASIC commands, statements, and functions. BASIC program samples then show you how each item works. Mix machine language with BASIC and use hi-res effectively with this easy-to-use guide.
ISBN: 0-672-22056-3, $19.95

☐ **Commodore 64®/128™ Assembly Language Programming** *Mark Andrews*
This step-by-step guide to programming the Commodore 64, Merlin 64™ and Panther C64™ shows you how to design your own character set, write action games, draw high-resolution graphics, create animated sprite graphics, convert numbers, mix BASIC and machine language, and program music and sound.
ISBN: 0-672-22444-5, $15.95

☐ **Commodore® 128™ Programmer's Reference Guide** *David L. Heiserman*
This excellent reference book gives you the keys to unlock the advanced features of the Commodore 128. Learn to master BASIC programming, machine language programming, the graphics system, sound system — including music — and much more. All hardware details are included too.
ISBN: 0-672-22479-8, $22.95

☐ **8088 Assembler Language Programming: The IBM® PC** *David C. Willen and Jeffrey I. Krantz*
This book is your comprehensive introduction to writing machine language software for the IBM PC. It functionally describes the 8088 microprocessor and furnishes detailed information about the PC's internal structure. Some programming experience is required.
ISBN: 0-672-22024-5, $16.95

SERVICE & REFERENCE DATA

☐ **Electronics: Circuits and Systems**
Swaminathan Madhu
Written specifically for engineers and scientists with non-electrical engineering degrees, this reference book promotes a basic understanding of electronic devices, circuits, and systems. The author highlights analog and digital systems, practical applications, signals, circuit devices, digital logic systems, and communications systems. In a concise, easy-to-understand style, he also provides completed examples, drill problems, and summary sheets containing formulas, graphics, and relationships. An invaluable self-study manual.
ISBN: 0-672-21984-0, $39.95

☐ **Gallium Arsenide Technology**
David K. Ferry, Editor-in-Chief
This comprehensive introduction to the structure and properties of this wonder compound also explores its application in analog and digital technology and examines the new band-gap engineering and the uses of gallium arsenide.
ISBN: 0-672-22375-9, $44.95

☐ **How to Read Schematics (4th Edition)**
Donald E. Herrington
More than 100,000 copies in print! This update of a standard reference features expanded coverage of logic diagrams and a chapter on flowcharts. Beginning with a general discussion of electronic diagrams, the book systematically covers the various components that comprise a circuit. It explains logic symbols and their use in digital circuits, interprets sample schematics, analyzes the operation of a radio receiver, and explains the various kinds of logic gates. Review questions end each chapter.
ISBN: 0-672-22457-7, $14.95

☐ **Image Tubes** *Illes P. Csorba*
This text provides a wealth of valuable, hard-to-find data on electron optics, imaging, and image intensification systems: image tube theory, design, construction, and components.
ISBN: 0-672-22023-7, $44.95

☐ **An Introduction to the Analysis and Processing of Signals (2nd Edition)**
Paul A. Lynn
Topics include periodic, aperiodic, sampled-data, random signals, signal comparisons, signals and systems, modulation and sampling, filters, signal recovery, detection, and prediction.
ISBN: 0-672-22253-1, $19.95

☐ **PHOTOFACT® Television Course (5th Edition)** *Howard W. Sams Engineering Staff*
This useful text contains hundreds of photographs, drawings, and schematics covering monochrome receiver design, construction, and operation.
ISBN: 0-672-21630-2, $14.95

SAMS/TI UNDERSTANDING SERIES™

☐ Understanding Artificial Intelligence
Henry C. Mishkoff

This book provides an introduction and basic understanding of this new technology. The book covers definitions, history, expert systems, natural language processing, and LISP machines.

ISBN: 0-672-27021-8, $14.95

☐ Understanding Automation Systems (2nd Edition)
Robert F. Farwell and Neil M. Schmitt

For the newcomer, here is an in-depth look at the functions that make up automation systems—open loop, closed loop, continuous and semi-continuous process, and discrete parts. This book explains programmable systems and how to use micro-computers and programmable controllers.

ISBN: 0-672-27014-5, $14.95

☐ Understanding Automotive Electronics (2nd Edition)
William B. Ribbens and Norman P. Mansour

This book begins with automotive and electronic fundamentals—prior knowledge is not necessary. It explains how the basic electronic functions, including programmable microprocessors and microcomputers, are applied for drive train control, motion control and instrumentation. Illustrations clarify mechanical and electrical principles.

ISBN: 0-672-27017-X, $14.95

☐ Understanding Communications Systems (2nd Edition)
Don L. Cannon and Gerald Luecke

This book explores many of the systems that are used every day—AM/FM radio, telephone, TV, data communications by computer, facsimile, and satellite. It explains how information is converted into electrical signals, transmitted to distant locations, and converted back to the original information.

ISBN: 0-672-27016-1, $14.95

☐ Understanding Computer Science (2nd Edition) *Roger S. Walker*

Here is an in-depth look at how people use computers to solve problems. This book covers the fundamentals of hardware and software, programs and languages, input and output, data structures and resource management.

ISBN: 0-672-27011-0, $14.95

☐ Understanding Computer Science Applications *Roger S. Walker*

This book discusses basic computer concepts and how computers communicate with their input/output units and with each other by using parallel communications, serial communications, and computer networking.

ISBN: 0-672-27020-X, $14.95

☐ Understanding Data Communications (2nd Edition) *John L. Fike et al.*

Understand the codes used for data communications, the types of messages, and the transmissions channels—including fiber optics and satellites. Learn how asynchronous modems work and how they interface to the terminal equipment. Find out about protocols, error control, local area and packet networks.

ISBN: 0-672-27019-6, $14.95

☐ Understanding Digital Electronics (2nd Edition) *Gene W. McWhorter*

Learn why digital circuits are used. Discover how AND, OR, and NOT digital circuits make decisions, store information, and convert information into electronic language. Find out how digital integrated circuits are made and how they are used in microwave ovens, gasoline pumps, video games, and cash registers.

ISBN: 0-672-27013-7, $14.95

☐ Understanding Solid State Electronics (4th Edition)
William E. Hafford and Gene W. McWhorter

This book explains complex concepts such as electricity, semiconductor theory, how electronic circuits make decisions, and how integrated circuits are made. It helps you develop a basic knowledge of semiconductors and solid-state electronics. A glossary simplifies technical terms.

ISBN: 0-672-27012-9, $14.95

☐ Understanding Telephone Electronics (2nd Edition) *John L. Fike and George E. Friend*

This book explains how the conventional telephone system works and how parts of the system are gradually being replaced by state-of-the-art electronics. Subjects include speech circuits, dialing, ringing, central office electronics, microcomputers, digital transmission, network transmission, modems, and new cellular phones.

ISBN: 0-672-27018-8, $14.95

☐ Understanding Digital Troubleshooting (2nd Edition) *Don L. Cannon*

This book presents the basic principles and troubleshooting techniques required to begin digital equipment repair and maintenance. The book begins with overviews of digital system fundamentals, digital system functions, and troubleshooting fundamentals. It continues with detecting problems in combinational logic, sequential logic, memory, and I/O.

ISBN: 0-672-27015-3, $14.95

☐ Understanding Microprocessors (2nd Edition) *Don L. Cannon and Gerald Luecke*

This book provides insight into basic concepts and fundamentals. It explains actual applications of 4-bit, 8-bit and 16-bit microcomputers, software, programs, programming concepts, and assembly language. The book provides an individualized learning format for the newcomer who wants to know what microprocessors are, what they do, and how they work.

ISBN: 0-672-27010-2, $14.95